APPLIED GEOPHYSICS

Respectfully dedicated
TO ALL PIONEERS

APPLIED GEOPHYSICS

IN THE SEARCH FOR MINERALS

BY THE LATE

A. S. EVE

Formerly Macdonald Professor of Physics,
McGill University

AND

D. A. KEYS, M.A., D.Sc., F.R.S.C.

Vice-President (Scientific) of the National Research Council of
Canada and formerly Macdonald Professor of
Physics, McGill University

FOURTH EDITION

CAMBRIDGE

AT THE UNIVERSITY PRESS

1956

CAMBRIDGE UNIVERSITY PRESS
Cambridge, New York, Melbourne, Madrid, Cape Town,
Singapore, São Paulo, Delhi, Tokyo, Mexico City

Cambridge University Press
The Edinburgh Building, Cambridge CB2 8RU, UK

Published in the United States of America by Cambridge University Press, New York

www.cambridge.org
Information on this title: www.cambridge.org/9781107600508

© Cambridge University Press 1929, 1954

This publication is in copyright. Subject to statutory exception
and to the provisions of relevant collective licensing agreements,
no reproduction of any part may take place without the written
permission of Cambridge University Press.

First edition 1929
Second edition 1933
Third edition 1938
Fourth edition 1954
Reprinted 1956
First paperback edition 2011

A catalogue record for this publication is available from the British Library

ISBN 978-1-107-60050-8 Paperback

Cambridge University Press has no responsibility for the persistence or
accuracy of URLs for external or third-party internet websites referred to in
this publication, and does not guarantee that any content on such websites is,
or will remain, accurate or appropriate.

CONTENTS

PREFACE

TO THE FOURTH EDITION

DEVELOPMENTS during the last fifteen years in geophysical exploration methods have been directed toward improvement in existing devices and the design of new ones utilizing modern advances in electronics. Examples of typical aerial survey instruments, gravimeters and seismic reflexion methods have been included. Special attention has been given to methods of locating deposits of radioactive ores, which are now of special interest. While the general character of the book has been maintained, experience gained in using it as an introductory textbook during more than twenty years at McGill University suggested the addition of a few problems, mostly based on field readings taken by the authors.

Many colleagues and friends have supplied information for this edition and our thanks are due to them. We acknowledge particularly the assistance received from C. S. Beals, W. J. Bennett, G. M. Brownell, H. Carmicheal, L. Gilchrist, S. Hammer, H. Lundberg, B. Pontecorvo, R. W. Pringle, H. G. I. Watson, J. T. Wilson, S. Worden, Houston Technical Laboratory, Imperial Oil Co., McCullough Tool Co., Schlumberger Well Surveying Corporation, McGraw-Hill Publishing Co. and the staff of the Cambridge University Press.

D.A.K.

DEEP RIVER, ONTARIO

February 1954

PREFACE
TO THE THIRD EDITION

S O M E minor modifications have been made, and a chapter has been added dealing with the chief advances during the past six years. Attention is directed to Schlumberger's 'electrical coring' for exploring uncased borings for oil; the ratiometer methods of electrical prospecting; and the improvements in the magnetic, resistivity and seismic methods of exploration.

A. S. EVE
D. A. KEYS

McGILL UNIVERSITY
MONTREAL, CANADA
September 1938

PREFACE
TO THE SECOND EDITION

W E have altered a little, and added a little, with an attempt to conserve the character, size, and price of the book. We are indebted to the Geological Survey of Canada, and to its Director, Dr W. H. Collins, for recent opportunities in the field. Our gratitude is due to the Cambridge University Press, to numerous reviewers and to H. G. Botset, L. Don Leet, E. F. Eckhardt, W. H. Fordham, E. A. Hodgson, J. W. Joyce, L. V. King, F. W. Lee, L. L. Nettleton and H. A. Wilson.

A. S. EVE
D. A. KEYS

McGILL UNIVERSITY
MONTREAL, CANADA
November 1932

ix

PREFACE
TO THE FIRST EDITION

The genesis and growth of a new branch of Science usually appeals to a large public, and fascinates those who are fortunate enough to take part in its development.

Nature holds in her hands infinite variety and boundless wealth, hence those who seek may find, but not always what they sought, nor when they deem it due. Her rewards are given to knowledge and to wisdom. Ignorance cannot discern, nor folly use, her bounties.

The search for hidden wealth has for ages urged men forward to high adventure, and no less has the search for truth its proud fellowship of gallant souls.

What shall we say then of the double quest where new knowledge flings wide the doors of fresh discoveries, and where science lays bare things invisible beneath the earth? Here, truly, should be fascination and adventure enough.

Is it not then strange that the modern application of sound physical principles to the discovery of hidden wealth was at first received with a little coldness by the physicist, some scepticism by the engineer, and even disfavour by some geologist?

No doubt the physicist has for some time been enthralled in his laboratory hunting the fugitive electrons within the elusive atom; a wise pursuit which has left mother earth overmuch in the dark and cold, so that 'earth physics' has arisen as a separate study, wherein some work for gain and some search for knowledge. Now these two are so bound together, that he who looks for one may find both.

For the most part the mining man admits the science but questions the results and benefits. Perhaps he unconsciously transfers to the new and scientific methods some of the vague but just misgivings associated with the use of the divining rod. If good results accrue from modern prospecting methods certainly the mining engineer will adopt them, but not, as history shows, hastily. In the meantime he will trust his diamond drill, which is indeed the ultimate appeal, but it is a laborious method, for the surface of the earth is vast.

As to geology, perhaps we await more men so well trained in mathematics and physics that they will eagerly seize the new torch with which to explore regions at present beyond sight or geological insight.

This book tries to set forth both means and end. It is founded on actual and varied experience in the field, for which we thank the Bureau of Mines at Washington.

Up to the present we know of no book in English which deals with the theoretical and practical sides of all the many schemes of exploration now available. Hence the difficulty to many engineers and geologists of estimating their relative merits.

If this book holds rather too much sound theory for the practical man, and some excess of field work for the man of pure science, then it will not have altogether missed its mark.

In the meantime, we shall welcome criticism and additions with a view to its enlargement and betterment.

A. S. EVE

D. A. KEYS

MCGILL UNIVERSITY
MONTREAL, CANADA
September 1928

CHAPTER I

INTRODUCTION

Modern advances in the detection of the unseen have been rapid. Bodies which were formerly regarded as opaque are now known to be penetrated readily by radiations of short wave-length, such as Röntgen and γ-rays, which can therefore be used to locate a flaw in a casting, or a foreign substance in the human body.

During the First World War the position of airships and of airplanes was found at night by direction finders, sometimes consisting of sound reflectors, sometimes of wireless detectors. Guns in France were located within a few yards, though miles away, by sound rangers employing fine wires cooled by puffs of air caused by the sound of gun fire, so that the electrical resistance of the wires was slightly but measurably affected. Today most ocean-going ships are fitted with wireless apparatus by which they can obtain their bearings in cloudy weather or fog. Directional wireless rays of a fairly narrow beam and radio pulses from two coordinated stations are employed to guide pilots on their path. During the Second World War many improvements in these devices were developed, such as radar in which short pulses of high-frequency electromagnetic waves are emitted from a transmitter and the reflexion from distant objects such as aircraft, ships or rainstorms recorded on a cathode-ray tube by which their position and distance are indicated. With suitable revolving or scanning antennae, maps of shore contours, the layout of towns and cities and the paths of aircraft and storms can be continuously viewed on the screen from aircraft, ships or ground stations. The detection of submarines during both wars afforded a difficult problem, which admitted of several solutions with varying effectiveness.

Light, sound, magnetism, electricity, and, in the last analysis, electricity, and perhaps electricity alone, brings to our consciousness all the available information both of ourselves and of the universe of which we form a part.

E & K

If then we can locate the hidden submarine by its noise, its oil patches, by its electrochemical properties, its magnetic effects, its electromagnetic effects, or by sound reflected or echoed from it, is there any reason why we should not detect an underground body by some of these methods? Indeed, a mineral vein has one great advantage: it is stationary, and does not move away like a submarine. Once located, we can remain above its neighbourhood and study its behaviour under stimulus, with variation of treatment.

Geophysics also offers indirect aids to geologist and miner, such as the discovery of faults where one region has moved up or down or sideways with respect to another region; the detection of dikes thrust up by igneous action; the location of the depth where a horizontal or sloping layer gives place to one of greater or less density or electrical conductivity; the finding of the concavity of an underground layer or syncline, or of its convexity in the anticline; in short, the elucidation of many of the complexities beneath the earth which the geologist has so patiently and skilfully deduced from observation of out-crop and section, or from the interpretation of fossil remains and of chemical constituents.

One restriction exists. The geophysicist can only detect a discontinuity, where one regions differs sufficiently from another. This is, however, a universal limitation, for man cannot perceive that which is homogeneous in nature, he can only discern that which has some variation with time or in space. Complete stagnation in time and space is to us inseparable from non-existence.

PURE GEOPHYSICS

There are two brands of every branch of science—pure and applied. Pure science is knowledge for its own sake with no material aim in view, not even a livelihood. Such a spirit of devotion is rare, and the enthusiasts who are guided by it sometimes look with scorn on those who work with ulterior motives. Pure and applied science by their very nature are, however, so bound together that the one always advances and helps the other.

Pure geophysics has to do with land, water and air, with all that constitutes the earth and its habits. As to the stars, sun, moon and planets, they give rise to astrophysics, perhaps always a pure science, which greatly enriches our knowledge of physics and chemistry.

For convenience geophysics may be divided into branches or departments. Strictly speaking, geology and also geography in its wider sense should both be included, but these venerable branches of knowledge have an established position which secures their independence. These departments of geophysics may perhaps be named as:

> Oceanography,
> Meteorology,
> Geodesy,
> Seismology,
> Gravitation and isostasy,
> Terrestrial magnetism and electricity,
> Atmospheric electricity and the ionosphere,
> Earth temperatures and heat,
> Radioactivity of earth, sea and air.

Some of these divisions are rather artificial, and the above list may not be exhaustive.

Those who wish to grasp the magnitude of the work already accomplished may read the large text-books by Heiland, Jakosky and others listed in the Bibliography. Others interested in the advanced methods of geophysical attack will find fascination in Jeffreys (1952) and Mathias (1924).

In these subdivisions of geophysics there must needs be much overlapping of fields. For example, oceanography will cover oceans and seas, but it may include estuaries, lakes and rivers. In that case the subject-matter includes not only tides and currents in the sea, problems of salinity and solution with their influence on plankton and fish food, but also questions relating to glaciers and to ice in rivers with their influence on electric power. So also it is not possible to investigate earthquakes without consideration of volcanoes. Yet again the earth's magnetism has variations closely linked with auroral displays, and these with earth currents and in turn with thunderstorms and with the perpetual circulation of electricity in the atmosphere. He who follows a lonely trail in science today is likely to end in a blind alley.

APPLIED GEOPHYSICS

The weather reports issued daily may be regarded as an example of applied geophysics. If earthquakes could be forecast with some precision, that result would again be listed under applied geophysics. The tides at leading ports are thus forecast with precision long in advance of their occurrences. When causes are many the future is obscure, when causes are few the outcome is simple to predict. Hence, an eclipse, or the return of a comet, may be foretold years in advance, but the weather a fortnight ahead cannot yet be calculated by the wisest. Toss a straw into the Niagara River and its every movement may depend upon the known laws of physics, but who shall say its path over and beyond the Falls? Nor can you persuade a golfer that every stroke in his round is predestined from an infinite past, or that it is the mere sport of electrons which are in and around him.

However fascinating the pursuit of applied geophysics may be, it must be frankly admitted that it has drawbacks. In the first place, its aims are purely material. Its unblushing object is to find and to make wealth. It looks for ore and for oil, and plenty of them, enough to pay! In the second place, it leads the searcher into the best and into the worst places of the world; in our experience to the splendour of the Central Rockies or the Adirondacks, or into the infested regions where the wildest stories of black fly and mosquito are rarely exaggerated. Lastly, if a discovery is made the geophysicist must expect the harvest to be reaped by another, for in mining truly 'to him that hath shall be given', and the rest of the quotation may prove applicable.

Now the branches of geophysics as applied to the search for
ore and oil are these: Magnetic,
                        Electrical,
                        Electromagnetic,
                        Gravitational,
                        Seismic.

These are all in the first rank, but in the second rank there
are also: Sound reflexion,
           Radioactive measurements,
           Temperature gradients,
           Micropalaeontology,

each of which may become prominent under special circumstances.

The magnetic method of prospecting has an interesting history,
which reveals the slowness of the spread of new ideas and
methods, and the passage from mere physics into ordinary
accepted engineering practice. Long ago the Chinese discovered
the remarkable properties of the 'lodestone' or leading stone,
and perhaps one of their ancient generals used it to guide his
troops through a fog to victory. Perhaps the Crusades spread
a knowledge of its properties, and navigators were then able to
guide their ships in fog, at night, or under a cloudy sky when
far from land. Its use enabled Columbus to venture on the
voyage to the West in search of China, so that it might be
claimed that the Crusades led to the discovery of America.

Dr Gilbert, physician to Queen Elizabeth, recognized that the
earth was a magnet and published his great scientific work *De
Magnete*, before the days of Galilei or Newton. But until about
the middle of the nineteenth century magnetic properties were
not used for the discovery of iron ores, when in 1843 von Wrede dis-
cerned their possibilities. In 1879 Thalén's book *On the Examination of Iron Ore Deposits by Magnetic Measurements* (see p. 16) was
written, yet in 1899 Nordenström expressed astonishment that
the great advantages of magnetic instruments to the practical
miner were not known outside of Sweden. Smyth, of Harvard,

made magnetic surveys in the United States early in this century. In 1904 Haanel's book was written *On the Location and Examination of Magnetic Ore Bodies by Magnetometric Measurements*; this good work was published by the Canadian Government, and it is now unhappily out of print and difficult to obtain.

Today the dip needle is ordinary equipment, and different types of variometers, as described in Chapter II, are available to the mining engineer.

The search for ore is usually direct, and for oil indirect. Pools of oil usually occur at too great a depth for detection by their intrinsic properties. Hence in the region near the Gulf of Mexico the search is made for salt domes, those remarkable upward intrusions of masses of salts crystallized from some ancient land-locked seas. This rock salt has high elasticity and low density, so that a sound shock travels faster through the salt than the surrounding strata. A high explosive will cause sound waves to reach a seismograph or quake-recorder more swiftly when salt domes are far below and yet intervene. Seismic waves also provide the means of measuring travel times from a source at the surface to a reflecting layer of different density thousands of feet beneath. Using adequate theory combined with experience, the presence of anticlines and synclines may be mapped and structures determined. Millions of dollars are being spent in such explorations, which have resulted in the discovery of immense new oil fields in Kuwait and in western Canada during the last few years. As a result of this increased tempo of geophysical exploration and development, production of oil in western Canada alone from such sources amounted to approximately a hundred million dollars in value in 1951, with known oil reserves well above a thousand million barrels. In addition, supplies of natural gas discovered by such means in Alberta are estimated to be billions of cubic feet. The reservoir of gas is so great that discussions are taking place on piping supplies to the Pacific coast and as far east as cities in Ontario. Here we have a case where pure geophysics reacts on applied geophysics, for the whole science of seismography, and the admirable instruments evolved for earthquake detection and measurement, all become available in this search for salt domes, and for the sulphur mines or oil deposits. A similar situation arose in the use of the torsion balance which Baron von Eötvös developed mainly for geological research.

In the use of electrical or electromagnetic methods we owe to the amplifier the possibility of magnifying effects a thousand-fold or more, and it is difficult to realize that, before the First World War, this powerful magnifier of small oscillations was not generally available to the physicist or engineer.

ON PROSPECTING

Homage is due to the old-time prospector who, facing the hardships of desert and wilderness, was buoyed up with per-pertual hope in the search for precious minerals, sometimes successful! Such discoverers have in the past contributed the lion's share, and the modern pioneer need not fear comparison with the past. The names of many famous mines perpetuate the memories of the more fortunate men: the heroic efforts of a greater number remain unrecorded.

THE DIVINING ROD

Sometimes men have sought aid from magic or enchantment by the use of a divining rod, in England sometimes called a 'dowsing rod', and in America a 'doodle bug', from a little beetle that the southern folk would call from its hole with a song.

This strange rod, usually a forked hazel twig, is grasped firmly by the two hands on each side of the fork, and the tip will, with some people only, point upwards or downwards, or become violently agitated when it approaches its quest. It has been used in the search for water, minerals, witches, criminals, Protestants, hidden treasure, lost animals and the points of the compass.

Rods and staffs have long been used by magicians, sooth-sayers, augurs and diviners. With a staff Moses struck the rock, and Aaron's rod budded; Cicero was an augur but scorned his own art. Marco Polo states that rods and arrows were used for divination throughout the Orient.

In 1556 Georgius Agricola wrote his fine work *De Re Metallica* (Fig. 1), and stated that 'a miner should be a good and serious man, and should not make use of an enchanted twig. If prudent and skilled he should follow the natural indications which he can see for himself and dig'.

Fig. 1. The divining rod. Agricola, *De Re Metallica*, 1556.

The divining rod appears to have come from Germany to Cornwall in the reign of Elizabeth. It is by no means extinct, and is frequently used for finding water. Its use for all other purposes appears to have waned, and certainly water is the easiest thing to find, and therefore increases the chance of apparent confirmation of magic or unknown influences.

The divining rod has been associated with pixies and with fairies, with sorcerers and with witchcraft. It was condemned by Luther as a breach of the first commandment. It has achieved sanctity by Christian baptism. It has provoked the wildest credulity and the uttermost scepticism. In 1674 a Jesuit priest, Dechales, inquired why the divining rod turns only in certain hands, and why it should discover ore and water. In 1692 Aymar used a divining rod to trace down a hunchback criminal, the last man to be broken on the wheel in Europe; while in 1701 the Inquisition forbade the use of the divining rod in criminal prosecution, yet in 1703 Aymar used it to point out Protestants for massacre. In 1854 Faraday declared 'table turning' to be due to involuntary muscular movements, and in the same year Chevreul was also advocating a scientific or psychological viewpoint, and showed that the magic pendulum depended on what one thought and not on external circumstances. Yet as late as 1914, in the United States, Henri Mager used divining rod and magic pendulum with much jargon on radioactivity and electromagnetism. The writers have come across three or four similar cases where bogus apparatus with a veneer of pseudo-science have deluded many credulous and superstitious people. Yet it must be admitted that the divining rod is still used by men of unquestionable integrity in search for water, frequently with marked success.

The rods or twigs are usually in a state of unstable equilibrium so that a small twist on each side of the fork will produce a large deflexion of the tip. This instability is essential for the possibility of the exercise of the psychic powers! But as the virtue is not wholly in the twig, but mainly in the man who holds it, we have to ask for the physiological or psychological causes of the violent paroxysms attendant upon the use of the twigs. To this no answer has been given, and we are reminded of the questions of how the mason bee finds its way back to its home, how the eel returns from rivers to the Caribbean Sea, and the salmon from ocean to the river of its birth, and of the ways of migratory birds and homing pigeons.

Inasmuch as man may possibly have some unknown physical,

mental and spiritual powers, it is not scientific to deny that which is not understood. It appears to be equally undesirable to remain a confirmed sceptic and to be a deluded believer, but perhaps the sceptic holds the safer position at present.

The divining rod and diviner have an immense literature. The views of the different writers are often influenced by their intellectual training and tendencies. Sir Ray Lankester may have found little, where Sir William Barrett might have seen much. A useful summary was written by Rasmond, of the Department of Geology at Washington, D.C., but it is unfortunately now out of print.

An engineer or mining man of any repute would not dream of calculating his expenditure, or of writing a report based on the findings of a divining rod. He generally disbelieves in its efficacy for ore hunting, yet when he wants water he may call in a dowser. A dowser was used in France for two years during the First World War and never failed to find a good source. Well, there was generally plenty of water to find in France!

The possibilities of conscious or of subconscious fraud must always be kept in mind by an investigator, who should be a trained scientific man, without bias or prejudice, and with some of the acumen of a sound lawyer. He will probably be faced with the profound difficulty that he cannot carry out the experiments himself, but that he will be forced to rely upon observations of another man who is therefore a medium. It is indeed desirable that sufficient indisputable facts should be collected in order to refute or to substantiate the position of divining, either as a notable but unusual power, or as a relic of medieval superstition and imposture.

In a letter to *Nature* (8 September 1928) Geddes states (1) that the faculty of water divining is possessed by some individuals, (2) that the individual responds to some, at present unknown, external stimuli, (3) that certain substances can prevent the arrival of those stimuli, in which cases the individual cannot respond. Maby and Franklin (1939) have described experiments investigating the divining rod and Tromp (1949) has attempted explanations in a recent book but gives little evidence of any real scientific basis for dowsing on accepted physical laws. Clearly we have not yet arrived at the bottom of the mystery and the end is not in sight.

SCIENTIFIC METHODS

Just as medical science has benefited mankind to an extent incomparably greater than the miracles at Lourdes and other shrines, so also sound scientific methods outstrip the dubious aid of the divining rod, and the chance guess of the wild cat. The tremendous output of metals and of oil during the present century has been mainly due to the combination of research and practice in chemistry, physics, geology, mineralogy and metallurgy, and of their use by competent mining engineers.

It is an open question whether applied geophysics should primarily be the handmaid of geology or of mining engineering. When the three sciences are represented by three sympathetic men who are capable of team work without jealousy or prejudice, there the best results will obtain. Certainly the prudent geophysicist will endeavour to acquire as much knowledge as possible of geology and of mining so that he will understand as far as possible the views and ideas of those with whom he comes into contact.

PRELIMINARY SURVEY

It is well to adopt the clothing and food of the men who have to work constantly in the region under inspection, and to guard reasonably against sun, weather and insects of all kinds. With a little caution it is easy to keep fit in work out-of-doors. One cheerful man is worth many grumblers.

The first step is to establish friendly relations with all the people in the neighbourhood, and to gather all local information from the inhabitants and from maps, plans and pamphlets, and to be observant.

A few runs to observe the natural currents due to ore bodies may quickly throw light on a region selected in consultation with geologist and mining man. The parallel wire electrical method is expeditious and easy to interpret. A brief magnetic reconnaissance may be useful. It is then possible to select the best method for more intensive investigations as set forth in later chapters of this book.

After the ground has been carefully selected by preliminary reconnaissance, it is necessary to prepare it by survey and line cutting.

When the area has been already carefully mapped by the government, or by the owners, it is possible to pick up a north and south line and to tie it in position with existing 'bench' marks or known stations. Otherwise it is necessary to select a well-marked station and to find the direction of true north by an observation of the pole star with a transit instrument, or theodolite.

It is not wise to put your faith in the magnetic compass, particularly where there are iron ores in the neighbourhood.

The selected north and south line must be measured with a good stainless steel tape or chain (200 ft.), and stations along the lines must be marked with stakes about 2 ft. long, firmly driven into the ground, with a nail in each, the position of which is determined with a plumb-bob and cord. These stations may be 100 or 200 ft. apart. The transit is set up at each in turn and east-west lines are obtained, which are then cleared of bush, sufficiently well for the workers to see along and to traverse quickly.

After these lines are ready, cross-lines can be cut and aligned rapidly, and the whole area is thus divided into squares with a stake at every corner, with the name of the station written upon it in indelible pencil. It is most convenient to number one set of parallel lines with capital letters, A, B, C, D, etc., and the other set of lines, 0, 100, 200, —ft.

The geophysicist is always in haste to get done with this preliminary survey, which must, however, be carried out thoroughly and correctly, otherwise there is a most undesirable uncertainty as to position and results.

Like everything else, such work is best done by experts—the survey by skilled surveyors, and the 'bush-whacking' by skilled bushmen, who have sharp axes, and know how to wield them.

Quickly the geophysicist will run into the vexed question of patent rights. Each new step in science may have evolved from a flash of genius, or from patient years of research work, and the discoverer is unquestionably entitled to his reward. Often the method patented is so obvious to a physicist that he is amazed that patent rights are sought or secured. Where much wealth can be obtained, much pains and much money are available for patent protection, and a small man may be unjustly extinguished by a powerful company. Moreover, an inventor is prone to believe that every analogous device is an infringement of his patent. A single method is not always efficacious in every locality, and because many different schemes must often be used, geophysicists have sometimes banded together into prospecting companies for mutual support and security. Investigation and advertisement are alike expensive, and the salaries of the ablest men are fortunately not small, central offices in large cities are expensive, and so the overhead costs are high.

Instruments are not cheap, and their repairs are not infrequent; transportation, salaries, food and lodging require money, specially in new regions under development. So it comes about that the mining engineer has to consider whether he shall be guided by his native wit, and by the geologist, and continue to use his diamond drill at a pound or two a foot, or whether he shall pay fifty pounds a day for the uncertain benefit of a geophysical survey.

Cases can be quoted where each of these decisions has been the more beneficial.

The question of costs is so complex that no safe or enduring estimate can be given. With seismic methods in Texas a few cents an acre will cover the cost, while in Northern Canada a few dollars an acre may not suffice. In 1949, 104 geophysical parties were prospecting for oil in the western provinces of Canada, 85 of these in Alberta. In 1952 the number of such parties in this province was 162, with a yearly expenditure of more than thirty million dollars.

The man with a sound general training in mathematics and physics, and with laboratory and research experience, will find no difficulty in quickly appreciating the theory, the instruments and field practice of the several schemes described in later chapters. Some insistence has been made on practical difficulties in the field because most physicists are trained today too exclusively in the laboratory, and many of them would benefit by a larger experience of surveying and of engineering generally.

The most striking differences occur in the variety of the methods used to suit various places and conditions. A skilled exponent of electrical methods might find himself a beginner when faced with torsion balance or seismic methods in the region of the Gulf of Mexico. A man who buried two hundred pounds of high explosive 20 ft. below the ground, who exploded it and left the crater unfilled, with effect comparable to war disasters, would certainly find it impossible to follow such procedure in the populous regions of Europe. For such reasons modern seismic prospecting has been developed by improvement in instrumentation, using recent advances in electronics so that moderate charges of only a fraction of a pound of explosive are required, leaving no undesirable evidence of disturbance on the surface.

A search for oil in the Gulf region is connected with salt domes, whereas faults and anticlines may be of main importance in south-west Persia or Alberta. The arid deserts of Western Australia are baffling from their high resistance, and the marshes and wet clay of Northern Quebec from their high conductivity. Thick sheets of magnetite give one type of indication, thin veins of pyrites another type. Open country may be swiftly surveyed, while bush or forest may present a serious problem in cutting lines with attendant cost and loss of time.

Geophysicists are sometimes asked why they do not go and find a mine of their own. The answer is that the cost of preparation and of maintenance is beyond the means of most men. As well build your own house and make your automobile! Moreover, a mine is not only found but made, and after prospecting there must be development before profits are made.

These introductory remarks may be to some extent superfluous, but it is hoped that they may be of service when the later chapters are read, wherein attention is paid not only to general methods and to instruments, but also to practical details and difficulties actually connected with work in the field. The aim has been to interest and assist geologists and mining men as well as practising geophysicists.

Is it permissible to close this chapter with a brief but true story?

In a speech at a meeting at Quebec in 1927 one of the authors declared that:

'The time may come when the geologist will go before, and the geophysicist follow after. Next will come the engineer with his diamond drill and, last of all, those men who with joy and with singing will gather up most of the dollars.'

At the end of the meeting a notable financier said to him: 'I liked the conclusion of your speech and the order you selected. Quite right! *Always send the cheapest man first!*'

How great the confidence in the application of geophysical methods as an aid in locating mineral and oil deposits has become during the last quarter century may be inferred from a statement by Hammer (1952) that world-wide investment in geophysical explorations for oil alone in 1951 amounted to $325,000,000. The additional expenditure on mineral exploration would greatly increase the total. How such modern scientific methods have improved the successful drilling for oil in the United States over the period 1947–51 is shown by the following table taken from Hammer's article:

Basis	Success ratio
Non-technical (random drilling)	1 in 30
Geological	1 in 10
Geophysical	1 in 6
Combined geological and geophysical	1 in 5

As a result of the use of modern exploration technique in this field of petroleum exploration, new reserves are being discovered at the average rate of six hundred million barrels of oil per year with an estimated potential world supply still undiscovered of about six hundred thousand million barrels, Moulton (1950). The amount of mineral wealth still undiscovered is even greater in prospect than oil. What physical principles are used in aiding such discoveries are described in the following chapters.

CHAPTER II

MAGNETIC METHODS

GENERAL DISCUSSION

The earth behaves as a great but rather irregular magnet. The magnetic properties of the earth have been studied for more than three centuries, but it was only as recently as 1843 that the first suggestion was made by von Wrede that the local variations in the earth's magnetic field might be used as a means of locating masses of magnetic ore. The science of applied geophysics may be said to have had its birth in 1879 when Thalén published his work *On the Examination of Iron Ore Deposits by Magnetic Measurements*. Since that time, and more especially in recent years, great strides both in technique and in theory have been made in this, the oldest of the geophysical methods of locating hidden ore.

A piece of hardened steel when magnetized retains its magnetism, hence we have the familiar compass needle and the dip circle or inclinator. Other alloys, such as permalloy and 'Mumetal', become readily magnetized in even a weak field and have very small retentivity. Faraday discovered that when a coil of wire is rotated in a magnetic field, a current of electricity is generated in the coil, hence the earth-inductor. These fundamental properties have been applied in a variety of ways, with the result that many types of magnetometers, variometers and earth-inductors have been developed for determining with extreme precision the 'elements' of the earth's magnetic field at any point.

The direction and intensity of the earth's magnetic field vary from point to point on the earth's surface. If a small compass needle capable of turning about a vertical axis is set up at any place, it will not point in general toward the geographical north pole, but rather toward a point called the magnetic north pole. The direction along which the true compass needle points defines the magnetic meridian at the point and d, the angle between the magnetic meridian and the geographical meridian, has two names. By navigators this angle is called the 'variation' of the compass, and by physicists the 'declination'. The declination will vary from place to place on the earth's surface. The strength of the horizontal component of the earth's magnetic field at any point along the direction of the magnetic meridian is called the horizontal component of the earth's magnetic field, and is designated by the letter H. The value of H is expressed in *oersteds* and is the horizontal component of the force, measured in dynes, exerted on unit magnetic pole at the given point. This is the customary physical method of expressing the strength of a magnetic field.

If now another type of magnetic needle, capable of rotating about a horizontal axis through its centre of gravity, is placed with its axis of rotation perpendicular to the plane of the magnetic meridian this needle will dip or point downwards. The angle which a true needle makes with the horizontal is called sometimes the 'dip' and sometimes the 'inclination' v. The intensity of the vertical component of the earth's magnetic field is usually designated by Z. The resultant intensity R of the magnetic field is obtained from H and Z by the well-known parallelogram law of combining forces,

$$R^2 = H^2 + Z^2.$$

The five quantities d, v, H, Z and R constitute at any point the 'elements' of the earth's magnetic field. The experimental determination of d and any two of the remaining four unknowns is clearly sufficient for the calculation of all five.

The declination and inclination and the corresponding values of H and Z have been determined for many years at various places over the earth's surface, over sea as well as on land. Places of equal declination may be joined on a map by curves called isogonic lines. In the same way charts are made showing lines of equal dip called isoclinic lines. So also another set of lines, passing through places where the resultant magnetic intensities are the same, give the so-called isodynamic lines. Moreover, it is often useful to draw lines of equal horizontal or of equal vertical intensity.

The declination is measured with a compass needle, the inclination with a dip circle, many varieties of both instruments being employed for the purpose. Some of these instruments are also used for geophysical work, and will be described later in that connexion. The values of H and Z are most frequently obtained by one of the following methods:

(1) The magnetometer and dip circle.
(2) The variometer, of which there are two distinct types:
  (a) Vertical variometer.
  (b) Horizontal variometer.
(3) The earth-inductor.

When the horizontal component H is measured, the vertical component Z is usually calculated from a knowledge of the inclination v.

The Kew magnetometer is still the standard instrument for finding the values of d and H in large-scale surveys. The angle between the magnetic and geographic meridians 'd' is deduced from the direction of the compass needle and observations made on the pole star or sun. The determination of H at the same point requires two experiments. In the one the period of oscillation T of a small cylindrical or bar magnet is observed, and this is known as the oscillation experiment. The other (deflexion experiment) consists of measuring the angular displacement α of a compass needle or other suspended small magnet. This deflexion must be produced by the same magnet (length $2l$) as that used in the oscillation method, which is now placed at a distance d, centre to centre, from the suspended magnet, at which it points, lying at right angles to the magnetic meridian.

From the oscillation experiment the value of the product of H and M, the magnetic moment of the magnet, is obtained in the form

$$HM = \frac{4\pi^2 I}{T^2},$$

I being the moment of inertia of the magnet, a constant depending upon its mass, shape and dimensions and also on the relative position of the axis about which it turns. The value of

M/H is found from the deflexion experiment and is given by

$$\frac{M}{H} = \frac{(d^2 - l^2)^2}{2d} \tan \alpha.$$

The solution of these two equations gives both H and M. The magnetic moment M of the magnet is the product of its pole strength and the distance between the poles, and this product is easily altered if the magnet receives any severe shocks or rough handling.

An earth-inductor is often used instead of the magnetometer; a certain type has been developed by the Carnegie Institution of Washington, and successfully used in such different places as Australia, South America and at sea. When a coil of known area, and with a large number of turns of wire, is revolved about a diameter at constant rate in a uniform magnetic field, a difference of potential between its ends will be generated which may be calculated from a knowledge of these factors. If the terminals of the rotating coil are attached to a sensitive measuring instrument of suitable type, a current may be detected and then measured. The strength of this current will depend upon the intensity of the magnetic field in which the coil rotates and also upon the orientation of the axis of rotation. When the axis of rotation of the coil is vertical, the current generated will be proportional to the horizontal component of the earth's magnetic field H. If the axis of rotation is horizontal, and in the magnetic meridian, the current in the coil will be proportional to Z.

In the compass form of the instrument, the alternating current produced in the earth-inductor is rectified by a commutator placed on the axis, so as to give a direct current which may be measured with a direct-current galvanometer. The orientation of the commutator brushes relative to the direction of the magnetic field will alter the value of the current obtained. Such an instrument may then be used for finding the magnetic meridian, the values of H and of Z or R, from which the angle of dip v may be found.

The earth-inductor has been used in aeronautics as a compass, with null-method indication. Lindbergh, in his lonely flight across the Atlantic, steered his course by means of such a compass.

Deposits of iron, nickel or cobalt will disturb considerably the earth's magnetic field in their vicinity. Some ores are like steel in that they are not only magnetic but tend to retain their magnetism unless they are unduly jarred or heated. Such magnetic permanence is usually shown by magnetite (Fe_3O_4) and pyrrhotite (Fe_7S_8). The magnetite at Caribou, Colorado, which contains titanium, retains its magnetism so strongly that a splinter of the ore when hung by a thread behaves like a compass needle, setting itself along the magnetic meridian. Other ores are like soft iron in that they can be readily magnetized in a magnetic field but do not retain their magnetism. Remove the field or rotate the body and their magnetism disappears or changes its polarity or direction. Thus the magnetite found at Mineville, N.Y., changed its polarity when simply rotated from a horizontal to a vertical position in the earth's magnetic field. In this respect, the Mineville magnetite resembled soft iron, or permalloy, which has a high permeability and is readily magnetized by induction in the earth's field.

Magnetic ores may thus possess permanent magnetic polarity, induced perhaps by the earth's field when the material was originally deposited, and in some cases this magnetism will be so permanent that subsequent upheavals, folds and displacements will not change the polarity the mass originally possessed. In other deposits the polarity will be determined entirely by the position of the ore in the earth's magnetic field. The presence of the magnetic material will then increase the strength of the earth's field within it by a factor μ, called the permeability of the substance. For most rocks and ores μ has a small value and another quality, called the susceptibility k, is usually measured, these two quantities being related by the equation

$$\mu = 1 + 4\pi k.$$

The following table gives the susceptibilities of a few of the more common minerals or rocks for low fields of the order of an oersted. Further measurements are indeed desirable for various values of H:

Mineral or rock	Susceptibility $= k \times 10^6$
Magnetite	97,300
Franklinite	35,600
Ilmenite	30,700
Pyrrhotite	7,020
Olivine-Gabbro	5,600
Gabbro	3,300
Specularite	3,200
Serpentine	1,270
Basalt	600
Chromite	240
Arsenopyrite	237
Hornblende	120
Zincblende	58
Pyrite	4

From this table it will be seen that only a few minerals, generally containing iron, are at all strongly magnetic; most rocks have small susceptibilities and a few have very low values. Often magnetite is found disseminated through other minerals, such as diabase dikes, and this fact is made use of in locating such intrusions. The relatively low susceptibility possessed by other minerals, such as rock salt, may serve as a guide in geophysical work over salt domes in particular regions. The low susceptibility possessed by other minerals may occasionally serve as a guide in geophysical work over particular regions.

We shall now consider what effect, on the horizontal and vertical components of the earth's magnetic field, a mass of magnetite buried below an overburden of rock would probably have. Consider Fig. 2, in which a magnetic ore-body NS is shown in profile below the surface AB. Let us for simplicity suppose that AB is the section along the magnetic meridian, and A towards the north. Then the vertical component Z will increase over the S pole and be less than normal over the N pole of the ore-body. The variation in the horizontal component will be rather more complicated, the approximate variations of Z and H being plotted as ordinates in the figure. The actual shape of either curve will depend of course upon a number of different factors, such as orientation and on the size, shape, depth below the surface and intensity of the magnetite. Some idea of the influence of inclination on the value of Z may be derived from the two cases shown in Fig. 3, in which the magnetic axis of the magnetite is shown as CD, and the value of Z plotted as ordinates along the magnetic meridian over the body. In drawing these figures the earth's field has been considered constant. In actual practice, however, the earth's magnetic field varies from place

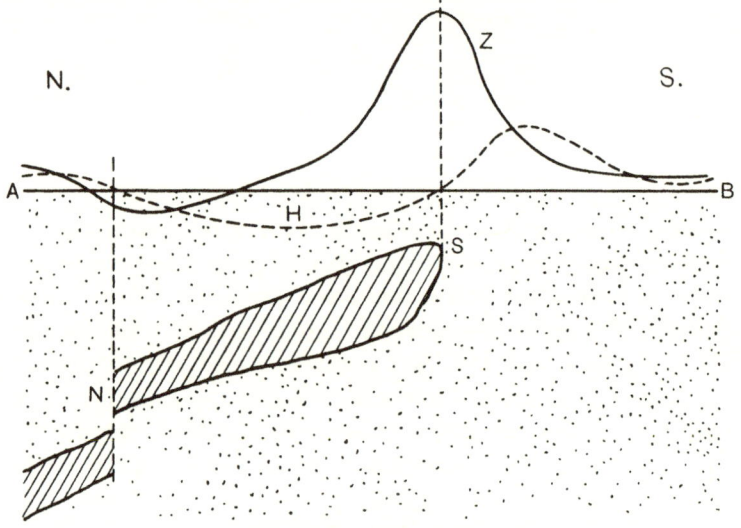

Fig. 2. Profile of magnetic ore-body and its effect on vertical and horizontal intensities.

to place and, indeed, is continually changing during the day. These minor daily variations have been given a special name —the diurnal variations. When accurate measurements are required, as in the search for salt domes, or for locating contacts between different rock formations, the magnitudes of the variation in H and Z are so small that the readings taken with the instruments have to be corrected for the diurnal variations as well as for any other changes caused by magnetic storms, auroral displays, etc. For this purpose it is customary to take continuous readings, usually photographically, of the variations in the horizontal and in the vertical component of the earth's field, at a given *stationary* point in the region of the survey. Investigations made in Ontario indicated that the continuous

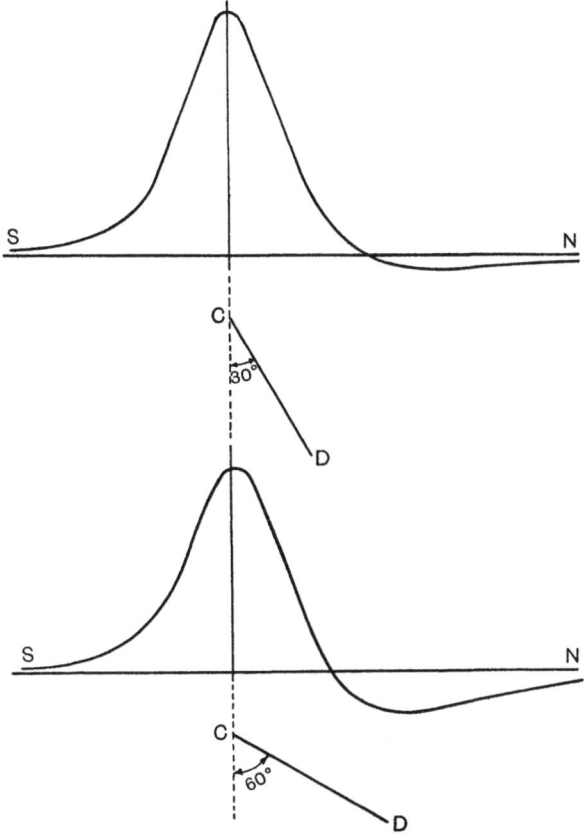

Fig. 3. The effect of the slope of a magnetic ore-body *CD* on the vertical intensity curve. (Upper figure slope 30°, lower 60°.)

readings taken at the Government magnetic observatory at Agincourt may be used for correcting diurnal variations up to distances of over 400 miles from the station, thus eliminating, within such a range, the necessity of taking readings in the field for such observations.

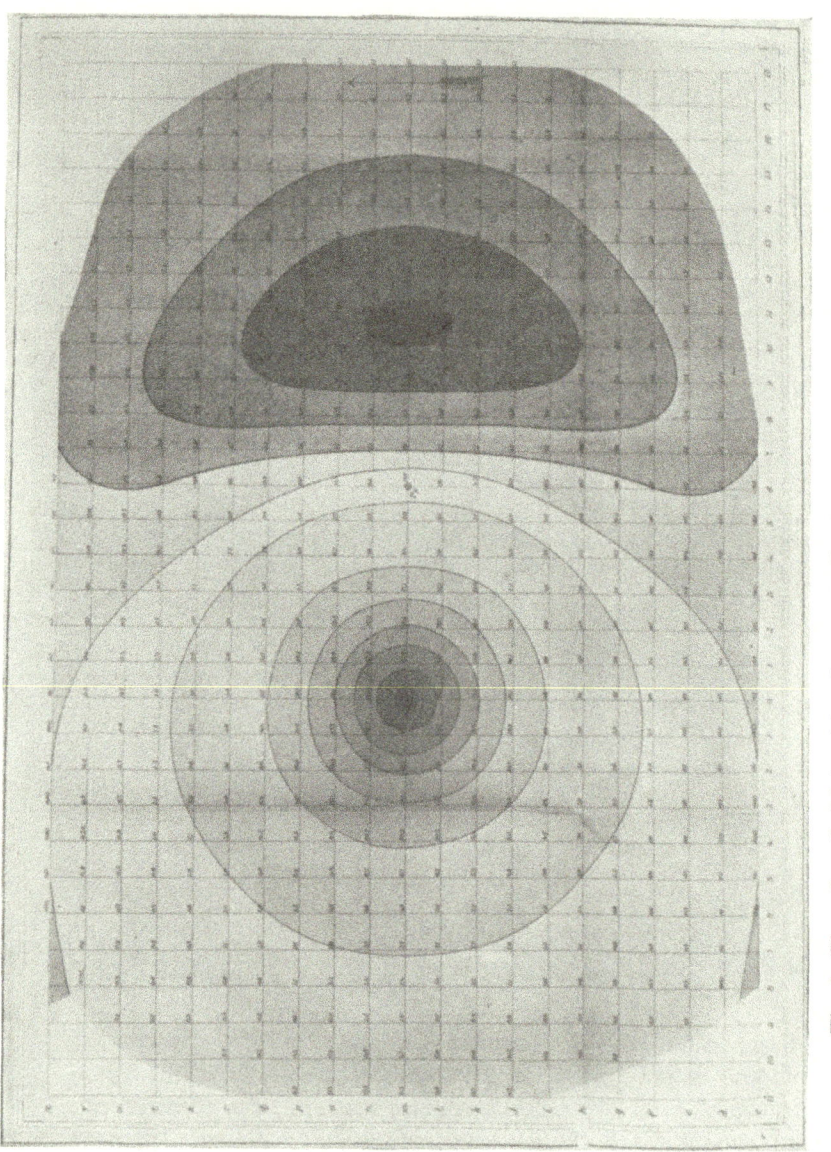

Fig. 4. Plan of vertical intensities above an inclined magnet, upper south pole on the left, lower north pole on the right. Isodynamic lines.

The value of H, Z, R, v and d may be found at a series of stations over the area investigated, and when the normal values are deducted from each reading we have a measure of the variation of these quantities at each station. Fig. 4 shows the distribution in a horizontal plane of the vertical intensity over an ideal magnet with its south pole on the left and its lower north pole on the right. The isodynamic lines are drawn and labelled with their proper values in arbitrary units. When such a map is obtained, the location of the poles of the magnetite deposit may readily be deduced. The principal methods by which such magnetic maps are obtained will be outlined in the following sections of this chapter.

THE MAGNETOMETER AND DIP-CIRCLE METHOD

The field is first laid out in squares of 100 ft. sides as already described in the introduction (p. 12). A survey may then be made with a magnetometer observing either (1) the angle of dip, or (2) the declination at each station. For the preliminary survey a dip needle is set up, and the inclination found at each station in turn. For this purpose the Tiberg inclinator shown in Fig. 5 is convenient, though any good type of mining dip needle will serve. The instrument is placed in a plane at right

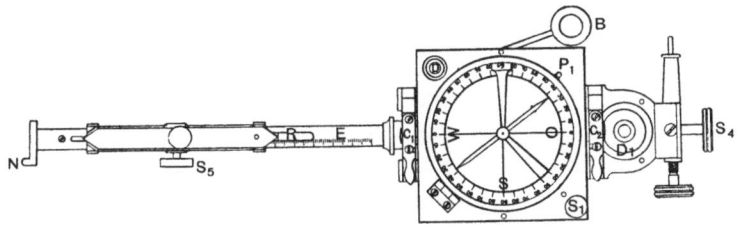

Fig. 5. Thalén-Tiberg magnetometer used as inclinator (to measure dip).

angles to the magnetic meridian and the angle of dip ϕ measured. The value of the vertical component due to the disturbed field will in this case be $Z = K \tan \phi$, where K is a constant which depends upon the weight of the needle, the position of the centre of gravity of the needle relative to the point of suspension, and the magnetic moment of the needle. In general

$$Z = (H \cos \theta + K) \tan \phi,$$

if the plane of the needle makes an angle θ with the magnetic meridian. The value of K may be determined experimentally as follows. The dip needle is set up in the magnetic meridian and the angle of dip ϕ_1 found. It is then turned at right angles and the angle of dip ϕ_2 read again. Then K will be given by

$$K = \frac{H \tan \phi_1}{\tan \phi_2 - \tan \phi_1},$$

where H is the value of the horizontal component of the earth's magnetic field at the point, which may be found by a Kew magnetometer or other method.

The values of Z found in this way for each station are recorded and plotted on the map of the field. Isoclinic lines are then drawn, giving magnetic contours from which the position of the axis of the ore-body is located.

To measure the horizontal component, the Thalén magnetometer (Fig. 6) is often used. In practice the dip needle is combined with the Thalén compass so that only one instrument is required which is called the Thalén-Tiberg magnetometer. The box containing the compass needle is shown at K in the figure; and an auxiliary cylindrical permanent magnet G, of known moment M, slides on a graduated arm E. The instrument is levelled, the magnet G removed, and the needle released on its bearings, when it will come to rest with its magnetic axis in the magnetic meridian. The arm E is then swung at right angles to the magnetic meridian, in which case the needle will read 90°. The bar magnet G is now replaced in its frame and its position along the arm adjusted until a suitable reading is obtained. If α is the deflexion, then

$$M = H \frac{(d^2 - l^2)^2}{2d} \tan \alpha,$$

in which equation M is the magnetic moment of the magnet, d is the distance from the centre of the magnet G to the centre of the needle, $2l$ is the length between the magnetic poles of the magnet G, numerically equal to five-sixths the length of the magnet, and H is the value of the horizontal component of the earth's field at the point. The value of the earth's field will thus be inversely proportional to the tangent of the angle of deflexion of the needle, provided the magnet G is kept at the same distance from the centre of the compass needle. Should the readings become too great at a station, then the magnet G must be shifted along the arm. The readings taken with this altered position of the magnet can be correlated with the previous readings by calibrating the scale on E. A simple way of making this calibration is to set the instrument up at a station where

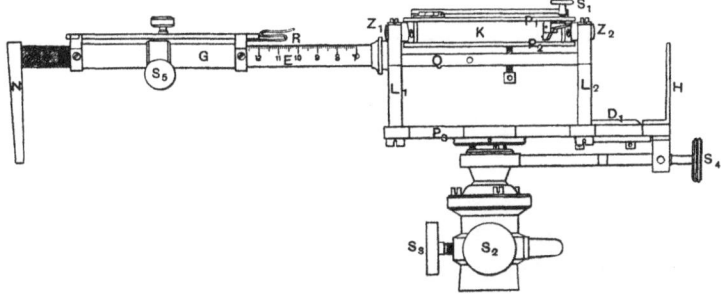

Fig. 6. Thalén-Tiberg magnetometer as used to measure
declination and horizontal intensity.

the value of H, the horizontal component of the earth's field, is known, and to observe how the deflexion varies with the distance of the deflecting magnet along the scale. When a curve is plotted, the corrections for the change in position of G may be readily made.

A sine method of determining the relative values of H at different stations is sometimes used. To apply this method, the apparatus is levelled as before and the magnet G placed in its holder. The compass box with its attached arm is now rotated about a vertical axis through the centre of suspension of the needle, until the needle is at right angles to the magnet G. When this is the case, the field due both to the earth and to the magnet G has a horizontal component parallel to the needle.

The magnet is now removed and the needle will at once move back through an angle θ until it lies in the magnetic meridian. Then

$$K = H \sin \theta,$$

in which K is a constant that can be determined experimentally by using the instrument in a magnetic field of known strength. The value of K will of course depend upon the distance of the deflecting magnet G from the compass needle, but a calibration curve may be drawn showing the relation between K and the distance. In practice the magnetometer is set up at one station, where the value of the horizontal component H_0 is known or taken as a standard for comparison, and the deflexion θ_0 observed. If the deflexion at another station is θ_1, then the corresponding value of the horizontal component H_1 may be readily found, since

$$K = H_0 \sin \theta_0 = H_1 \sin \theta_1, \quad \text{whence} \quad H_1 = H_0 \frac{\sin \theta_0}{\sin \theta_1}.$$

Most magnetometers are subject to instrumental errors which require a special procedure to correct, the nature of which depends upon the type of apparatus used. To quote as examples a few sources of error common to most instruments—the magnetic axis of the needle may not coincide with the geometric axis— the needle may be supported eccentrically or there may be friction on the bearings. Care should be exercised by the observer that no magnetic objects, such as penknives, iron keys, metallic pencils, iron-rim spectacles and magnetized watches are carried about his person when taking measurements, which should be made at some distance from iron pipes, posts, rails and hydrants.

The values for H and Z are thus obtained for every station and are charted on the map. Contours of equal intensity are then drawn from which the location of the magnetic deposit is deduced. A typical chart of the vertical intensities determined with a Thalén-Tiberg magnetometer in the Bow Lake district is shown in Fig. 7.

It is interesting to note the rapid changes of dip, from positive

Fig. 7. Magnetic survey (Bow Lake) showing lines of equal vertical intensity. Double-hatch area, dip positive > 60°; single-hatch area, dip negative > 60°; due to changes in horizontal intensity. (Department of Mines, Canada.)

to negative, shown in the diagram by double- and by single-hatched shadings, in each case denoting regions where the dip exceeds sixty degrees. This variation along a line of equal vertical intensity must of course be due to rapid changes in the horizontal intensity caused by iron ore beneath. It is found by experience that charts constructed from vertical intensities are more easily interpreted, and consequently the dip needle is the most popular form of magnetic detector, at least in high magnetic latitudes.

The Hotchkiss Superdip Magnetometer (Stearn, 1930), developed in the U.S.A., is on the principle of the Tiberg dip circle, over which it has many advantages, particularly in the measurement of the vertical magnetic field. A light brass arm is associated with the magnetic needle at an adjustable angle. These two are perfectly balanced before magnetization, and a small control weight is placed on the brass arm, so that the final centre of gravity is slightly below the axis.

When the needle is perpendicular to the resultant magnetic field, and the counterweight arm horizontal, an infinite sensitivity is obtained, and by varying the angle between the two, the instrument may be adjusted to any desired sensitivity.

THE MAGNETIC VARIOMETER METHOD

In many cases it is only necessary to measure the variations in the intensity and direction of the earth's magnetic lines of force, rather than their actual values. For such purposes variometers may be used, which are actually very sensitive balanced magnetic needles with which small variations of the order of 10^{-5} oersted may readily be observed and measured. The unit 10^{-5} oersted has been called *one gamma* (1γ), and charts are made showing the variations in terms of these units.

Variometers may be classified in two groups. The instruments of one group are of the type called horizontal variometers, because with them the variations in the horizontal component H are measured over the area under investigation. Those of the other group are known as vertical variometers, because they measure the variation in the vertical component Z. These devices depend upon the fact that as the intensity of the magnetic field increases or decreases, the direction of the magnetic needle (or needles) changes slightly, and by observing these alterations the variations in Z and H are found. The diurnal variations caused by other sources, such as magnetic storms, must be allowed for by obtaining from a neighbouring observatory the variations noted there, or by keeping one variometer stationary while others are moved about the region being surveyed, and by then using the stationary instrument for recording any variations in the earth's magnetic field during the survey. In this way only those variations that are due to the nature of the formation underground are measured, and a map, constructed from such readings, indicates the location of any disturbing field due to mineralization.

One of the oldest types of horizontal variometers is that devised by Kohlrausch, shown in Fig. 8. It consists essentially of a good compass needle, enclosed in a box with glass covering, and mounted at the top of a vertical rod. This rod is supported on a three-legged base, which may be levelled by means of screws and a small bubble level, so that the rod carrying the compass box is exactly vertical. A horizontal bar magnet slides up and down the vertical rod, which passes exactly through its centre. The magnet may be clamped at any point on the rod, its position below the magnetic needle being determined by the extent of the variations in horizontal intensity that are to be measured. The bar magnet may also be turned in any azimuth, and two adjustable stops are provided so that the magnet may be turned to the same positions right and left of its zero.

The method of procedure in using this type of horizontal variometer is as follows. The apparatus is set up and levelled at a station where H, the horizontal component of the earth's magnetic field, either is known or is considered as an arbitrary standard which we shall call H_0. The bar magnet is brought to its zero mark when it will have its magnetic axis directly under the 0–180° diameter of the magnetic needle scale. The vertical support, with magnet and compass box, is now turned until the needle reads zero, and when this is the case, the needle is in the magnetic meridian and the south pole of the bar magnet being

Fig. 8. Kohlrausch horizontal variometer used to measure
local changes in horizontal intensity H.

north, the magnetic axis of the bar magnet will also be in the magnetic meridian. The bar magnet is now turned to the right until the compass needle reads 90°, and one of the adjustable stops is clamped at this position. The magnet is now rotated in the opposite direction until the compass needle is again deflected through 90° also in the reverse direction, when it will read 270°. The second stop is then clamped in position. The magnet must have been turned through an angle greater than 90° in each case, and the mean value of this angle is read and the value of its supplement called θ. If M is the strength of the magnetic field produced at the needle by the bar magnet M, then we have from Fig. 9 (in which the needle is represented in the resultant field direction R)

$$H_0 = M \cos \theta. \tag{2·1}$$

The apparatus is now moved to a new station and is levelled with the needle and magnet in the magnetic meridian. Let us suppose that at the new station the intensity of the horizontal component of the earth's field is weaker than H_0, and let its new value be H. When the bar magnet M is rotated to its stop, the compass needle instead of reading 90° will be pulled round further and read $90 + \phi$. If H had been greater than H_0 the needle would have come to rest in a position less than 90°. We have then, as indicated in Fig. 10, by resolving the fields in a plane perpendicular to R,

$$H \cos \phi = M \cos (\theta + \phi). \tag{2·2}$$

Substituting for M from (2·1) in (2·2) we have

$$H = H_0 (1 - \tan \theta \tan \phi).$$

Thus, knowing H_0 and θ for the first station, we can calculate the variation in the horizontal field by observing ϕ. The magnetic moment of the magnet will change with time, but if the apparatus is carefully handled so as to receive no violent shocks, no alteration in M will occur during a survey.

The angle ϕ in an ore-free region will only change by one or two degrees, but the variations over a magnetite deposit may be considerable. The change in H from H_0 is called the magnetic anomaly, and it is this quantity which is determined for each station. When the anomalies are charted, we have a a map of the variations in the horizontal component of the earth's field, and these variations will be greatest over ore and least over regions possessing less than the ordinary amount of magnetic material. The location of the ore is deduced from such maps, as already explained.

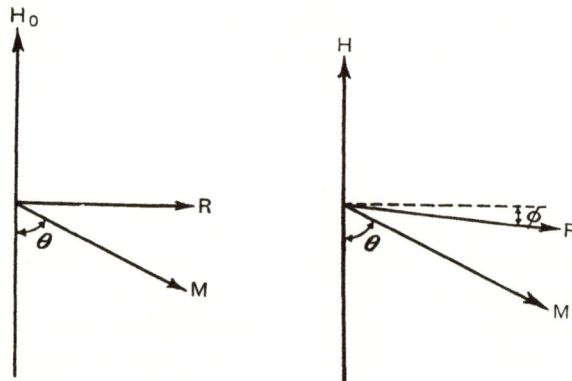

Fig. 9. Plan showing the horizontal forces on unit magnetic pole. (Kohlrausch variometer.)

Fig. 10. Plan showing the horizontal forces on unit magnetic pole, when H_0 is decreased to H.

When the magnetic anomalies are small, such as those sometimes found over salt domes, more sensitive and delicate types of variometers have been developed. These instruments will read to 1γ, but with such sensitivity accurate corrections must be made for temperature changes as well as for the diurnal variations. Several types of modern horizontal and vertical variometers of this kind are now in use, each requiring special methods of adjustment and corrections peculiar to the particular instrument used. For our purpose we shall select the well-known Askania instruments as typical of this class of magnetic balance.

The vertical variometer (Fig. 11) is used to measure the magnetic anomalies in the vertical field Z only. The magnetic needle

Fig. 11. Vertical variometer as used in the field.

consists of two thin magnets of tungsten steel, elliptical in shape, connected together by a cubical block of aluminium. This aluminium cube serves to support the quartz knife edges, shown in Fig. 12, so that the system resembles somewhat the beam of a chemical balance. The apparatus may be adjusted so as to be horizontal with the centre of gravity of the needles directly below the knife edges. This adjustment is made by two brass screws, each fitted with a small german silver screw for tightening purposes. As in the case of the chemical balance, the sensitivity depends upon the distance of the centre of gravity of the system below the point of support. In this instrument the sensitivity

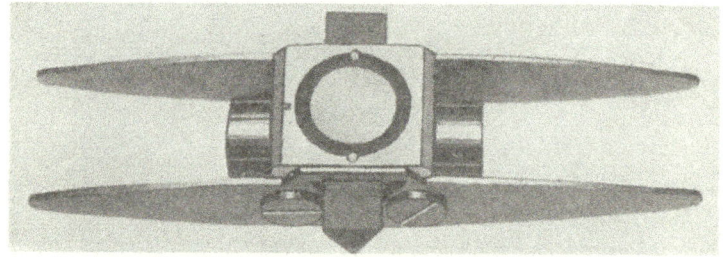

Fig. 12. Long twin magnets connected with aluminium block, showing mirror above and knife edges on either side.

may be altered by screwing in, or out, a large brass screw located directly below the knife edges on the lower side of the aluminium block. This screw is also fixed in position by a smaller german silver tightening screw, and a special type of screwdriver is used for adjusting and tightening these screws.

A circular plane mirror is attached on the upper side of the balance which reflects the light used for observing the slight variations in the position of the needle. The optical arrangement for measuring the deflexion is shown in Fig. 13. A little plane mirror outside and near the gauss eyepiece may be adjusted so as to reflect the skylight through the frosted glass window of the eyepiece on to the clear glass reflecting plate arranged at 45°. The beam of light then passes through the plate with a scale etched on it, down through the optical system, being reflected from the mirror on the magnetic balance, and returns back through the eyepiece into the observer's eye, or on to a photographic recorder, depending upon whether the instrument is to be used for surveying or for obtaining a continuous record of the diurnal variations. When the balance is horizontal, the image of the scale reflected by the mirror will appear coincident with the direct image seen through the eyepiece. In this case the scale will appear as in Fig. 14 A and the reading is taken as + 20·0. If at the next station the field is weaker, the scale will move to the left as shown in Fig. 14 B, where the reading is − 7·2. Fig 14 C indicates an increase in the field strength, the reading being + 49·4. The range is thus − 20 to + 60 divisions. A small handle on the outside of the case serves to raise the magnetic balance off its bearings and lock it so that the system may be transported readily. The case is double in order to lessen effects due to radiation and is fitted with thermometers for reading the temperature.* The whole apparatus is fitted to a convenient form of tripod provided with levels, which serves as a support. Directly below the centre of gravity of the balance and the knife edges, a cylindrical bar

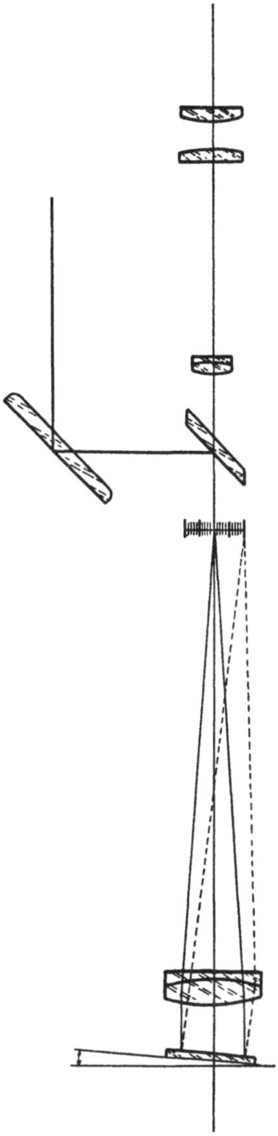

Fig. 13. Optical system of the Askania variometer· The light enters on the left, and after reflexion from the bottom mirror on the magnetic needles, passes to the eye at the top.

Fig. 13.

* An improved type of apparatus largely reduces temperature corrections.

magnet of known moment may be inserted with its axis vertical. This magnet, shown in Fig. 11 below the box containing the balance, serves to produce an auxiliary field at the needle. The strength of this artificial field will depend upon the distance of the magnet from the needle, the distance being adjustable

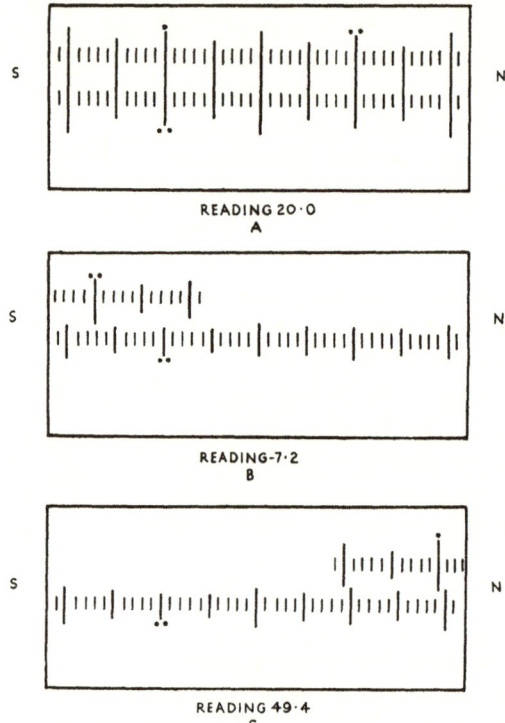

Fig. 14. Scale of Askania variometer.

and easily read from the scale on the support. The Watts* magnetic force variometers, now used quite extensively in Canada and elsewhere, are similar in principle to the Askania instruments.

The apparatus is first set up at a given station which is chosen as the standard for reference—all anomalies being read in terms

* Manufactured by Hilger and Watts, Ltd., London.

of the value of the earth's field at this point. The magnetic meridian is found by a compass needle which is provided with the instrument. This compass box will only fit on the tripod head in one position, and when the needle is released it is necessary to rotate the whole head in a clockwise direction until the compass needle points to the magnetic north as indicated on the scale below the needle. In case the needle does not swing freely, a small aluminium weight must be adjusted until the needle rests horizontally on its pivot. The tripod head is now correctly orientated and its position is fixed by the two clamps provided. The box containing the variometer needle will now fit the head in such a way that the needle lies in the magnetic meridian. The instrument must then be turned through one quadrant of $90°$, which is easily done by observing the reading on the scale of the tripod head and bringing the next quadrant mark to the same position. A simple plan, however, is to set the compass needle at the west point when adjusting the tripod head, for under this condition the north of the compass will be really to the magnetic east. When the variometer box is now placed on the tripod, the apparatus will be in such a position that the axis of oscillation of the magnetic balance is in the meridian and no further adjustment other than the levelling of the apparatus is required. When the apparatus is level, as shown by the two levels provided, the system is released and the scale deflexion is observed. If the deflexion is too great the brass screws on the magnetic needles are adjusted until the deflexion is brought into the field of view. In case the vertical field is still too strong, the auxiliary cylindrical magnet is inserted below with its appropriate pole upward and its distance from the needle is adjusted until the scale deflexion is nearly zero, say with some value D_0. Three auxiliary magnets of different strengths are usually carried in order to provide for the measurement of large magnetic anomalies and thus extend the range of measurement. The sensitivity ϵ of the instrument is next found by observing the distance R of the auxiliary magnet below the needle with its north pole up. Let us suppose that the reading is now D_1, a value less than D_0. When the south pole is upward the reading will be D_2, a little greater than D_0, if the distance of the centre of the magnet below the needle is still R. We then have

$$(D_1 - D_0)\, \epsilon = -F,$$

and
$$(D_2 - D_0)\, \epsilon = F,$$

in which equations F is the field at the balance due to the cylindrical magnet of magnetic moment M. Subtracting, we obtain
$$2F = (D_2 - D_1)\, \epsilon,$$

and the sensitivity ϵ, which is a measure of the deflexion of the balance, caused by an auxiliary field of 1γ, will be given by

$$\epsilon = \frac{2F}{D_2 - D_1} = \frac{4M}{(D_2 - D_1)}\,\frac{1}{R^3}.$$

As a check on the readings, we may use the fact that

$$D_0 = \frac{D_1 + D_2}{2}.$$

If the sensitivity is not of the proper order, it may be altered by screwing in, or out, the brass screw under the pivot. The deflexion of the balance is very nearly directly proportional to the number of turns of the screw.

It is often simpler to calibrate the instrument by placing it in the centre of a large horizontal circular loop of ten turns of wire, 100 ft. in diameter, laid out on a level piece of ground free from any disturbing magnetic ore. A series of different currents are then passed through the loop and the resulting deflexions of the instrument read, the instrument being set at right angles to the magnetic meridian. The deflexions are plotted as abscissae and the corresponding values of the currents in amperes as ordinates. The resulting curve will be a straight line through the origin, the reading when no current is flowing in the wire being taken as zero. The slope of the line gives the mean value of field due to the current I and the deflexion D. But the magnetic field in oersteds at a point x cm. above the centre of a large loop of radius a cm., and of n turns, produced by a current of I amperes in the coil, is

$$F = \frac{2\pi n I a^2}{10(a^2 + x^2)^{\frac{3}{2}}} \text{ oersteds.}$$

Hence in the present case, with a equal to 50 ft., x to 3 ft. and n equal to 10, we find

$$F = \frac{I}{243 \cdot 8}.$$

If $\tan \theta$ is the slope of the curve in the graph (Fig. 15) it is clear that

$$D = I \tan \theta,$$

so that

$$F = \frac{D}{243 \cdot 8 \tan \theta},$$

and the sensitivity, or field in oersteds to produce a deflexion of one division, is given by $\dfrac{1}{243 \cdot 8 \tan \theta}$. With the help of Fig. 15,

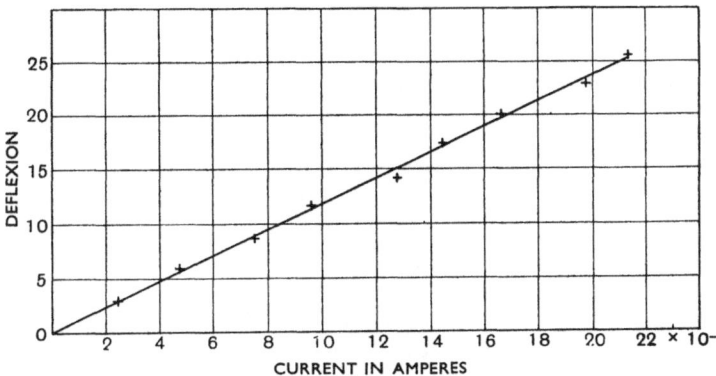

Fig. 15. Calibration curve of Askania vertical magnetometer.

the numerical value arrived at is $34 \cdot 4 \times 10^{-5}$ oersted or $34 \cdot 4\gamma$, for one division on the Askania vertical magnetometer.* A special attachment, consisting of two identical circular Helmholtz coils which produce a uniform magnetic field at the centre, can be obtained to fit on either instrument as shown in Fig. 16. The field is proportional to the current through the coils, the constant depending upon the radius of the coils and the number of turns of wire on each. This constant is provided for each coil by

* For a more detailed account of the theory and methods of calibrating Askania magnetic variometers, see Joyce (1937).

Fig. 16. Calibrating coil.

the manufacturer and the current I read on a milliammeter, the source being a good storage battery. The field $H = kI$, where k is the constant of the coils and I the current in milliamperes.

The instrument when adjusted is read at the standard reference station. The needle is usually clamped and released three times and the average of the three readings used as the proper value at the point. The apparatus is now moved from station to station, and the values of the deflexions at all the various stations are recorded. When the anomalies are small, one instrument is kept stationary and continuous readings are taken of the variations with time. The time is noted at which each observation in the field is made, and the proper corrections for diurnal variations are applied later. As stated earlier in this chapter, it is often sufficient to use the readings from a neighbouring observatory, if the location is within two degrees of latitude and longitude of the area surveyed.

When the anomalies for each station are all calculated and corrected, their values are plotted on a chart and the position of the magnetite or other deposit finally deduced.

The horizontal variometers are similar to the vertical, except that the pair of magnetic needles is mounted so as to hang vertically with the north pole downward. A photograph of such an instrument as used in the field at Caribou, Colorado, is shown in Fig. 17. The optical system, adjustments and corrections are made in precisely the same manner for each variometer. The auxiliary magnet of known moment is supported below, as in the case of the horizontal variometer, but instead of being vertical, it is now placed horizontally in the magnetic meridian with the north pole either toward the north, or south, depending upon whether the earth's field is too strong or too weak. The sensitivity will be given by

$$\epsilon = \frac{2M}{(D_2 - D_1) R^3}$$

in the case of the horizontal variometer, the symbols having the same meaning as in the previous case of the vertical variometer.

Both the vertical and horizontal anomalies are found at each station. The corrected readings of the instruments are plotted and contour lines showing equal anomalies in H and Z are obtained. The anomaly in H and Z may also be plotted in profile, or the resultant anomaly R may be shown in direction and

Fig. 17. Horizontal variometer, showing auxiliary
horizontal magnet beneath it.

magnitude, this disturbance vector being obtained from the relations

$$\Delta R = \sqrt{\{(\Delta H)^2 + (\Delta Z)^2\}} \quad \text{and} \quad \tan \lambda = \frac{\Delta H}{\Delta Z}.$$

A typical example of a survey made with such instruments is shown in Fig. 18 and also in Fig. 19, where the vertical anomalies are given over the magnetite deposit at Caribou. The magnetic variometer is used for locating salt domes and the oil which may lie on the flanks of such deposits. Fig. 20 gives the results of a survey over a salt deposit.

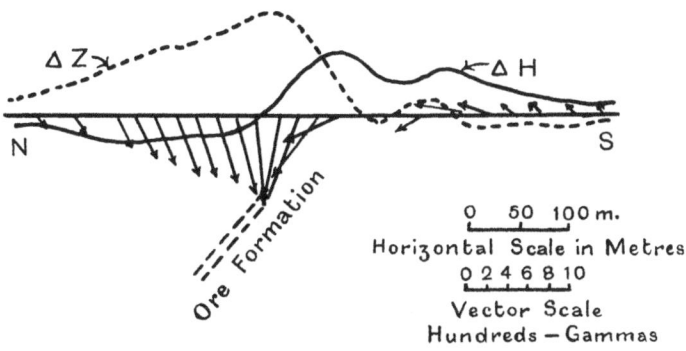

Fig. 18. Vertical, horizontal and resultant intensity differences
in a vertical plane across a magnetic vein.

Deposits of magnetic ore often occur in long veins which are nearly vertical, as, for example, in the pyrrhotite-nickel deposits in the Sudbury district of Ontario. Many diabase dikes also contain disseminated magnetite, and the geophysicist may obtain some information as to the approximate depth of overburden over such veins, from magnetic observations with the Askania variometers, although other methods will be described later by which the dip and the thickness of overburden may be found by electrical schemes. For locating pyrrhotite-nickel veins, the Askania vertical variometer is very convenient and large areas may be surveyed rapidly with such an instrument. The indications, when found are often so large and conclusive that no allowance need be made for diurnal variations, or even

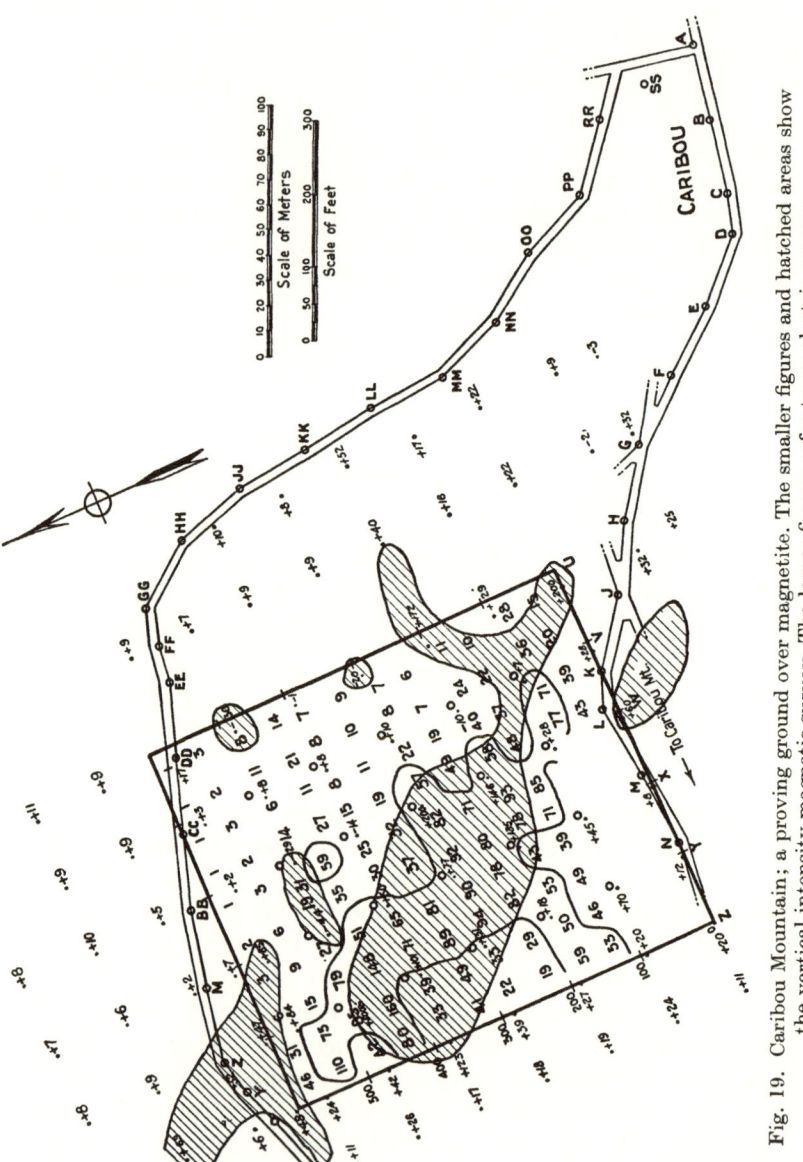

Fig. 19. Caribou Mountain; a proving ground over magnetite. The smaller figures and hatched areas show the vertical intensity magnetic survey. The larger figures refer to an electric survey.

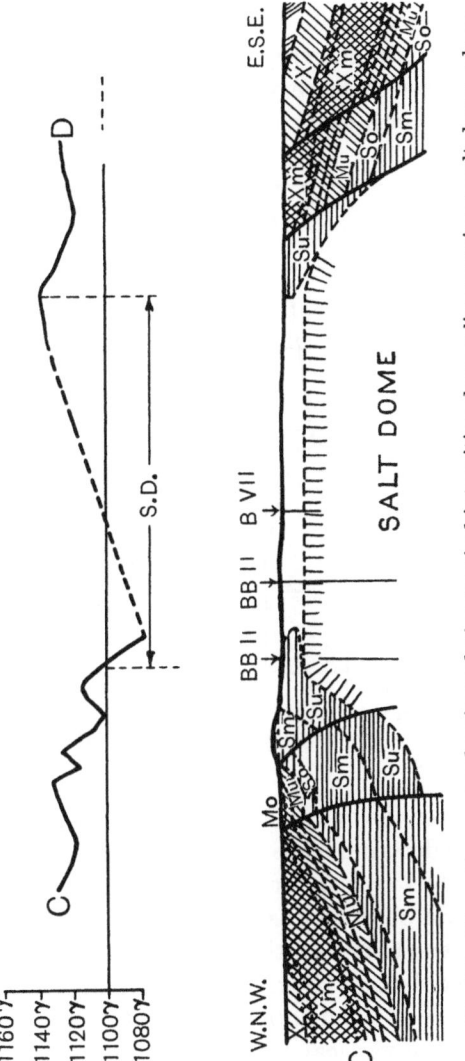

Fig. 20. Magnetic survey showing relative vertical intensities along a line crossing a salt dome, where the salt is deficient in the small magnetic content as compared with the surrounding rocks.

temperature changes. The method followed by the authors was to have the bush fairly well cleared along lines running approximately north and south and perpendicular to the known direction of the quartzite-norite contact in the region surveyed. Readings were then taken every hundred feet until an indication was found and after that every fifty feet, or less, thus obtaining when plotted a smooth curve of the variations in vertical mag-

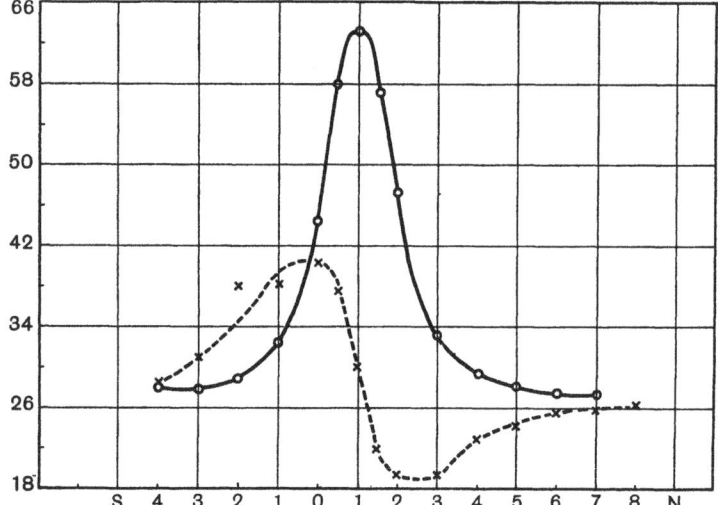

Fig. 21. The variation in the vertical component (continuous curve) and the horizontal component (broken curve) of the magnetic field along a line crossing a pyrrhotite-nickel vein.

netic intensity over the vein. Other parallel lines were similarly surveyed so as to determine the strike, until the vein or magnetic dike had been thus traced for a mile or more, until it either thinned out, or an outcrop was found. In Fig. 21 the results of such a survey over a north-south line across the pyrrhotite-nickel vein at the Falconbridge Nickel Mine, Sudbury, is given, from which the presence of the magnetic ore is at once evident by the rise in the magnetic vertical attraction.

The direction of the dip of the vein may also be obtained from such a curve. An inspection of the theoretical curves, Fig. 3,

p. 23, will indicate that the vertical intensity falls to a lower value on the right towards the north than towards the south, when the magnetic dip of CD is to the north. A similar asymmetry occurs in the case of readings taken with the horizontal variometer.

It is helpful in this connexion to indicate the type of curves obtained when both the horizontal and vertical Askania variometers are passed over a small bar magnet orientated at a given dip. For this purpose the needle of the variometer was set up as usual, either in the magnetic meridian (horizontal variometer) or perpendicular to it (vertical variometer). A wooden plank was placed horizontally on the ground underneath the instrument, so that its length was along the direction of the magnetic meridian. Distances in feet were marked off north and south from the centre, which was directly below the needle of the variometer. The small bar magnet was then placed successively at the different points, with its south pole uppermost and the readings taken on the instrument. The resulting curves will be identical in shape with the effect obtained by moving the variometer across the magnet but with this difference, that when the magnet is moved from north to south, it has the same effect as if the variometer had been moved south to north with the magnet stationary.

Readings were taken with both the horizontal and vertical variometers for each position of the bar magnet. The readings are plotted as if the variometer had been moved and are shown in Fig. 22 for the case of the magnet dipping about 50° to the north and in Fig. 23 for the case of the same magnet dipping 50° to the south. The readings taken with the vertical variometer are shown as continuous lines and those taken with the horizontal as broken lines. An inspection of the vertical variometer curves will show that the ordinate is lowest on that side towards which the magnet, or dike, dips. It will further be noted that the side of the curve which has the steeper slope is parallel to the direction of the dip.

In the case of the curves obtained with the horizontal variometer it will be quite apparent that there is again a definite difference between the two sides of the curves. If we draw a line horizontally through the normal reading of the instrument, when the variometer is not disturbed by the magnet (or magnetic vein), we shall see that the maximum rise is not equal to the maximum depression. Thus in Fig. 22 the mean zero reading is 23·7, giving the maximum increase to be 1·8 divisions and the maximum depression 3·5 divisions. Hence the side on which

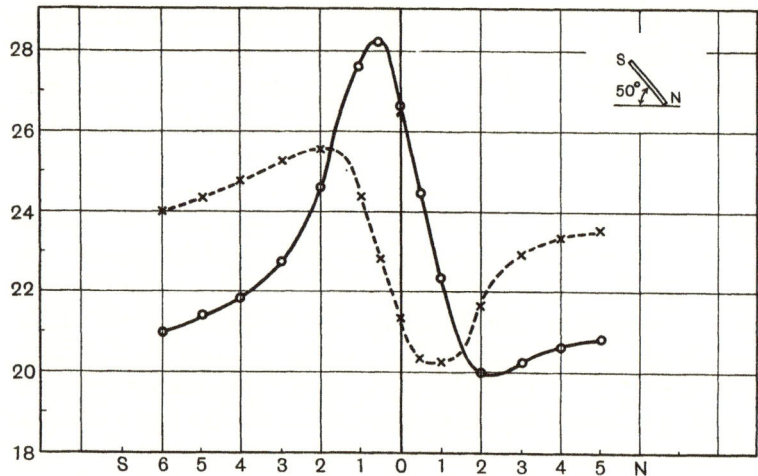

Fig. 22. The variation in the vertical component (continuous line) and the horizontal component (broken line) of the magnetic field due to a small bar magnet, dipping to the north.

the maximum deviation from the normal occurs gives the direction of the dip; for example, in Fig. 22 the dip is to the north, while an inspection of the field results shown in Fig. 21 indicates a slight dip to the south. Since a magnetic diabase dike gives similar curves, the dip of the dike may be determined by the same means.

Some idea of the amount of overburden may be obtained by taking readings with a horizontal and with a vertical variometer at each station on a survey over the dike or vein. The changes in the horizontal and vertical intensity may be measured in divisions of the instrument and lines showing the *direction* of the resultant magnetic pull at each station plotted. In order that these vectors should represent the resultant direction, the horizontal and vertical variations must be measured in the same units, but since a change of one division on the horizontal variometer is caused usually by a change in the magnetic field which is less than that required to produce a deflexion of one division on the vertical variometer, the sensitivity of the two instruments

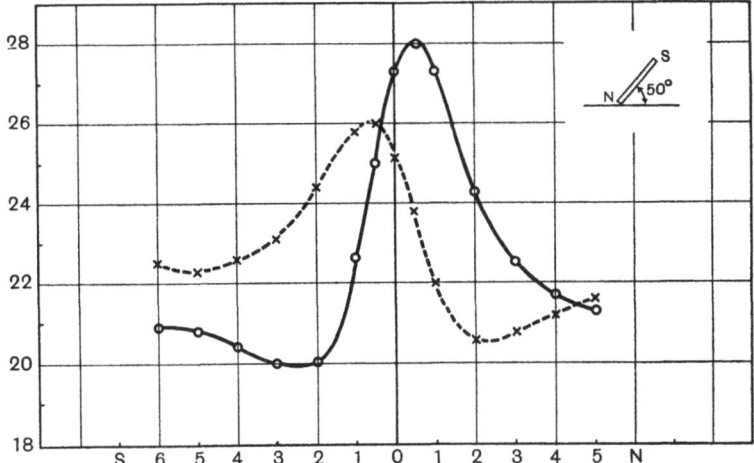

Fig. 23. The variation in the vertical component (continuous line) and the horizontal component (broken line) of the magnetic field due to a small bar magnet, dipping to the south.

must be compared. Thus it is necessary to compare the values of the magnetic fields which cause a deflexion of one division on the two instruments. This may be carried out in the following simple manner. The Askania instruments are provided with a long vertical brass rod, on which there is a sliding holder for an auxiliary bar magnet. The rod is attached directly beneath the point of support of the needles of the instrument, in exactly the same place as that of the ordinary short holder for the auxiliary magnet. The rod is graduated with two scales, the one giving the distance beneath the point of support in the case of the horizontal variometer and the other scale giving the distance to the sliding holder from the point of support of the magnetic needles in the case of the vertical variometer. Both instruments fit the same tripod, so that the vertical variometer is first placed in position and a given bar magnet screwed into the holder with say its north pole up. The reading is then taken, after which the magnet holder is lowered about 1 cm. Let R_1 and R_2 be the readings on the vertical variometer when the distance to the centre of the auxiliary magnet, as read on the scale, is d_1 and d_2 respectively. If K is the magnetic field required to produce a deflexion of one division, we shall have

$$2M \left[\frac{1}{d_1^3} - \frac{1}{d_2^3}\right] = K(R_1 - R_2),$$

where M is the magnetic moment of the small bar magnet, and when the length of the magnet is small compared with the distances d_1 and d_2.

The vertical variometer is now replaced by the horizontal variometer which is set up along the magnetic meridian in the usual manner. The auxiliary magnet is now turned to a horizontal position with its north pole north and is set at the same distance d_1 below the point of support of the needle, reading the other scale on the vertical brass rod. The reading is taken, the distance is increased again to d_2, and the final reading is made. Suppose that the readings in this case are R_1' and R_2'. Then we shall have

$$M \left[\frac{1}{d_1^3} - \frac{1}{d_2^3}\right] = K'(R_1' - R_2'),$$

where K' is the constant of the horizontal variometer. Hence we can express the sensitivity of the horizontal instrument in terms of that of the vertical variometer by the relation

$$K' = \frac{1}{2} \frac{R_1 - R_2}{R_1' - R_2'} K.$$

In one pair of instruments used by the writers, the horizontal variometer was found to have a sensitivity of 2·0 times that of the vertical variometer. In order to obtain the vectors, the horizontal readings were divided by 2·0 and the results tabulated. From the joint readings the directions of the resultant magnetic fields due to a pyrrhotite-nickel vein were drawn and are shown in Fig. 24. The depth indicated is about 150 ft., whereas drilling records indicate a depth of overburden to the top of the vein to be about 120 ft. The magnetic vectors will of course always point to a depth rather greater than the amount of overburden and only accumulated field experience will permit the geophysicist to make a fair estimate. The procedure for estimating

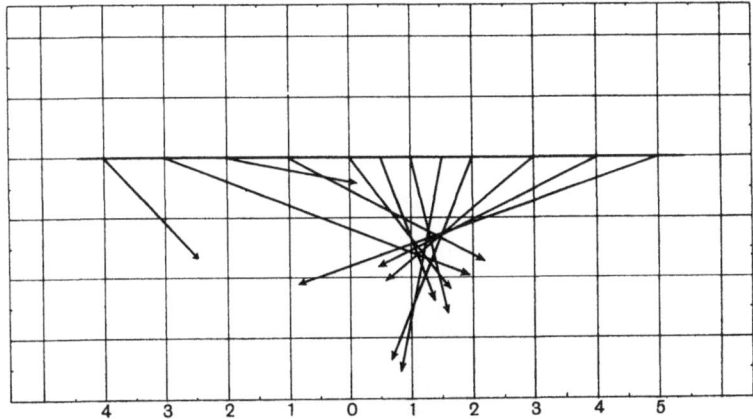

Fig. 24. The direction of the resultant magnetic field at different points on a line crossing a pyrrhotite vein 120 ft. below the point 100 ft. north of the zero station, indicating a depth of about 140 ft. to the north pole of the orebody.

depth of overburden from magnetic observations made on the surface will depend upon the form of the deposit. The accuracy of such determinations will naturally vary with the extent to which the shape of the body conforms to the mathematical premises on which the theory is based. Laboratory model experiments are helpful, but field experience, when checked by drilling, is invaluable for correct interpretation of the survey.

In the summer of 1930 the authors were asked if they could determine approximately the vertical extent of the main orebody at the Falconbridge Mine, and this inquiry led to the following investigations which are of a preliminary or tentative character.

When the dike or vein has been discovered, a few readings are taken at stations on either side of the centre of the vein in a line perpendicular to the strike. A rough, light, rigid wooden platform was made about 10 ft. high and 4 × 4 ft. in area on

Fig. 25. A light wooden platform, 4 × 4 ft., and 10 ft. high, on which the Askania vertical variometer is being read.

which an observer with the vertical Askania instrument could stand and take readings (Fig. 25). A series of readings at the same stations but 10 ft. above the ground are taken. If D is the depth of overburden, H the height of the platform above the ground, h the vertical thickness of the dike and r the ratio of the reading when the instrument is on the ground to the reading on the platform at a point over the centre of the vein, then it may be shown that*

$$(D + H) (D + H + h) = rD(D + h).$$

In obtaining r, the normal value of the vertical field is subtracted from each reading, so that only the deviations caused by the presence of the thin magnetic dike are used. For a first approximation, one may suppose that h is very much greater than D, in which case the above equation simplifies to

$$h = \frac{H}{r - 1}.$$

It is next possible to make an estimate of the vertical height h of the ore-body, but slight variations in the ratio r make large variations in h, and this means that an effort should be made to increase H, the height of the platform, in order to get reliable results.

LaCour (1930) has perfected small portable types of horizontal and vertical magnetometers. The horizontal instrument consists of a light rectangular magnet about 15 mm. long and 1 sq.mm. cross-section with a magnetic moment of about 1 c.g.s. unit. The magnet is suspended horizontally at the end of a quartz fibre inside a brass cylinder, provided with a glass window through which the position of the magnet may be ascertained by means of a spot of light reflected from a small mirror attached to the magnet and observed with a telescope provided with a cross-hair. The whole instrument fits on the top of a Kew magnetometer circle, and measurements of the value of H are made by rotating the entire instrument round the vertical axis until the image of the cross-hair reflected from the mirror is again in coincidence, so that a torsion of $2\pi + \alpha$ degrees is produced by the fibre. The restraining torque, $c(2\pi + \alpha)$, produced by the fibre is balanced by the opposing torque, $HM \sin \alpha$, of the magnet, where c is the torsion constant of the suspension, M the magnetic moment of the magnet and H the horizontal

* See Appendix III.

component of the earth's magnetic field. The small angles α are read at each station, and thus the values of H are computed in reference to some value at a given place. The instrument is provided with an ingenious clamping device that enables it to be transported readily without danger of breaking the delicate fibre.

The vertical magnetometer is similar to the Schmidt type made by Askania. It consists of a small magnet like that used in the horizontal instrument, provided with a quartz knife edge and resting horizontally on an agate support. An auxiliary magnet of known magnetic moment is moved up or down a vertical rod beneath the centre of the magnetometer until the small magnet is horizontal as determined by an eyepiece and the reflected image of a cross-hair from the mirror on the small magnet. The position of the auxiliary magnet is read and the change in Z may be computed. The axis of rotation of the small magnet is placed in the magnetic meridian. Dr E. H. Vestine informs us that this type of magnetometer has been used in field observations in China.

Many types of magnetic variometers manufactured by different firms are available, constructed on the same general principles as the Askania instruments. The essential features of all such instruments is a permanent magnet of high retentivity, capable of being deflected by small variations in the earth's magnetic field against a restoring force which may be either gravitational, magnetic or torsional. One recent portable Arvela magnetometer (L. A. Levant & Co., 1939) is a vertical type suitable for use in northern latitudes, which does not require orientation with respect to the magnetic meridian. Its sensitivity is not high, but in prospecting for pyrrhotite and other highly magnetic ores, the rapidity with which readings may be taken and the ease of transport offer special advantages.

METHODS OF ESTIMATING DEPTH OF OVERBURDEN

The actual procedure in estimating the depth of overburden from magnetic observations made on the surface will vary with the form of the deposit, and any theoretical discussion will apply only so far as the form of the deposit conforms to the mathematical premises on which the theory is based. Since the shapes of natural deposits in the field cannot be expressed rigorously in mathematical form, the best that can be done is to assume some simple form and deduce results. Fortunately many small variations in form produce only slight changes in the estimated depth of overburden, therefore the methods have a definite practical value.

For theoretical treatment, magnetic-ore deposits may be divided roughly into three classes, depending upon whether the magnetic anomalies may be considered to arise from (1) a single magnetic pole, (2) a dike, or (3) a magnetic doublet. In general, it may be stated that when the magnetic anomalies as observed on the surface are due only to one pole of the body, the other pole being so remote that its effect may be neglected, the resultant anomaly F at any point will vary as the inverse square of the distance r of the point from the pole. When the deposit occurs in the form of a long thin dike, the resultant maximum anomaly will vary inversely as the depth of the dike below the point of observation,* and when the deposit is in the shape of a lens, so that both poles have an appreciable effect on the surface, the resultant maximum anomaly will be similar to that caused by a magnetic doublet and will vary as the inverse cube of the distance to the centre of the deposit.

The particular class under which a deposit falls may be deduced from the results of the magnetic survey. The second class is readily found, but there may be difficulty in deciding between the first and third. Isodynamic contours of both vertical and horizontal anomalies will often be a good guide in making a correct decision. The following simple methods of estimating the amount of overburden in the three cases mentioned above have been found to be useful by the authors in the field.

(a) *Magnetic anomalies arising from a single pole*

Such cases occur when the deposit is near the surface and extends nearly vertically to so great a depth that the other pole is remote.

* See Appendix III.

The position of the pole is found from the survey. The following methods may then be adopted.

Method 1. If ΔZ_m be the maximum value of the vertical anomaly over such a deposit and x the distance measured along the horizontal surface to a point P, where the vertical anomaly $\Delta Z_p = 0 \cdot 353 \Delta Z_m$, the depth of overburden will be x. In practice the contours of equi-anomalies will not be circles and the mean distance in different directions is taken as x. The above relation follows from the fact that if the line joining the point P to the pole makes an angle α with the vertical, then

$$\Delta Z_p / \Delta Z_m = \cos^3 \alpha \quad \text{and} \quad x = d \tan \alpha,$$

where d is the depth of overburden. If $\alpha = 45°$, $\cos^3 \alpha = 0 \cdot 353$.

Method 2. The horizontal ΔH and vertical ΔZ anomalies near the point M of maximum vertical anomaly are determined for various points. These may be combined vectorially at each point, corrections being made for any difference in sensitivity of the instruments, and the direction of the resultant field due to the ore-body found. When the vector diagram is drawn to scale, the depth of the pole may be determined.

Method 3. Suppose ΔZ_m is the value of the maximum vertical anomaly on the surface. Erect a platform and measure the value of the vertical anomaly ΔZ_h on the platform at a vertical height h above the ground at the same point as ΔZ_m was determined. It follows from the inverse square law of force from a single pole that $\dfrac{\Delta Z_m}{\Delta Z_h} = \dfrac{(h+d)^2}{d^2}$, where d is the depth to the pole. Hence d may be found from

$$d = \frac{h}{\sqrt{(\Delta Z_m / \Delta Z_h) - 1}}.$$

(b) *Deposit in the form of a long narrow dike*

This form of deposit is revealed both from the vertical and horizontal anomalies. The strike of the dike may be ascertained and its magnetic axis. A survey is then made along a line at right angles to the strike, and the vertical and horizontal anomalies are determined at small intervals along this line, so as to obtain a smooth curve when the anomalies are plotted as profiles across the dike.

Method 1. If the dike is vertical or nearly so and extends to such a depth that the effect of the other pole may be neglected, it may be shown that the value of the maximum vertical anomaly ΔZ_m at M on the surface will be*

$$\Delta Z_m = \frac{2mb}{d},$$

where m is the magnetic pole strength per unit area, b the width of the dike and d the depth from the surface to the top of the dike. Then the vertical anomaly due to the dike at a point P at a distance x from M along the line at right angles to the strike will be given by

$$\Delta Z_p = \frac{2mbd}{r^2} = \frac{2mbd}{x^2 + d^2},$$

where r is the distance from P to a point on the ore-body vertically below M. If the point P be so chosen that $\Delta Z_p = \frac{1}{2}\Delta Z_m$, then it follows that $d = x$. Thus the depth of the overburden is found by plotting the vertical anomalies as a profile along the line perpendicular to the axis of the strike of the dike and finding the distance at which the anomaly is one-half the maximum value.

Method 2. The depth of overburden may be determined by drawing the vector diagram of resultant anomalies near the point of maximum vertical anomaly. The procedure is the same as that given in case (a), method 2.

Method 3. This method is similar to that described in case (a), method 3, but the law of force in the case of a dike is different. If ΔZ_m is the value of the maximum vertical anomaly on the surface and ΔZ_h the value on the platform at a height h above this point, it follows that

$$\frac{\Delta Z_m}{\Delta Z_h} = \frac{2mb}{d} \bigg/ \frac{2mb}{h+d} = \frac{h+d}{d}$$

or

$$d = \frac{h \,\Delta Z_h}{\Delta Z_m - \Delta Z_h}.$$

* See Appendix III.

Method 4. When the profiles of the vertical and horizontal anomalies along the traverse perpendicular to the strike of the dike are examined, a point P may be found on this line at a distance x from the point M of maximum vertical anomaly, at which the horizontal anomaly will be equal to the vertical. If the two profiles be plotted on the same axis, the point where the horizontal anomaly cuts the vertical anomaly will be above the point P. The distance x from this point to M will be equal to the depth of overburden d.

How well field practice agrees with theory may be illustrated by applying these methods to the particular case shown in Fig. 21. The horizontal scale is 100 ft. to a division. The maximum vertical anomaly at station 1N is $\Delta Z_m = 63 \cdot 0 - 27 \cdot 0 = 36$ divisions. Hence $\Delta Z_p = \frac{1}{2}\Delta Z_m = 18$ divisions, which corresponds to a reading of $27 \cdot 0 + 18 = 45 \cdot 0$. The mean width of the vertical anomaly at this reading is about 218 ft., so that the distance from the point of maximum vertical anomaly to the point where the anomaly is half this value is approximately 109 ft. Using the vector method, a value of approximately 140 ft. was obtained. A survey made along an adjacent line crossing the dike gave a maximum anomaly of 33·2 divisions on the ground, and on a platform 10 ft. high the maximum anomaly was 30·6 divisions. Applying method 3, one obtains a depth of overburden

$$d = \frac{10 \times 30 \cdot 6}{33 \cdot 2 - 30 \cdot 6} = 118 \text{ ft.}$$

The fourth method, where $\Delta Z = \Delta H$, gives a value of approximately 120 ft. The actual depth of overburden as determined by diamond drilling was 120 ft., in very good agreement with the mean of the values obtained by these four methods. These readings were made on the Falconbridge property, and it is not to be expected that field observations will usually give such good results.

(c) *Deposit in the form of a lens*

When the deposit is a lens, both poles of the equivalent magnet may produce appreciable anomalies on the surface. By plotting contours of isodynamic lines, the position of the maximum anomaly ΔZ_m may be found. The application of this method will be most accurate when the axis of the equivalent magnet is either vertical or horizontal, and the inverse cube law can be applied when the anomaly on a platform at a height h above the point of maximum anomaly is determined. If this value is ΔZ_h, then

$$\frac{\Delta Z_m}{\Delta Z_h} = \frac{(h+d)^3}{d^3},$$

from which

$$d = \frac{h}{\sqrt[3]{(\Delta Z_m/\Delta Z_h)} - 1},$$

where d is the depth to the centre of the lens.

THE COMPASS VARIOMETER

The compass variometer was designed for measuring variations in H, the horizontal component of the earth's magnetic field. Many forms have been developed; one portable type was the result of a series of experiments by the Carnegie Institution of Washington (Bauer *et al.* 1926, pp. 339-57). It is shown in diagram in Fig. 26. The apparatus consists essentially of two equal magnets, which are circular disks so magnetized that the poles are at opposite ends of a diameter. These circular magnets are supported on pivots, the one vertically above the other at a distance e cm. apart. The magnetic moment of each magnet is M and the magnets are placed on their pivots so that like poles are at the same end. Hence the magnets repel one another but are pulled together by the force of the earth's field. The sensitiveness of the scheme lies in the action of these balanced fields. The two needles will stand at an angle ψ to each other, ψ being the angle between the two imaginary vertical planes through the magnetic axes of the two magnetic needles. The stronger H is, the smaller will the angle ψ be, and the more nearly will the axes of the two magnets be in the same direction. Also the nearer the two magnets are together, that is, the smaller e, the greater the repulsion between them, so that the angle ψ for two given magnets in a given field may be altered by increasing or decreasing the distance e. The variation of ψ with H and e is obtained from the following equation:

$$H = \frac{2MD}{e^3} \cos \frac{\psi}{2},$$

where D is a constant of the instrument depending upon the distribution of magnetism in the two magnets. D is found experimentally by placing the compass variometer in a known uniform field such as that of a Helmholtz-Gaugain coil arrangement, and thus measuring ψ, while M and e are known. The values of D for two instruments used by Bauer, Peters and Fleming were found to be 0·935 and 0·928 respectively. The change in H may be measured in two ways with this type of variometer. In one method the distance apart of the magnets e is kept constant and the angle ψ is observed at each station. This angle is easily measured with the help of a large magnifying glass which serves as a cover for the cylindrical box containing the magnets and their supports. To each magnet a quartz pointer is attached, and the position of each pointer is read from a circular scale provided with a plane mirror beneath to avoid parallax. This is shown in the figure. The pointers are usually coloured to render them readily visible, and also to identify which pointer belongs to each magnet. The north pole of the upper needle is painted red, and the south pole green. Both ends of the lower pointer are painted black. The bottom of the cylindrical container is closed with a piece of ground glass, so that a lamp placed underneath will illuminate the interior sufficiently to allow the instrument to be read at night. To reduce the time required for the magnets to come to rest after the instrument has been moved to a new station, the interior of the container is filled with gasoline or acetylene tetrabromide, a liquid which quickly damps the oscillations. The sensitivity of the apparatus may be determined from the equation

$$\Delta H = -\frac{DM}{e^3} \sin \frac{\psi}{2} \Delta\psi.$$

With an instrument of this type, a deflexion of one degree is equivalent to a change in H of about 21γ.

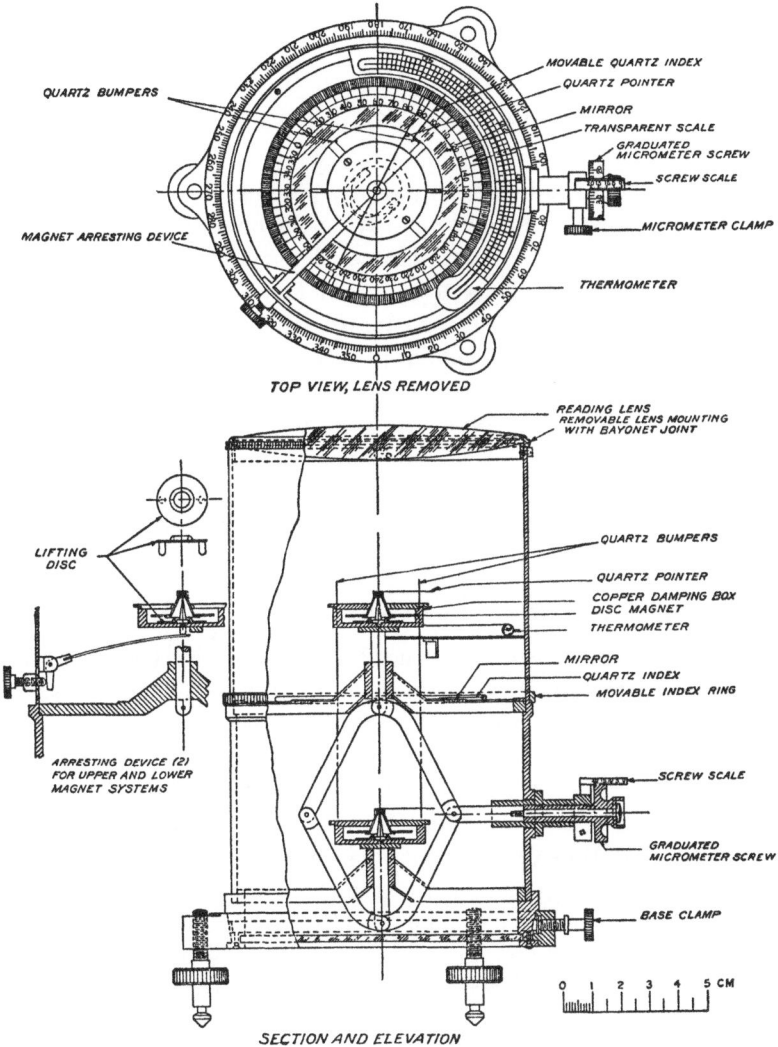

QUARTZ BUMPERS

MOVABLE QUARTZ INDEX
QUARTZ POINTER
MIRROR
TRANSPARENT SCALE
GRADUATED MICROMETER SCREW
SCREW SCALE
MICROMETER CLAMP

MAGNET ARRESTING DEVICE

THERMOMETER

TOP VIEW, LENS REMOVED

READING LENS
REMOVABLE LENS MOUNTING WITH BAYONET JOINT

LIFTING DISC

QUARTZ BUMPERS
QUARTZ POINTER
COPPER DAMPING BOX
DISC MAGNET
THERMOMETER
MIRROR
QUARTZ INDEX
MOVABLE INDEX RING

ARRESTING DEVICE (2)
FOR UPPER AND LOWER MAGNET SYSTEMS

SCREW SCALE

GRADUATED MICROMETER SCREW

BASE CLAMP

0 1 2 3 4 5 CM

SECTION AND ELEVATION

Fig. 26. Compass variometer. (Department of Terrestrial Magnetism, Carnegie Institution, Washington.)

The second way of using the instrument is to keep a fixed angle ψ between the two magnets by altering their distance e apart as the field changes. In this case the sensitivity is determined by the equation

$$\Delta H = -\frac{6MD}{e^4} \cos \frac{\psi}{2} \Delta e.$$

A calibration curve is determined by placing the instrument in a series of known fields and measuring the changes in e required to keep the angle ψ constant. The sensitivity for ordinary values of H ($= 0 \cdot 18$ c.g.s. unit) will be about 80γ for one division.

This type of variometer has many advantages for field use. It does not have to be levelled, it is light to carry (8–10 lb.), it is easily read and its sensitivity is quickly and readily altered in the field. It does not appear to have been applied in the field for geophysical work as yet, but it has been used at sea, and for making surveys around ships. Fig. 27 shows the results of a survey made at Sandy's Parish, Bermuda, with such an instrument.

THE EARTH-INDUCTOR METHOD

The earth-inductor was first used for determining the strength of the terrestrial magnetic field at any point by Weber in 1837, and since that time instruments have been improved and developed to such an extent that the modern type is perhaps the most accurate of all types of inclinators for determining the dip. A photograph of the form of earth-inductor used by the Carnegie Institution of Washington for their work on terrestrial magnetism both on land and over sea is shown in Fig. 28. Such an instrument may equally well be used for geophysical exploration work. It consists of a coil of about two or three thousand turns of no. 30 insulated copper wire wound on a hard rubber spool, 24 mm. thick with an inner diameter of 26 mm. and an outer diameter of 74 mm. The ends of the coil are brought to a commutator on the shaft by which the coil is rotated. The phosphor-bronze axle, which is along a diameter of the coil, is supported on double gimbles in such a way that the coil may be set with its axis in any direction. Screws for clamping the gimbles when in position are provided, and micrometer attachments for facilitating slow-motion adjustments permit accurate setting of the coil. The coil is rotated on its axle at a constant rate by means

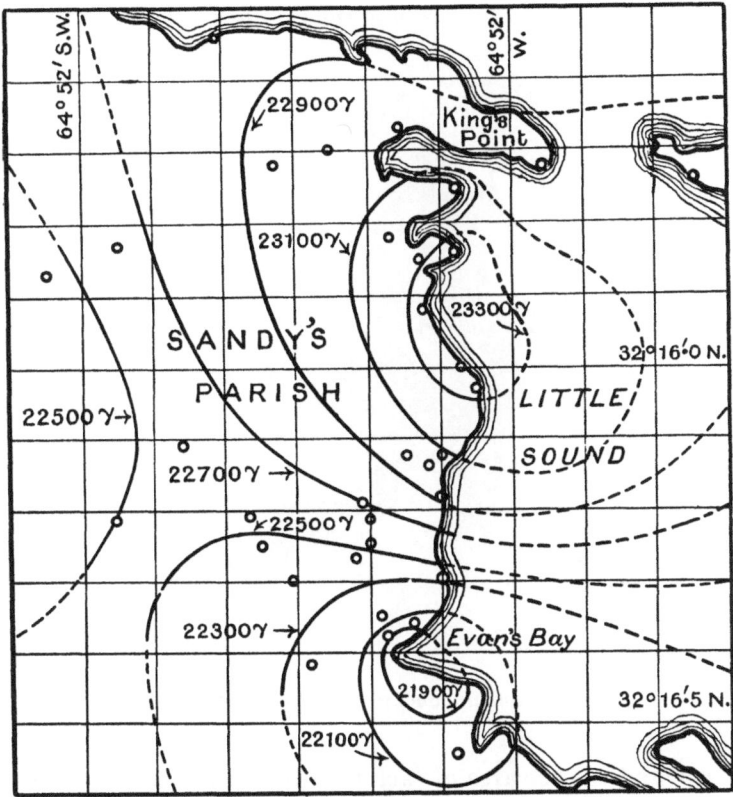

Fig. 27. Magnetic survey, Bermuda, showing horizontal
intensities in gammas, with isodynamic lines.

of a handle fitted with gears and a flexible transmission shaft, which is connected to the coil by the universal type of joint and gearing shown in the illustration. The current is measured with a sensitive galvanometer of either the Kelvin or moving-coil type, such as a unipivot microammeter, and when absolute values are desired, a ballistic galvanometer may be used, though this is not so practical in field work.

The theory of the earth-inductor is a simple illustration of Faraday's laws of induction. Let R be the resultant field strength and A the effective area of the turns of wire on the spool, so that $A = na$, where n is the number of turns of wire in

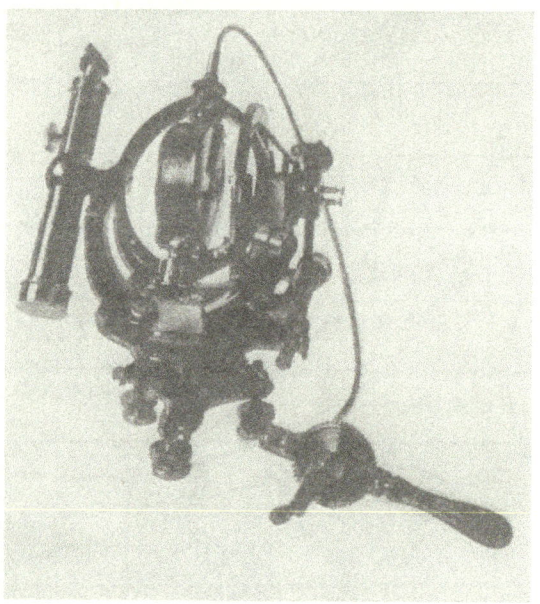

Fig. 28. Earth-inductor. (Carnegie Institution.)

the coil and a is the area of each turn. When the coil, with commutators, is rotated so that the axle of the rotating coil is perpendicular to the direction of R, then the charge which will pass through the ballistic galvanometer for a single turn is $Q = \dfrac{2RA}{r}$, where r is the total resistance of the circuit, i.e. of the coil and galvanometer. In the field and at sea it is more convenient to use a galvanometer for measuring the current. The sensitivity of the galvanometer should be about 1 mm. scale division per

10^{-8} amp. The current induced in the coil will be proportional
to the speed of rotation, but for practical purposes the rotation
can be kept sufficiently constant by turning with one hand and
using a watch or metronome for indicating the time. Of course
a driving mechanism and governor may be used, but they must
all be constructed of non-magnetic materials.

The inclination is determined by rotating the coil with its
axis of rotation in the magnetic meridian (determined usually
with a compass needle) at various inclinations, until a position
is found when there is no current recorded by the galvanometer.
In practice the coil is rotated first in the one direction and then
in the other at measured angles on opposite sides of the line of
the magnetic inclination, so as to get an accurate reading of the
angle of dip. If the dip is just half-way between the two
positions in which the coil was rotated, the readings obtained on
the galvanometer will be equal. If the readings differ slightly,
the proper angle may be determined from the readings. When
the magnetic dip is found in this way, the axis of rotation of the
coil is turned through 90° and the coil rotated at the usual constant rate. The current indicated by the galvanometer will then
be proportional to the strength of the earth's field R. For prospecting work, either R or the angle of dip v or both may be
determined.

The inclinator may be used as a null instrument in the way
described, but another method is to rotate the coil first with its
axis horizontal and along the magnetic meridian, which gives
a current proportional to the vertical component Z, and then
with its axis vertical which gives a current proportional to H.
From these two readings the resultant field may be calculated
as indicated on p. 17 and the angle of dip v determined.

In such instruments there are, of course, sources of error and
fine adjustments which must be considered. The position of the
coil at which commutation takes place, the width or space
separating the commutator segments and the consequent time
interval of commutation, the position of the brushes and thermo-
electric effects all require attention and adjustment. The extent of
these errors will naturally depend upon the type of instrument

considered. Such sources of error may usually be either eliminated,* calibrated or compensated in special ways depending
on the apparatus. The use of slip rings instead of a commutator, thus using the alternating current produced by the
coil rotating in the earth's magnetic field, may be advantageous in rough preliminary work. The inclinator is then used
with telephones as a null instrument for detecting the dip. When
there is no current in the telephones, then the axle of the rotating
coil lies along the line of magnetic inclination. Intensity measurements may be made as already indicated, but an amplifier with
vacuum thermocouple and microammeter are used as the
measuring instruments.

For accurate work corrections for diurnal variations and magnetic storms must be made in the manner referred to on p. 31.
The routine for such surveys then becomes rather involved, as
all sources of error in the instrument must be corrected. For
a description of the method of making observations with an
earth-inductor of the type described, the reader is referred to
the papers mentioned above.*

MAGNETIC VECTOR METHODS

Jenny (1935) has pointed out the advantage of using a vector
method of plotting the magnetic anomalies. From the declination at any station, the north-south horizontal anomaly, the
east-west anomaly and the vertical anomaly, a resultant magnetic anomaly may be formed which will be a vector in space.
The size and direction of this vector is then plotted at each
point and the map thus formed will consist of a number of
vector triangles. From such a map much more regional and
local information may be obtained than from maps of horizontal
and vertical magnetic anomalies alone.

* Details of the method of making corrections for one type of earth-
inductor are given by Dorsey (1913).

THE MAGNETIC AIRBORNE DETECTOR

This instrument, developed in the United States (Balsley, 1946) during the Second World War, is now being applied in the United States, Canada, Australia and elsewhere for making rapid regional geophysical surveys from a low-flying aircraft. The instrument consists essentially of three magnetometer elements of the fluxgate type, two of which maintain the third in the direction of the earth's total or *resultant* magnetic field. The variations in intensity of this field are traced on a moving strip of paper by a recording milliammeter of the Esterline Angus type, which may be adjusted to have a sensitivity of 100γ for a full-scale deflexion. The detecting unit is towed sufficiently far behind the plane so as not to be affected by the magnetic field of the aircraft. A picture of the instrument in flight is shown in Fig. 29. The magnetometer itself, showing the orienting gimbals and slip-ring contacts, may be seen in Fig. 30. The details of its construction and the electronic recording equipment with theory of operation and calibration will be found in an article by R. Bailey (1948). The general principle of the apparatus is that the variation in the value of the total resultant earth's magnetic field is continuously recorded as the plane flies at a constant recorded height along a traverse. A continuous photograph of the terrain is taken as the plane flies, and this is tied in with the magnetic record, thus providing easy identification of any strong anomaly. Detailed ground surveys can then be made over these regions.

Many thousands of miles of traverses have been made with such equipment both in the United States and Canada. The magnetic contours drawn from a series of profiles obtained in a survey in eastern Pennsylvania (Jensen, 1951) are shown in Fig. 31. A survey made by the Geophysics Section, Geological Survey of Canada, showing isomagnetic lines compiled from flights at 1000 ft. above ground-level, with airborne magnetometer along the straight lines shown, is given in Fig. 32. The anomalies are due largely to variations in the composition of the rocks making up the basement, but in some cases may be due to

Fig. 29. Magnetic airborne detector in flight.

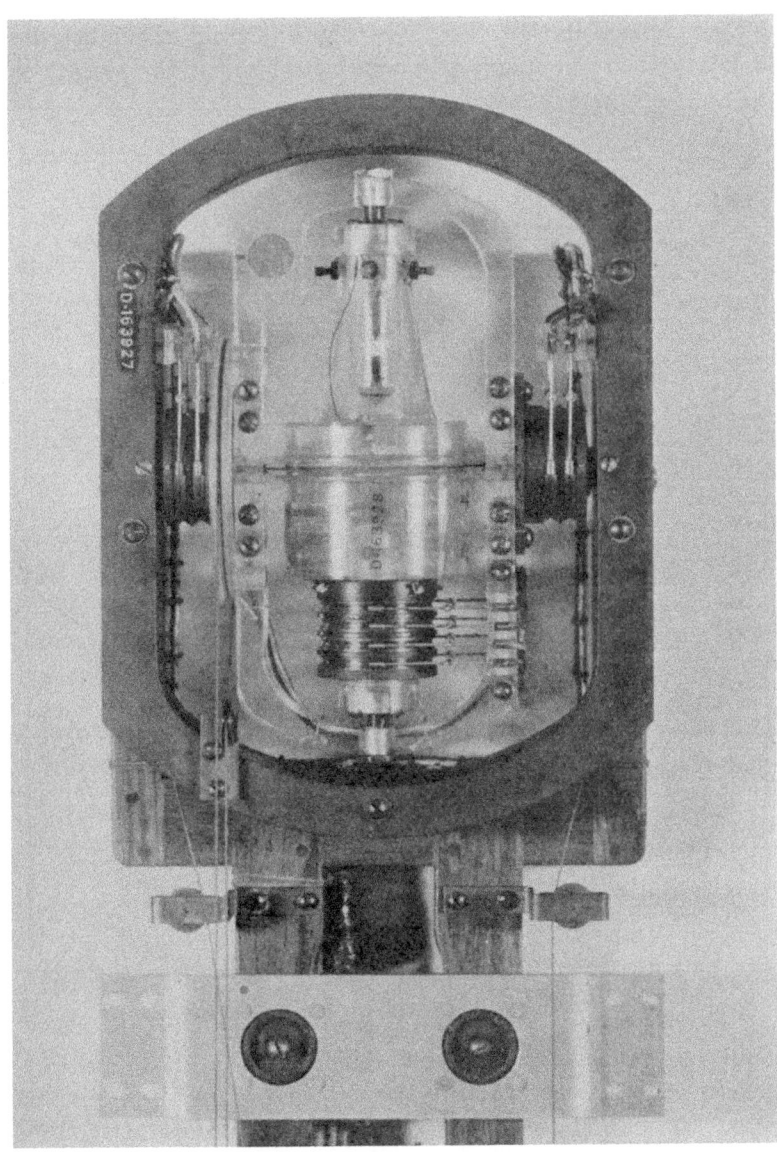

Fig. 30. The airborne magnetometer.

changes in altitude of the basement according to members of the geophysics section. They also state 'strong anomalies are probably due to an increased magnetic content in the rocks, but

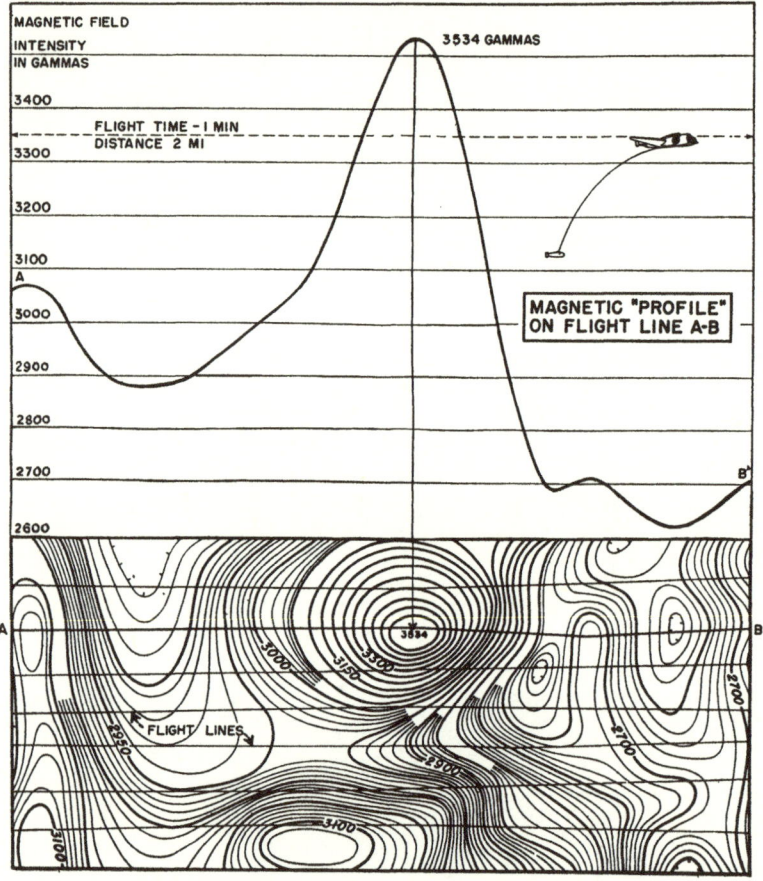

Fig. 31. Magnetic airborne survey (Pennsylvania).

small anomalies may be due to either of the above causes'. Results of flights at two different altitudes along the same line might assist in the interpretation using methods described on pp. 56–61.

Earth-inductors have been receiving more attention in recent

years, and new varieties with more simple design and operating
mechanism are appearing in the literature (Jakosky, 1950,
p. 100). The mining engineer, however, still retains his magnetic

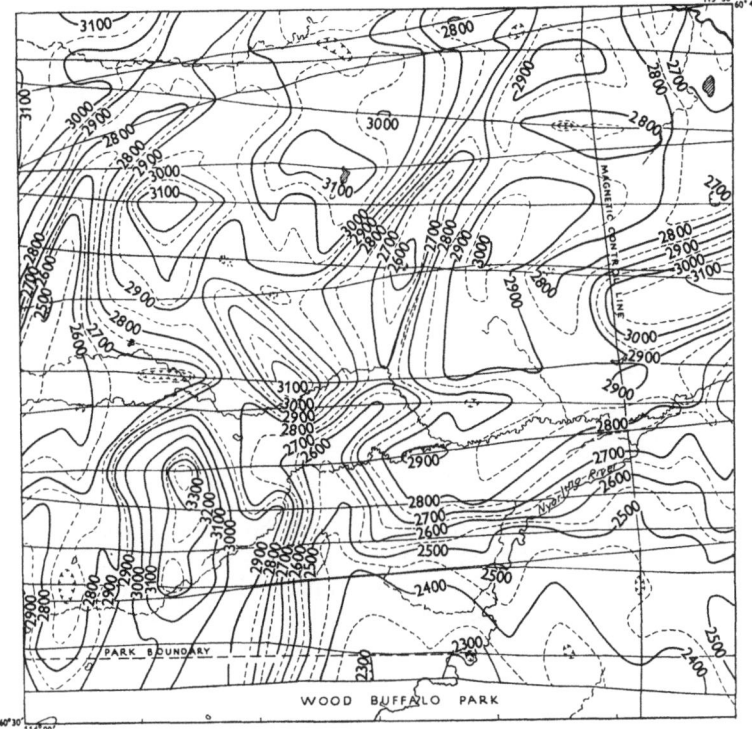

Fig. 32. Magnetic airborne survey (N.W.T.).

needles and variometers which have proved of such great value
in the past. Magnetic methods have become an established
adjunct to prospecting for magnetic ores and for surveying salt
domes and oil structures. But there are many valuable minerals
which possess such weak magnetic properties that these methods
are entirely useless for their detection. It is for such ores, which
constitute indeed the majority, that the more recent and different
geophysical methods have been evolved and developed from
known principles of physics. These methods will now be discussed and described in the following chapters.

PROBLEMS

1. The following readings were taken with a vertical and horizontal magnetic variometer at stations 25 ft. apart on a line at right angles to the strike of a pyrrhotite-nickel vein in the Sudbury area. The sensitivities of the two instruments were adjusted to be the same. The base station was at the south end of the line. Find the position of the vein, its approximate depth and the amount of overburden by three methods.

Station	Vertical	Horizontal	Station	Vertical	Horizontal
S. 0	20·1	21·1	10½	27·0	21·6
1	20·0	21·1	11	26·3	19·0
2	20·1	21·0	11½	25·0	17·9
3	20·0	20·7	12	23·3	17·6
4	19·9	20·0	13	21·5	18·0
5	19·8	19·4	14	20·7	18·4
6	19·4	19·5	15	20·3	18·9
7	19·0	20·1	16	20·1	19·4
8	19·0	21·5	17	19·9	20·0
9	20·6	23·4	18	20·0	20·5
9½	24·0	23·6	19	19·9	20·9
10	26·3	23·0	N. 20	20·0	21·0

2. A magnetic survey was made using the Askania vertical and magnetic variometers at stations along a traverse running 16° east of north on the Falconbridge mine property. The sensitivity of the horizontal instrument was twice that of the vertical variometer. Locate the vein, its probable direction of dip and estimate the amount of overburden.

Station (ft.)	Vertical	Horizontal	Station (ft.)	Vertical	Horizontal
0	− 12·1	29·0	700	− 6·7	10·6
100	− 12·2	33·8	800	− 10·8	17·5
200	− 11·4	41·8	900	− 13·0	20·2
300	− 7·8	48·0	1000	− 12·0	20·4
400	− 5·2	53·0	1100	− 13·2	23·5
450	− 17·8	47·1	1200	− 13·1	24·0
500	+ 22·9	31·8	1300	− 13·0	24·9
550	+ 17·2	16·2	1400	− 12·6	28·6
600	+ 7·0	10·4	1500	− 12·2	28·8

3. A preliminary magnetic survey indicated that there is a long magnetic vein anomaly beneath overburden extending in an east-west direction. To estimate the depth of overburden and dip of the vein, a survey was made with a vertical and horizontal magnetic

variometer along a line at right angles to the strike of the vein. The anomalies in gammas at stations 25 ft. apart are given below. What conclusions may be deduced from the readings?

Station	ΔH	ΔZ	Station	ΔH	ΔZ
0	0	0	14	−1200	5500
1	60	30	15	−2200	2700
2	150	60	16	−3100	1700
3	210	80	17	−2800	560
4	260	100	18	−2150	310
5	400	125	19	−1600	250
6	610	155	20	− 850	190
7	800	220	21	− 600	160
8	1600	290	22	− 400	120
9	2050	400	23	− 300	100
10	2650	700	24	− 180	90
11	2900	1720	25	− 100	80
12	2500	3240	26	− 60	50
13	1000	5400	27	− 10	30
$13\frac{1}{2}$	0	5850	28	0	0

4. The following readings were taken with a Tiberg dip needle along a line across a dike. Estimate the depth of overburden.

Station (ft.)	Dip in degrees	Station (ft.)	Dip in degrees
0	2·2	450	5·0
50	2·3	500	4·25
100	2·8	550	3·60
150	3·4	600	3·15
200	4·0	650	2·71
250	5·0	700	2·5
300	5·75	750	2·25
350	5·9	800	2·12
400	5·7	850	2·1

5. Chromite is known to occur in serpentine in the Colerine district of Quebec. A magnetic analysis of some specimens indicated that the ore contained less magnetite than the serpentine. The following readings were taken along a line at stations 25 ft. apart on the Reed Belanger property with a vertical magnetic variometer. Determine the depth of overburden.

Station	Reading	Station	Reading	Station	Reading
1	11·0	6	− 0·4	11	9·3
2	11·0	7	− 2·0	12	11·2
3	7·5	8	1·8	13	11·1
4	5·0	9	7·8	14	11·0
5	− 0·6	10	9·2	15	11·1

6. On a claim in the Little Long Lac area, the following readings were taken along a line at stations 25 ft apart with a vertical magnetic variometer. Determine the position of the contact and estimate the overburden.

Station	Reading (divisions)	Station	Reading (divisions)
0	12·2	15	13·2
1	12·3	16	35·1
2	12·3	17	25·8
3	12·0	18	23·7
4	10·0	19	63·9
5	10·3	20	19·8
6	11·0	21	9·8
7	11·1	22	9·1
8	10·8	23	8·0
9	9·8	24	0
10	11·7	25	−12·2
11	18·1	26	3·2
12	15·1	27	10·0
13	10·0	28	10·7
14	7·2		

7. The table below gives the results of a magnetic survey across a dike with the object of estimating the dip of the dike and the amount of overburden. The anomalies in vertical and horizontal intensities at the stations 40 ft. apart are given in gammas. Estimate the dip and the depth of overburden.

Station	ΔZ	ΔH	Station	ΔZ	ΔH	Station	ΔZ	ΔH
0	42	41	14	380	535	26	0	−245
1	38	38	15	700	540	27	−15	−232
2	50	39	16	1120	460	28	−20	−222
3	76	45	16½	1255	400	29	−27	−220
4	84	42	17	1370	230	30	−20	−220
5	82	50	17½	1390	−30	31	−10	−208
6	63	54	18	1280	−250	32	0	−180
7	50	70	19	1020	−482	33	15	−160
8	42	100	20	710	−580	34	12	−135
9	48	148	21	480	−545	35	5	−110
10	52	192	22	300	−515	36	10	−90
11	86	250	23	138	−420	37	8	−70
12	141	314	24	90	−340	38	0	−40
13	320	425	25	20	−280			

CHAPTER III

ELECTRICAL METHODS

INTRODUCTION

It is well known that metals and many metallic ores conduct electricity many thousands of times as well as such rocks as gneiss, porphyry and quartz. Most sulphides, especially those which possess a high metallic lustre, are excellent conductors and have a low specific resistance. The outstanding exception is zincblende or sphalerite (mostly ZnS), which is a poor conductor. The guiding principle underlying all the electrical methods of locating ore deposits depends upon this marked difference in electrical conductivity between the metalliferous minerals and the surrounding rocks. Just as the magnetic properties of iron ores are used as an indication of their presence, so the low resistance of such minerals as chalcopyrite, galena and pyrites may serve as a means of their detection.

In order to find regions or deposits of good conductivity, it is usually necessary to introduce electrical currents artificially into the ground and to trace their behaviour. From the knowledge thus acquired as to how the current distributes itself under the various arrangements by which it is introduced into the ground, the effects of local regions of good conducting ore are noted and the presence of the ore-bodies revealed. When the resistance between metal stakes driven into the ground is measured, the main resistance is near the stakes; the vast mass of the earth between offers but little resistance.

In some cases, especially when looking for sulphide ores, chemical action is continually going on, owing to the oxidation of the sulphides due to the action of surface or rain water which seeps through the ground. This chemical action sets up electrical currents in much the same manner as currents are produced by voltaic cells. Such natural currents may often be detected and traced on the surface of the ground, revealing to the geophysicist the presence of hidden sulphides many feet below. Natural currents due to ore-bodies are strictly a local effect and have high values compared with the normal earth currents, which are large-scale phenomena, accentuated in times of magnetic storms or auroral displays. Earth currents have been the subject of numerous investigations in various parts of the world, and as a result of these experiments, a simple and direct method of locating conductors and of determining their depth has been evolved, which will be described on pp. 112 et seq. of this chapter.

Several well-tested methods of locating ore have been developed, which depend upon the factors mentioned above. In this chapter the principles underlying each system will be given with full descriptions of the various types of apparatus convenient for use in the field. Typical results of surveys actually made in the way described will serve as an indication of what may be expected in each case. In such field work unexpected difficulties arise, both instrumental and natural, and these will be discussed in their appropriate places.

THE SELF-POTENTIAL METHOD

The natural current, or self-potential, method of locating ore is unique among the electrical methods, in that no currents are artificially introduced into the ground. The ore itself produces the supply of electrical energy by means of which its detection is made. Let us consider the simple case of a vein of sulphide ore BC, as in Fig. 33, covered with an overburden of from 20 to 50 ft., for at such a depth a sulphide ore may usually be located with assurance. The rain and surface water sinking into the ground reaches the sulphides and brings about oxidation. The upper portions of the vein are usually more active chemically than the lower or deeper levels. As a result of such chemical action, differences in electrical potential exist and hence the term 'self-potential' is often applied to this method. The upper portion B, where the activity is greatest, becomes the negative electrode, the lower C becomes the positive electrode. As a result of this difference in potential, which may amount to more than a half volt, currents of electricity flow down through the good conducting ore and return through the surrounding poorly conducting ground, making a complete circuit as in a battery. The return currents will spread out to large distances from the ore, owing to the high resistivity of the ground. The approximate lines of flow are indicated by the dotted lines in Fig. 33. On the surface of the earth it will be found that the currents tend to flow into the region D, which is for this reason called a negative

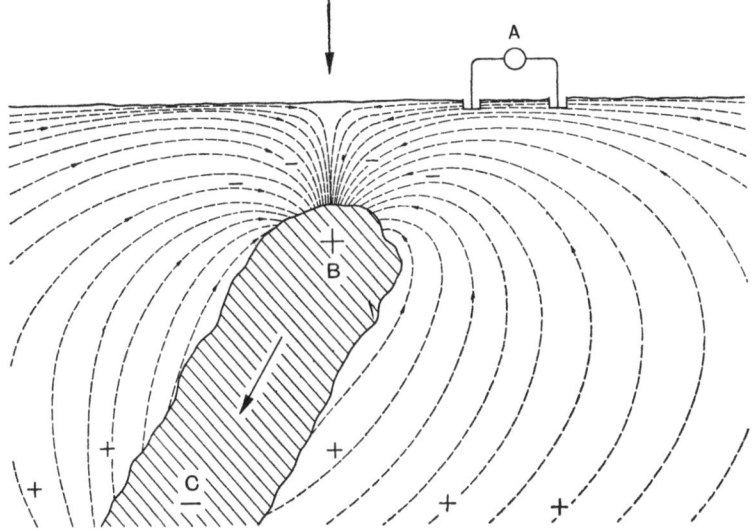

Fig. 33. A diagram illustrating the natural currents produced by an oxidizing sulphide vein.

centre. In Fig. 34 the direction of the lines of current flow are shown by lines with arrows. As a result of the flow of current into D, there will exist imaginary lines, always at right angles to the direction of current flow, and these 'equipotentials' are represented by the closed curves in Fig. 34. Hence the negative centre may be found by determining in the field either the equipotential lines or the direction and magnitude of the currents. The oxidizing sulphide deposit will then usually be discovered directly below the negative centre as located by such experiments.

It is important to note that over graphite deposits a positive centre will exist instead of the negative centre which occurs over oxidizing ores. This difference is sometimes of value in distinguishing a sulphide ore from a deposit of graphite.

The question at once arises as to how the direction and magnitude of the natural currents, produced near an oxidizing sulphide body, may most conveniently be detected and measured on the surface. As long ago as 1830 Fox worked inside a Cornish

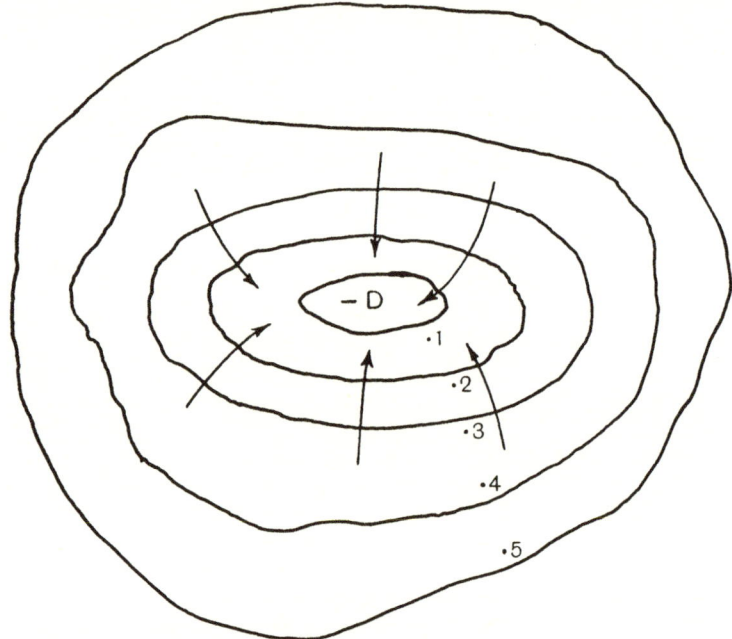

Fig. 34. Plan showing the direction of the currents flowing towards a negative centre, and the equipotential lines.

mine and discovered copper sulphide deposits, using large copper plates placed against the rock as his electrodes, and a galvanometer for measuring the natural current. Even with such a simple arrangement he had considerable success in locating the deposits, an account of which he gave in the *Philosophical Transactions of the Royal Society* (1830). It might easily be supposed that two iron stakes driven well into the ground, several feet apart, would serve as efficient electrodes, so that when joined by insulated wires in series with a sensitive micro-ammeter, or similar galvanometer, they would be sufficient to detect the natural currents. But a very serious error arises when two such metal rods are used as electrodes. Chemical action at once commences at the surface of such electrodes, especially when the ground is moist, and there result large fluctuating potentials between the two stakes, due to what is termed in elec-trolysis 'polarization'. Such polarization effectually masks the more feeble natural currents that are produced by the ore. Hence it is desirable to employ some form of non-polarizable electrodes at which there will be no disturbing influences when they are placed in or on the ground. For this purpose, porous pot electrodes have been used with considerable success. Thus Carl Barus used a zinc plate im-mersed in a solution of zinc sulphate which was enclosed in a semipermeable animal membrane to prevent the solution flowing away. With these electrodes Barus made a survey of the Comstock Lode in 1882. The metal electrode in such cells is always in contact with a concentrated solution of its salt, so that any currents of electricity passing through the cell will either deposit or remove some of the metal from the electrode, the solution and metal remaining in uniform con-centration and therefore in voltaic equi-

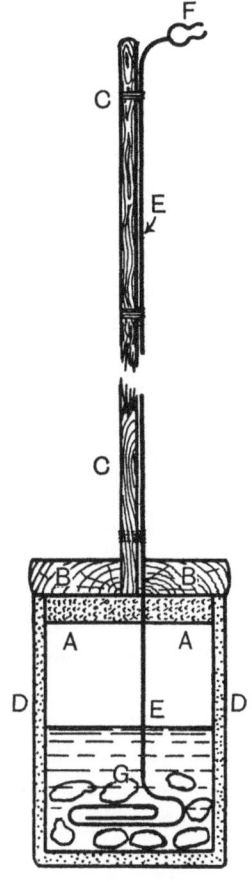

Fig. 35. A non-polarizable porous pot electrode, con-taining copper sulphate solution and crystals.

librium. By an arrangement of this kind, all the spurious effects of polarization are removed, and yet the semipermeable pot with its electrode assumes the potential of the ground. For example, if one of the two non-polarizable electrodes is placed

in a galvanized iron pail containing some water, then a difference of potential is at once evident, in fact the pail is detected
in much the same way as the ore-body! The explanation of non-
polarizable cells as considered by the physical-chemist, in terms
of the solution pressures of the various ions, may be found in
text-books on the subject (for example, Nernst, *Theoretical
Chemistry*, 6th ed. p. 771).

To construct a non-polarizable electrode which will be suitable
for field use, a form similar to that shown in Fig. 35 has been
found useful and is simple to make. An ordinary unglazed or
porous pot DD, such as that used in a Daniell cell, is fitted with
a cork stopper AA, which is cemented to a wooden top BB.
A wooden stopper will not serve, as the swelling due to the
absorption of moisture by the wood is found to break the pot.
The plug must fit snugly so as to prevent the solution inside
from splashing or leaking out, and also to reduce evaporation.
A few turns of electric friction tape, wound round the top and
over the wood BB, will hold the porous pot securely to the
stopper. A wooden rod CC, about 3 ft. long, like a broom handle,
is attached to BB and this serves as a convenient handle.
A piece of stout insulated copper wire EE, with a terminal or
clip at the top F, passes down the handle through the stopper,
and reaches to the bottom of the porous pot. The insulation is
removed from the lower part of the wire. A saturated solution
of copper sulphate, with an excess of crystals, G, fills the porous
pot about two-thirds full. Such electrodes are non-polarizable,
and, when two such pots are placed in the same watered hole
in the ground, the difference in potential between them seldom
amounts to more than a fraction of a millivolt. An iron nail
thrust into the water will at once give rise to fluctuating
potential of the order of 100 mV. or even more. When not in use,
the porous pots should both be placed in the same saturated
copper sulphate solution contained in an earthenware vessel.

Schlumberger (1930), who has been largely responsible for the
practical development of the natural current method of locating
sulphide ores, has used porous pot electrodes similar to the one
described above. Frail as such a piece of apparatus may appear,

we have used a pair of porous pots for more than two months in rough mountainous country without mishap. It is always prudent, however, to have 'spares' when working far from a source of supply.

Surveys of an area by the natural current method may be made in two distinct ways. In the one, a microammeter is used for measuring the direction and magnitude of the natural current between two points about 100 ft. apart. In the other, a sensitive and portable potentiometer serves to determine the direction and magnitude of the potential difference, or is used to plot out equipotentials. The potentiometer gives more accurate results, since its use is independent of the stake resistance, the name given to the resistance between the ground and the non-polarizable electrode. This resistance may vary considerably, depending upon the quality of the contact between the electrode and the ground. In the case of an iron stake driven about 1 ft. into the ground, the resistance may vary between 50 and 10,000 ohms or more, according to whether it is driven into marshy land on the one hand, or into sand or stones on the other. In the case of rocks, Rooney secures a good contact by using several turns of bare copper wire and a pad of felt, or other material, soaked in copper sulphate solution, and by pressing the whole against the rock with heavy stones. In the case of sand a copious supply of water will temporarily improve the contact.

The microammeter can be obtained in such a handy portable form that it requires no levelling when placed on the ground, whereas the potentiometer, as supplied by the trade, needs a plane table for levelling purposes. For rapid and rough survey work, we have found the Weston microammeter most convenient and quite satisfactory. No doubt Paul unipivot instruments would be equally effective.

To survey a region using the microammeter and porous pots, a definite line is chosen in a given direction crossing the field. Two or three yards of insulated flexible wire are attached to one porous pot, and 100 ft. or more of similar flex to the other non-polarizable electrode. A small hole is scooped out of the ground with a short piece of angle-iron sharpened at one end so as to serve as a convenient form of stout trowel. A little water, usually a cup full, is poured into the hole and one pot is placed in this little watered cavity. The microammeter is attached with its negative terminal to this electrode. A hundred feet along the line of the survey, a second similar hole is made in the ground, and well watered. The other electrode is placed in it, and connected to the positive terminal of the microammeter with the 100 ft. of wire. If the natural current flows in the same direction as that in which the instrument has been connected, that is, with the forward pot positive relative to the other, the needle will be deflected in the proper direction. The reading of the microammeter is then recorded. Should the natural current be in the opposite direction, the connexions to the terminals of the microammeter must be reversed and the reading is recorded as negative.* Both pots are then advanced 100 ft., a fresh hole being made and watered for the forward electrode. Proceeding in this manner, a survey of the direction and magnitude of the natural currents may rapidly be made along a given line. Such a preliminary survey is extremely useful. The microammeter usually has a small resistance of about 30 ohms. We have found in actual field work that the resistance of each pot, when placed in a watered hole of the type described, was about 500 ohms. This means that the total resistance in the circuit is about 1000 ohms, the earth's resistance being usually only a few ohms. Hence a reading of $1\,\mu$A. corresponds to a difference of potential between the pots of about 1 mV. That this relation remains fairly constant will be seen from the comparative tests made on Fisher Hill, Mineville, N.Y., which are given on p. 87.

When the readings are completed for a given series of points along a line, they are plotted with the plus and minus values of the currents as ordinates, and with the distances from the starting-point as abscissae. The readings will be small and variable in sign when passing over ground free from ore. On traversing a single vein, however, the values will gradually in-

---

* It is more convenient to have an instrument which will read right and left from a central zero and the scale should read up to $100\,\mu$A. each way.

crease to a maximum over the vein and then abruptly reverse
to an approximately equal value as the centre of the vein is
passed. The readings gradually decrease and resume their minor
fluctuations as the vein is left behind. If a survey line is thus
run and the readings plotted, then the presence of the ore is
indicated by the large values of the ordinates, and the deposit
will be located almost directly below the point on the ground
where the reversal takes place.

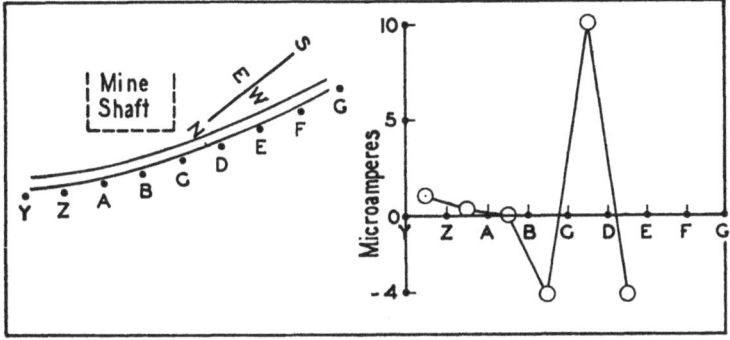

Fig. 36. The natural current method indicating a vein,
Baxter Mine, Ward, Colorado.

Fig. 36 illustrates the results obtained by the authors in a
survey along an old disused road near the Baxter Mine at Ward,
Colorado. The strong indications and reversal are evidence of
a vein crossing the road at or near the station C. Another very
striking example is given in Fig. 37, where a survey near the
O Mine at Newmarket gave strong indications of ore at OP.

A single line is often sufficient to locate the approximate
position of a vein, but, in many cases where an area is to be
investigated, several lines are followed in turn. Fig. 38 illustrates
the results obtained along four directions over a piece of meadow
land on Fisher Hill, N.Y. Ore was crossed on each line as
indicated by the large currents, of potential differences, and
their reversals. From the positions of the deposit thus found on
these lines the general direction of the strike was deduced. A fifth
line, just beyond the X point on the diagram, was investigated,

but the results gave no indications. It will be noted that the potentiometer was used in some of these measurements (readings given in volts), while the microammeter was used in others, when currents were recorded. The microammeter method, though simple, is merely *qualitative* owing to the variable contact resistances of the porous pots with the ground. It is therefore more precise to use a potentiometer, and in practice this instrument is now usually used.

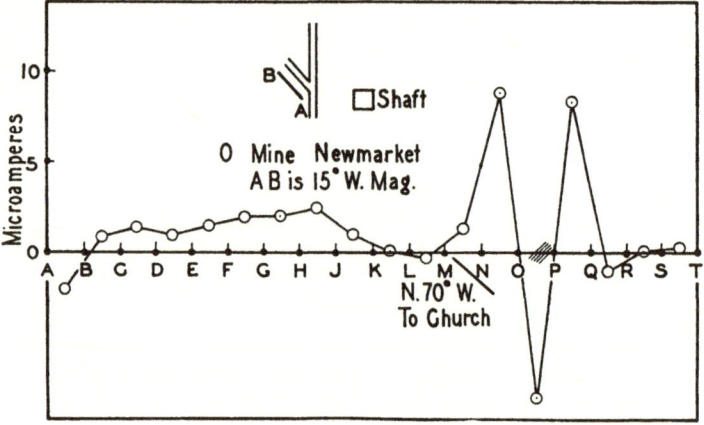

Fig. 37. The natural current method indicating a sulphide vein at Ward, Colorado.

In the portable potentiometer a dry cell contained in the box is used as a source for comparing potential differences. The dry cell is connected to a long piece of uniform wire, wound on a cylinder, and to an adjustable resistance in series with it. One end of the wire on the cylinder is brought to a potential terminal, and the other potential terminal is connected to a sliding contact that makes an adjustable connexion with the wire on the cylinder. The potential terminals thus give the difference in potential along a length of the uniform wire on the cylinder. This potential difference will be directly proportional to the length of wire between the two contacts. The wire is calibrated by using a standard cell, also contained in the box, which may be connected between the ends of the wire on the cylinder by turning a key and then altering the variable resistance in series with the dry cell, until the current from the dry cell gives one volt difference in potential between the ends of the uniform wire on the cylinder. When the instrument is thus calibrated, the standard cell is disconnected, and the sensitive galvanometer

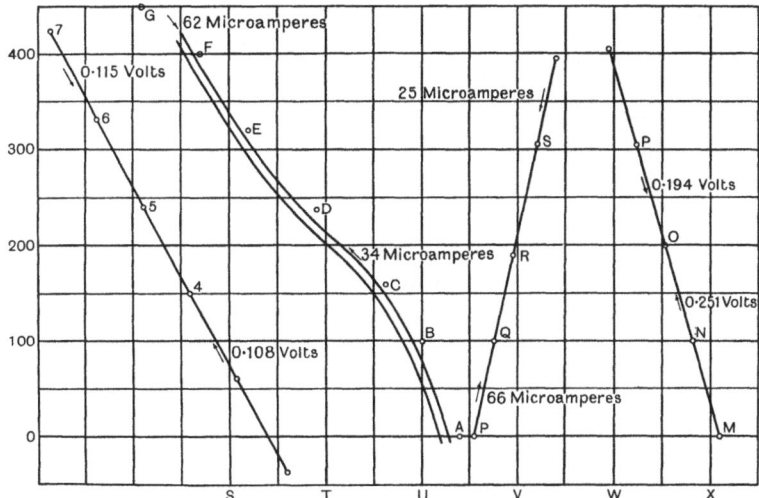

Fig. 38. A preliminary survey over an area at Fisher Hill, N.Y., with the natural current method employing microammeter and potentiometer.

of the apparatus will read from 0·0001 to 1 V. Portable potentiometers of this type are available, but, for geophysical work in the field, such instruments should be provided with unipivot galvanometers.

The potential difference between the two electrodes, for each 100 ft. interval on the ground, is found with the instrument, and the results, when plotted, are interpreted as already described in the case of the microammeter readings. The potentiometer is often used for determining equipotential lines from which the negative centre over the ore is found. To carry out such a survey, one pot is placed at a given point near the potentiometer. The other porous pot is moved about at a distance of approximately 100 ft. until a point is found which is at the

same potential, as indicated by a zero reading of the potentiometer. The two pots are then on the same equipotential line.
The first pot may then be advanced to the new position, and
the operation repeated. Equipotentials are closed curves, and
if the path found by experiment in this way closes within a few
feet, the result may be considered satisfactory.

When an equipotential has been determined, a point at right
angles to the equipotential line is located, which has a potential
about one-tenth of a volt less. This is easily done by setting the
potentiometer at 0·1 V. and finding the position of the porous
pot for which the needle reads zero. Care must be exercised in
making the connexions so as to find a point of *lower* potential.
A second equipotential curve is then found which passes through
the new point. The work is then repeated until the negative
centre is located. In plotting the results it is advantageous to
label each curve with its potential, choosing any one line as an
arbitrary zero, as shown in Fig. 34.

We have found that in field work a survey is most easily made
by taking measurements along a set of parallel lines 100 ft.
apart, and then a second series of measurements along directions
at right angles, the whole field being laid out in 100 ft. squares
for this purpose. The result of a survey using a potentiometer is
shown in Fig. 39. The large values of the readings at T, 100, also
between W and V on the 200 line, and again in the north-west
corner, are indications of ore at these points. These results were
later confirmed by other methods, and they serve as a typical
example of the ease and success of the natural current method,
when it can be used.

The details given in the previous paragraphs illustrate the
general principles of the self-potential method, but in present
practice the potentiometer is used to eliminate the contact
resistance between ground and pots. The area to be surveyed
is laid out in the usual grid form, parallel lines marked out
100 ft. apart. One non-polarizable pot is placed on the ground,
at the mid-point of the central line and attached to one terminal
of the potentiometer. The other pot is connected to one end of
a reel of about 1000 ft. of insulated flexible wire, which remains

beside the potentiometer and the pot carried out 50 ft. along the line and placed on the ground, assuring a good contact. The reel is provided with a terminal which is readily connected to the potentiometer. The potential differences between the mid-point and the different points 50 ft. apart along the line in both directions from the central station are thus measured. If there is heavy brush or trees, communication between the observer reading the potentiometer and the assistant carrying the pot is

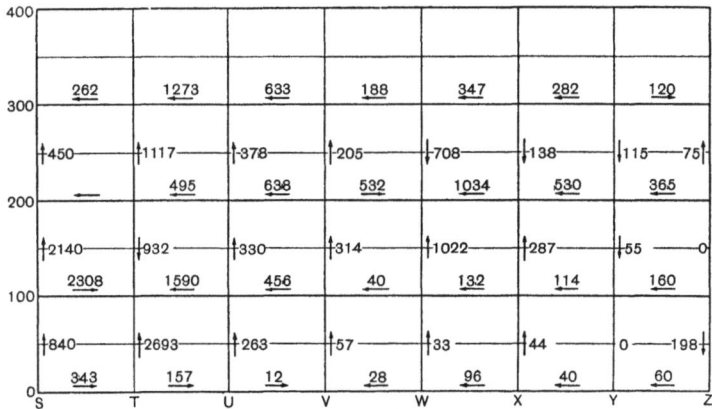

Fig. 39. The natural currents over Fisher Hill area; potential differences in tenths of a millivolt per 100 ft.

accomplished by means of whistle or radio walkie-talkie. When surveys have been made along each line, the potentials of the mid-points are tied in by comparing their respective potentials in terms of the original station. The relative potentials of each point on the grid having been determined, contours of equi-potentials are drawn and negative centres located. The method is very rapid, two experienced operators being able to survey an area of three to four acres in a day.

THE PARALLEL-WIRE METHOD

A simple method of locating good conducting ore-bodies has been developed and applied by Lundberg (1922, 1928). The arrangement resembles the well-known experiment in elementary physics for plotting equipotential lines between two electrodes placed in a tray of conducting water. In its field application, two bare copper wires, about 3000 ft. long, are laid on the ground parallel to each other, 2000 ft. or more apart, the actual dimensions depending upon the size of the area and upon the nature of the terrain investigated. The wires A, B (Fig. 40) are connected to the ground every 100 ft. of their length by pegs well driven into the earth. A very useful type of peg for making good ground connexions is shown in Fig. 41. It is made of quarter-inch angle-iron, sharpened at one end and fitted with a rod firmly welded to it at the other. Such a stake grips the

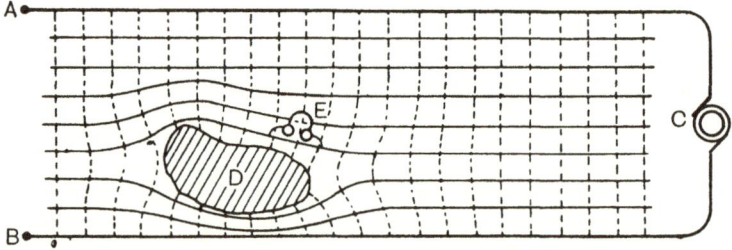

Fig. 40. The directions of current flow (dotted lines) and of the equipotentials (full lines) between two parallel long bare copper electrodes pegged to the ground every 100 ft.

earth and forces a good contact in the angle part, so that it is much superior in this respect to the circular pegs sometimes used. The ends of the wires A, B are connected to a generator C, which supplies alternating current of frequency 500 cycles per second. We have found that the best results are obtained when the generator is specially designed so as to have few harmonics, and then to reduce the intensity of any harmonics which are present by filters or choke coils. The presence of a strong harmonic in the current from the generator is often sufficient to prevent entirely the use of some electrical methods, and the reason is this: In using a null method, the harmonics, as in the older gramophones, are wont to produce the illusion of the fundamental note being present.

When an alternating current flows between the two wire

electrodes, the lines of current flow, on the surface, will be perpendicular to the wires if the ground is homogeneous. In vertical cross-section the lines of current flow are arcs of circles passing through both wires. Hence the current reaches to all depths, but the bulk of it is confined to a region near the surface, whose depth is comparable to the distance between the wires. If the ground is not uniform, but conceals a large mass of conducting ore, as at D (Fig. 40), then the currents will crowd together into D as indicated by the dotted lines. Had D been a region of quite bad conducting material, surrounded by better conducting ground, then the current near D would be less dense, and the lines of flow would have diverged from it.

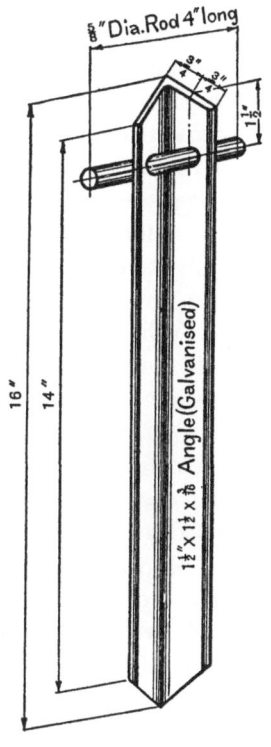

Fig. 41. A useful type of ground stake.

The equipotential lines on the surface of the ground are always perpendicular to the direction of current flow, and so they will be parallel to the wire electrodes in regions undisturbed by conductors. But they will diverge from conductors, such as ore-bodies, as shown by the continuous lines in Fig. 40. Though this figure is drawn in plan, the reader must endeavour to form a three-dimensional picture in his mind. The distortion of the equipotentials on the ground makes possible the detection of the ore-body beneath the surface. All that is necessary is to trace the equipotentials between the wires, to find how they are deviated and to interpret the results.

To trace the equipotentials two prodders or sounding electrodes are made, each consisting of a round wooden rod about 4 ft. long, fitted at one end with a sharp metal spike 6 or 7 in. long, to which a short piece of insulated copper wire is welded.

This wire ends in a convenient form of clip, at the top of the rod, which serves as a terminal for making necessary connexions. The two prodders are connected together by a piece of well-insulated flexible wire about 150 ft. long, with a pair of good radio headphones in the circuit as at E (Fig. 40). An assistant thrusts one prodder into the ground at a given surveyed point, and the observer with the phones on his head walks away after the fashion of a blind man walking with a cane, sticking the prodder into the ground as he proceeds. When the two prodders are not on the same equipotential, the note from the alternating current generator will be heard in the headphones. The nearer the equipotential is approached, the less intense does the sound become, until finally when on the same equipotential there is silence. When the currents between the two sounds are too feeble to be heard properly, they may be amplified by a good three-stage audio-frequency amplifier. Here we have one good reason for the use of alternating current as the exciting source on the electrodes. Sometimes complete silence is not attainable, in which case the sound is only reduced to a minimum. This may be caused by a change in phase, which is not constant over large regions of ground, but usually such troubles are small, and may be neglected in this method.*

When two points about 100 ft. apart are found on an equipotential, the asistant brings his prodder forward, and places it in the ground at the point where the observer stood. The observer with his headphones then goes forward tracing the equipotential for another 100 ft., and so the process is repeated until sufficient equipotentials over the area are found, surveyed and plotted to allow of an adequate interpretation of the position of the ore-body. Fig. 42 gives the actual results of a survey made in the way described, and the positions of the ore-bodies are shown by the divergence of the equipotential lines.

* But indeed at Caribou, Colarado, using a buzzer, we were forced to abandon the metal prodders, and to substitute a bare metal wire wound round one foot only, wearing rubber boots; whereas at Mineville points were found with no difficulty with the prodders, but there we used a 500-cycle generator.

From a first inspection of Fig. 42 the presence of conductors near the wire electrodes might easily be inferred, since the equi-potentials are drawn away from the wires near their centres. But Fig. 43 indicates how good conductors, situated at A and B with respect to the parallel wires CD and EF, will produce a distortion of the lines of current flow (shown as dotted lines) and the consequent deviation of the equipotentials away from the wires, as represented by the continuous lines. A second

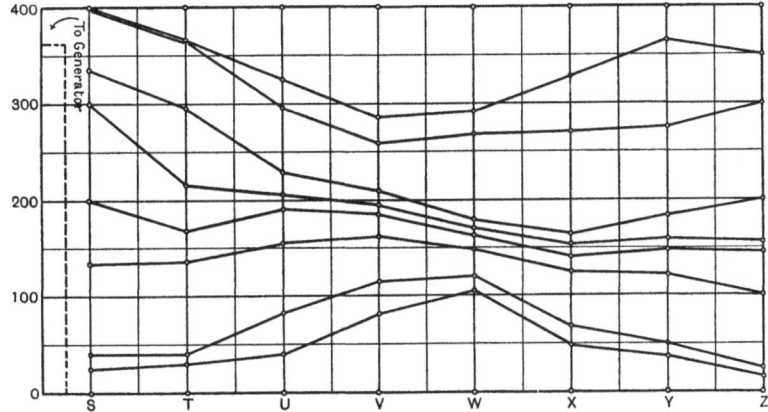

Fig. 42. The results of a survey at Fisher Hill, N.Y., using parallel wires.

survey made with the electrodes CD and EF at a different angle to A and B would confirm such an interpretation.

A good example of field work carried out by Lundberg (1928) using the equipotential method is shown in Fig. 44. The Lucky Strike Mine in Newfoundland is a group of wide lenses of a lead-zinc-copper ore containing silver and small traces of gold. The shallow overburden of gravel and of muskeg made the conditions of the survey ideal for the parallel-wire method, and recent results of diamond drilling based upon the predictions of the electrical surveys have shown an estimated volume of ore ex-ceeding 3,000,000 tons.

Direct, instead of alternating, current may be used to excite the wires, and a sensitive direct-current galvanometer, such as a portable microammeter, may be employed as the detecting instrument in place of the headphones. The use of direct current, however, introduces polarization effects, also the microammeter will be affected by any natural currents which may be present in the ground. The difference of potential applied to the parallel wires is usually more than 100 V., so as to ensure a sufficient flow of current through the ground between the two electrodes. On account of polarization effects at the prodders, these sounding rods are replaced by the non-polarizable porous pots. In the absence of natural currents, equipotentials are found exactly as already explained, a zero reading on the microammeter indicating that the two pots are on the same equipotential line. Whether one pot is at a higher or lower potential than the other is readily shown by the direction of the motion of the needle of the micro-ammeter, and to this extent the method is superior to that using the headphones, which gives no indication as to which electrode is at the higher potential.

The presence of natural currents complicates the method, for even when the pots are on an equipotential due to the parallel wires, yet the galvanometer will indicate a current. The authors, when using this method in Colorado, where natural currents were present, overcame this difficulty in the following simple manner. The exciting current was applied to the parallel wires by means of a switch which could readily be turned off or on. When the two porous pots were on an equipotential, the reading of the microammeter was *not* changed by connecting the battery or dynamo to the wires, whereas if the two exploring pots were not on the same equipotential, the switching on or off of the current to the parallel wires did change the reading of the micro-ammeter. From an inspection of whether the current increased or decreased, the direction in which the observer was to move so as to get on the equipotential lines was at once evident. The method was successfully applied in a survey of a region at Caribou, and the results agreed extremely well with the equi-potentials found when alternating current and headphones were used over the same area.

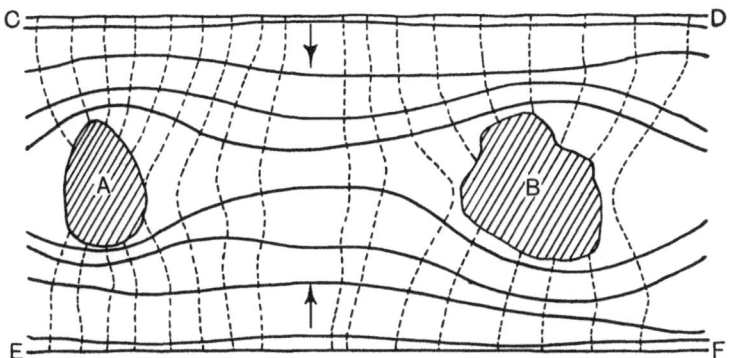

Fig. 43. A diagram illustrating the distortion of the equipotential
lines caused by two conducting ore-bodies *A* and *B*.

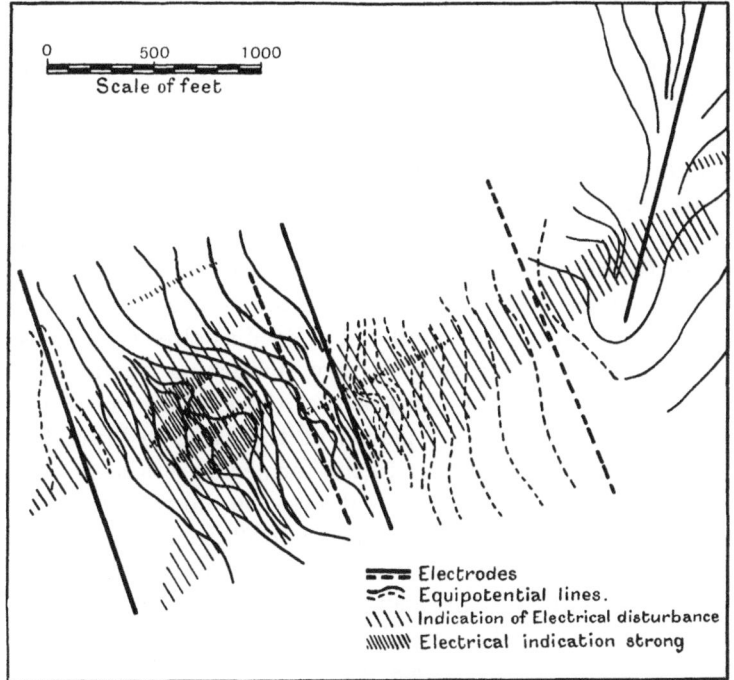

0   500   1000
Scale of feet

▬▬▬ Electrodes
≈≈≈ Equipotential lines.
\\\\\ Indication of Electrical disturbance
\\\\\\\ Electrical indication strong

Fig. 44. A survey of the Lucky Strike Mine, Newfoundland, by
Lundberg, using the parallel-wire method.

THE POINT ELECTRODE METHOD

Instead of using long parallel wires as electrodes for introducing the current into the ground, two point electrodes may
serve (Gella, 1921, 1925; Krahmann, 1926). A single stake placed
in the ground is found to have too high a stake resistance, so in
practice about a dozen iron rods, placed a few feet apart in the
form of a square or of a circle about 10 ft. in radius, are driven
well into the ground, and these serve as a single electrode when
connected together with bare copper wire. Two such groups of
stakes when placed several hundred feet apart behave as two
points. The two electrodes are connected to a source of alternating or direct current which flows through the ground between
and around them. The reader is again invited to imagine the
directions of current flow in three dimensions, extending to large
distances but with the major portion confined to a region reaching
to a depth comparable with the distance between the two groups
of stakes. The left part (Fig. 45a) illustrates in diagram the
direction of the currents (dotted lines) between the electrodes
AB, and also the resulting equipotentials (continuous lines) as
existing on the surface of a region free from the disturbing
influence of a good conducting ore-body. In the right half,
Fig. 45b, the altered directions of current flow and equipotentials
due to the presence of the ore-body C are shown.

The equipotentials are found and plotted as already described
on p. 91, and the location of any disturbing conductor is
deduced from the results. In practice a second survey is often
made with the electrode B placed at the same distance from
A but in a line perpendicular to the former direction AB. This
is done in order to avoid errors due to the body C being, by
chance, symmetrically placed with respect to A and B. The
interpretation of the deviation of one set of curved lines from
another, necessary in this method, is a much more difficult
problem than the detection of the deviation of curved equipotentials from straight lines when long parallel wire electrodes are used. For this reason the authors have found, by
their experience in the field, that the long parallel-wire

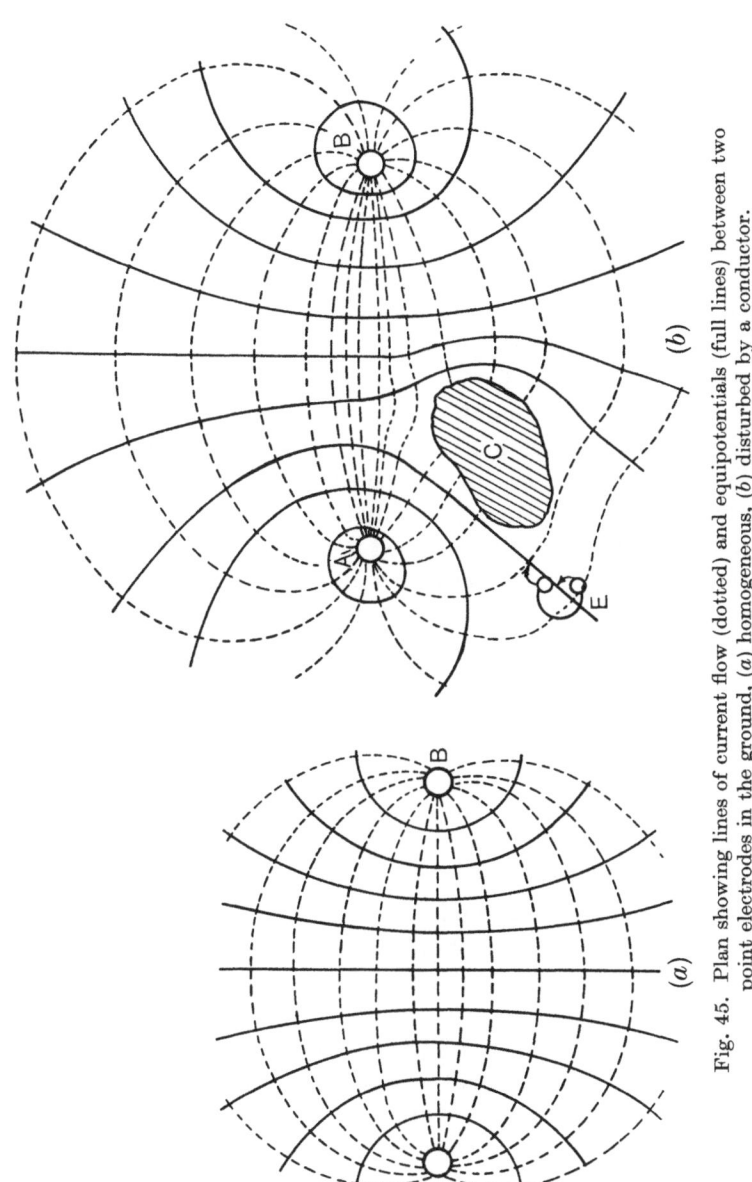

Fig. 45. Plan showing lines of current flow (dotted) and equipotentials (full lines) between two point electrodes in the ground, (a) homogeneous, (b) disturbed by a conductor.

method is both simpler and more reliable (Eve and Keys, 1928, pp. 8–16).

THE IMPORTANCE OF STAKE-RESISTANCE

Many geophysical electrical methods depend upon the introduction of electrical currents into the ground. In such cases the resistance in the vicinity of the electrodes may be of great importance. An investigation carried out in 1927 (Eve and Keys, 1928, pp. 24–30), will now be described because it throws some light on the general theories of resistance, both of stakes and of the ground, which indeed are not separable.

When the Caribou field was investigated with the parallel-wire method, we thought it desirable to measure the fall of potential over the region between the wires on the earth's surface. Two porous pots were joined together by a piece of flex and a high-resistance voltmeter was inserted in the circuit. The two parallel bare copper wires, as used in the method given on p. 90, were pegged to the ground every 100 ft. and they were 600 ft. long and 500 ft. apart. The current was supplied to the long electrodes by three 'B' batteries, and the potential difference between the wires was 122 V. when the circuit was closed. When one porous pot was placed on the one wire and the other inserted in a small well-watered hole in the ground a few feet away, then the drop in potential between the two pots as measured with a voltmeter was about 26 V. When the same experiment was repeated on the other wire, the reading was about 65 V. The drop in potential per 100 ft. in regions between the wires was found to be only a volt, and often so small that it could not be read on a voltmeter. We should expect this result, for most of the fall of potential will be near the electrodes. But why this difference between the two regions near the electrodes? On inspecting the area, it was found that the west side, where the reading was 26 V., contained a large amount of magnetite, which is an excellent conductor, while on the eastern side there was little ore. The low stake-resistance was due to the presence of good conducting material. Experiments were therefore made to find how the potential difference increased with distance from the wires, one porous pot being kept on the wire and the other moved to points where there were well-watered holes at the stated distances from the wire. The results are shown below.

It will be seen from these results that the drop in potential is mainly near the electrodes, but it will also be evident that the difference in potential for a given distance is consistently less on the western side, where the ore is, than on the eastern side. The minor fluctuations are due to local variations in conductivity near the electrodes:

Eastern wire		Western wire	
Distance in feet	Voltage difference	Distance in feet	Voltage difference
5	65	5	25
10	66	10	25
20	70	20	25
30	69	30	23
40	66	40	24
50	67	50	25
60	67	60	24
70	71	70	23
80	67	80	23
90	64	90	20
100	70	100	21·5

The theory of this method is as follows. Let A be an electrode, of a stake in a well-watered hole, thus forming a good conducting hemisphere of radius a, in good contact with the surrounding ground of resistivity ρ (Fig. 46). The resistance of this shell is proportional to its thickness and inversely proportional to its area. Hence the resistance of the shell of thickness dx is equal to

$$dR = \frac{\rho \, dx}{2\pi x^2}.$$

To find the resistance of the earth or ground outward from the electrode we sum these shells up from a to infinity and obtain

$$R = \int_a^\infty \frac{\rho \, dx}{2\pi x^2}$$

$$= \frac{\rho}{2\pi a}.$$

If the two electrodes are similar and at some distance apart, then the resistance of the ground between them is a measure of the conductivity of the rocks and earth of a large region between

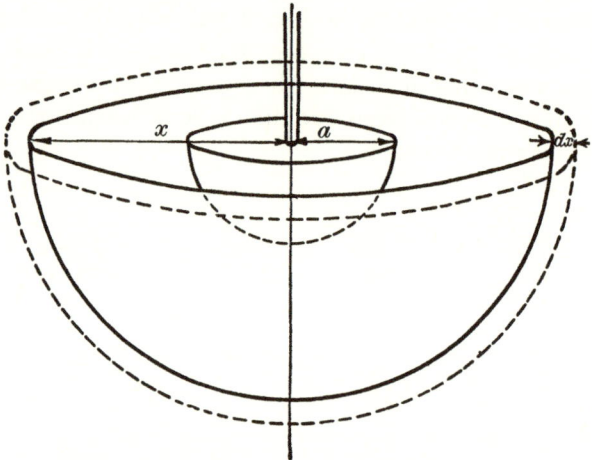

Fig. 46. Theory of ground resistance near an electrode.

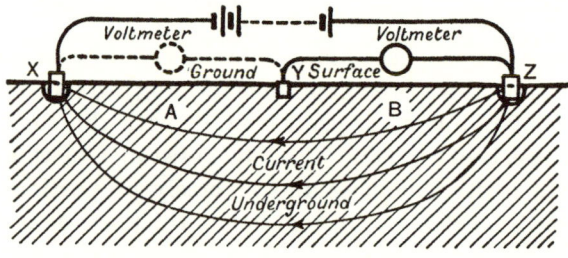

Fig. 47. Circuit for comparing relative conductivities of adjacent regions.

and around them.* So that, if Y is half-way between X and Z (Fig. 47), and if V_1 is the potential difference between X and Y, and V_2 that between Y and Z, and if ρ_1, ρ_2 are the mean

---

* Where there is much variability in conductivity, the stake resistances will greatly predominate over the resistance of the ground between the stakes and at some distance from them.

resistivities (specific resistances) of the left- and right-hand regions respectively, we then have

$$V_1/V_2 = \rho_2/\rho_1.$$

By measuring V_1 and V_2, using three stakes or electrodes, it may be possible, under very favourable conditions, to compare the relative conductivities of the two adjacent regions. Any such scheme, however, will usually be nothing more than a comparison of the stake-resistances. When the ore comes to the surface and the stake-resistances are small, the outline of the deposit may be found in this way.

In the experiments described in the previous paragraph, the electrodes were points. Linear electrodes are used in other geophysical methods which will be described later. At this point it is of interest to investigate how the resistance varies between two such linear electrodes, for in this case the resistance will be that between two semi-cylindrical surfaces instead of between two semi-spherical surfaces as described on p. 99. The problem is then to calculate the resistance between two semi-cylindrical surfaces in the ground when the region between has a mean resistivity ρ. Let A, Fig. 48, be a long wire electrode of length l, pegged to the ground, and suppose a current of electricity to enter the ground by this electrode and leave by some other electrode at some distance away. The resistance of the thin section of the ground between C, D is proportional to the thickness dx and inversely proportional to the area $\pi l x$. Hence the resistance of this section is given by

$$dR = \frac{\rho \, dx}{\pi l x}.$$

To find the resistance of the ground between the cylinders passing through the points A and B at distances a and b from the centre of the electrode, we must sum up the resistances of all the thin semi-cylinders between A and B, obtaining

$$R = \int_a^b \frac{\rho \, dx}{\pi l x}$$

$$= \frac{\rho}{\pi l} \log_e \frac{b}{a}.$$

It will be noted that this result is quite different from that obtained in the case of the point electrode given on p. 99, where the summation was taken from a to infinity. In the case of the line electrode, it is the resistance between two surfaces that is always used. Thus if V_0 is the potential difference between the two points A and B on the surface, measured with a potentiometer using non-polarizable electrodes (see p. 81), and if I is the current flowing through the ground from the electrode O, then

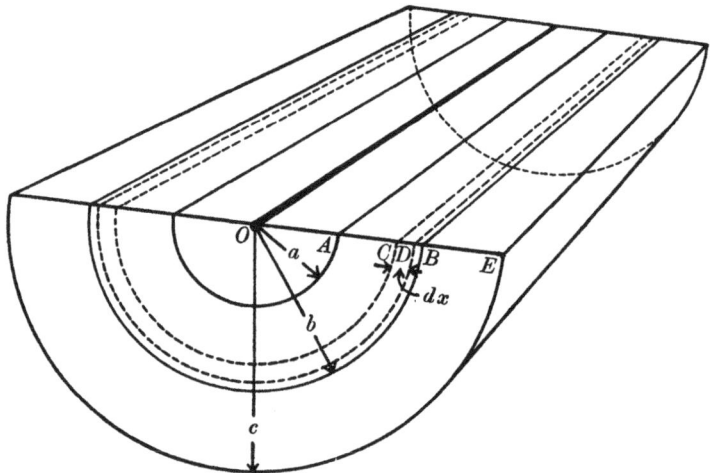

Fig. 48. Theory of ground resistance near a line electrode.

the resistance of the ground between semi-cylindrical surfaces passing through A and B will be $R_0 = V_0/I$. The mean resistivity of the earth in this region may be found from the relation

$$R_0 = \frac{\rho_0}{\pi l} \log_e b/a$$

or

$$\rho_0 = \frac{\pi l R_0}{\log_e b/a}.$$

Suppose the potential difference between B and some other point E at a distance c from the electrode is V_1, then the resistance between the cylindrical surfaces passing through these points will be $R_1 = V_1/I$. If ρ_1 is the mean resistivity of this region, we obtain as before

$$\rho_1 = \frac{\pi l R_1}{\log_e c/b}.$$

By division we can compare the mean resistivities of the two regions

$$\frac{\rho_1}{\rho_0} = \frac{R_1}{R_0} \frac{\log_e b/a}{\log_e c/b} = \frac{V_1}{V_0} \frac{\log_e b/a}{\log_e c/b}.$$

It will be evident that by choosing appropriate values of the ratios a/b, b/c, c/d and so forth, for a number of points, it is possible to compare the mean resistivities of different regions by merely comparing the potential differences between the given points. For example, let us make $b = 2a$, $c = 2b = 4a$ and so on. Then ρ_1/ρ_0 becomes directly proportional to V_1/V_0. Such a system of parallel wires and method of comparing mean resistivities of adjacent regions form the basis of one system of electrical prospecting, which will be described later.

EARTH-RESISTIVITY METHODS

The Department of Terrestrial Magnetism of the Carnegie Institution of Washington has for many years investigated earth currents, earth-current storms, the potentials to which they give rise, their daily variations and the resistances of the 'grounds' or stakes which are used as electrodes. This valuable sequence of investigations has given rise to measurements of the mean resistivities of the ground. Variations of the conductivity of successive layers of soil, or of water, were easily recognizable by the method employed, and test cases under quite varied conditions were made by Gish and Rooney near Washington, in Spain and in Western Australia.

It has already been pointed out that most metal sulphides are good electrical conductors, and it is this fact which has led to the use of the earth-resistivity methods for the location of ore. The resistance of a uniform rod or wire varies directly as its length and inversely as the area of cross-section, provided the temperature remains constant. We may express this fact by the equation

$$R = \rho \, \frac{L}{A},$$

in which R is the resistance in ohms of a wire of length L cm. and of cross-section A cm.^{2}. Then the constant ρ, called the specific resistance, or resistivity, of the material, is measured in

Table 1*

Material	Resistivity in ohm-cm.
Granite	10^9 to 10^{11}
Sandstone	$5 \cdot 0 \times 10^9$ to 10^{11}
Porphyry	10^9
Zincblende	$1 \cdot 5 \times 10^8$
Diabase (dry)	$3 \cdot 0 \times 10^5$
Diabase (moist)	$2 \cdot 0 \times 10^4$
Serpentine	$3 \cdot 0 \times 10^5$ to 2×10^6
Limestone (dry)	$6 \cdot 8 \times 10^4$
Limestone (moist)	$4 \cdot 0 \times 10^4$
Blende-galena ore, Galena, Ill., U.S.A. (dry)	$1 \cdot 6 \times 10^4$
Blende-galena ore, Galena, Ill., U.S.A. (moist)	$1 \cdot 5 \times 10^3$
Blende-galena ore, Sweden (dry)	$1 \cdot 3 \times 10^3$
Blende-galena ore, Sweden (moist)	$3 \cdot 0 \times 10^2$
Pyrite ore (dry)	$1 \cdot 0 \times 10^4$
Pyrite ore (moist)	$6 \cdot 0 \times 10^3$
Magnetite	$0 \cdot 6$
Chalcopyrite	$1 \cdot 0$
Pyrrhotite	$0 \cdot 01$
Mercury	$0 \cdot 96 \times 10^{-4}$
Lead	$2 \cdot 1 \times 10^{-5}$
Copper	$1 \cdot 7 \times 10^{-6}$
Silver	$1 \cdot 6 \times 10^{-6}$
Yellow river sand 0·86 % water	830
Yellow river sand 1·52 % water	380
Yellow river sand 9·5 % water	95
Garden soil 3·3 % water	1670
Garden soil 17·3 % water	60
Clay 4·4 % water	1450
Clay 16·1 % water	50
Clay 28·0 % water	16

* Koenigsberger (1928) gives for sandstone 400,000 to 51,000; lime-stone 350,000 to 250,000; wet humus 4000; rock salt 10^6 to 10^7 ohm-cm.

ohm-cm. and will be different for every kind of rock, soil or metal.

Great divergence exists among the values given for the resistivity of rocks, minerals and soils by different authors. Such disagreement may in part be due to variation in the conditions under which the measurements are made, for example, whether the specimens are dry or moist. Also a difference is found in the resistivities of similar minerals from different localities. As the available data refer mostly to European investigations, it is desirable that measurements should be made on rocks and

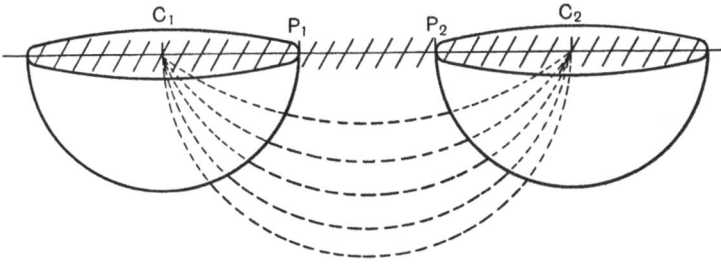

Fig. 49. Profile showing current flow (dotted lines) and two equipotentials (full lines), with current electrodes C_1, C_2 and potential electrodes P_1, P_2.

minerals found in the new mine fields of Canada and of the United States, where at present geophysical methods are being employed so extensively. A table giving the resistivities of some of the more common substances is given above, in which the limits of the values obtained from various sources and under different conditions are listed (Ambronn, 1926, p. 110; Sundberg et al. 1925, pp. 11–12; Gutenberg, 1927, p. 571; Heiland, 1940, pp. 657–65; Jakosky, 1950, pp. 441–2).

In order to measure the mean resistivity of the earth, four stakes or electrodes (Fig. 49) are placed in a straight line, at equal intervals, of length A cm. The stakes are driven into the soil to a depth of B cm. A current of electricity enters and leaves the ground at the two outer stakes C_1 and C_2. This current is measured by a milliammeter and is written down as I amp. The difference of potential between the two inner electrodes P_1, P_2 is measured by a potentiometer and is recorded as E V. Wenner

(1916) has shown that the average resistivity of the soil is given by the equation

$$\rho = \frac{4\pi A R}{1 + \dfrac{2A}{(A^2 + 4B^2)^{\frac{1}{2}}} - \dfrac{A}{(A^2 + B^2)^{\frac{1}{2}}}}.$$

If B is small compared to A, as is usually the case when the electrodes are at the surface of the earth, then this equation simplifies into

$$\rho = 2\pi A \frac{E}{I}$$

(for proof, see Appendix I).

The lines of current flow are shown in Fig. 45 a as dotted lines, and the equipotentials are drawn in full lines. This is merely a picture of the conditions at the *surface* of the earth. We are more interested in the three-dimensional flow in the unknown and invisible region beneath the earth. Indeed, the reader is invited to use his imagination and to regard the equipotentials as oval bowls, not quite hemispheres of course, with their top rims on the earth's surface. Reference may be made to Fig. 49, where such equipotentials are drawn. Instead of thinking of the current as passing between C_1 and C_2 we may consider the whole current I as flowing between the two 'bowls' which are there shown. Nor need we think of anything at all inside these bowls, but we may think of a current of I amp. flowing outside and between the two bowls, which have a difference of potential E V., and we may talk about $R = E/I$ as the resistance between these two imaginary bowls.

For quite homogeneous earth, since $\rho = 2\pi A E/I$, it is clear that, since ρ is constant, as A increases, so does E/I get less. This value E/I is of the nature of a resistance, and may be denoted by R and measured in ohms, but we must be careful not to think of it as quite such a definite thing as the resistance of an ordinary piece of wire. When A is measured in cm. then ρ will be in ohm-cm.* The effect of a conducting ore-body,

* It is convenient to remember that if A is written in *feet* and E/I or R in ohms, then $\rho = 191 A \times R$ and the result is in ohm-cm.

such as a vein of chalcopyrite lying between the two bowls, may be illustrated in the following simple way. Suppose the potential terminals are 50 ft. apart, and that the ore-body which is short-circuiting the two potential bowls is 5 ft. in radius. The ratio of the area of a hemispherical bowl is to the area of the cross-section of the chalcopyrite as

$$2\pi . 50^2/\pi . 5^2 = 200.$$

But the conductivity of the ore is a million times that of the surrounding rock. Hence the resistance between the two potential stakes will be decreased by a factor of 10^4, which will of course be at once apparent in lowering the resistivity considerably.

So far we have regarded the earth as homogeneous, which is an ideal and rarely obtained condition. What would be the effect if the right side were twice as conducting as the left side? Well, the equipotentials would crowd together where the resistance is high, and spread out where the resistance is low, so that corresponding equipotentials would no longer pass through P_1 and P_2.

What is the effect of stake (or electrode) resistance on this work? Clearly high resistance at the current electrodes will diminish the amount of current entering the ground from the battery or other source of current used. Their combined resistance is effective in this respect. If very great, it will diminish the accuracy of the reading, for if I, and hence E, are both small the value of their ratio E/I tends to become indeterminate or uncertain.

Rooney found on Barton Hill, Mineville, N.Y., that stake resistances varied from 50 ohms in the marsh to 3000 ohms in sandy patches on the hill-side. In natural current work we have always watered our electrodes and then found the resistance to be of the order of about 500 ohms apiece.

In some earth-resistivity methods it does not appear to be necessary to water the stakes because the 'grounds' or stake resistances are either bad and require inspection, or good enough to give a reliable reading. In the Gish-Rooney method, the stake resistances are always tested and measured before a reading is taken.

There are then five different methods of measuring the earth-resistivity which will be considered in the following order:

(a) Porous Pot, or direct method.

(b) Gish-Rooney method.

(c) 'Megger' method.

(d) Lee method.

(e) The Single Probe method.

Each of these schemes has definite advantages in geophysical work. One may be of use for rapid but rough reconnoitring work, another useful when accurate and thorough surveying is required, while yet a third may have the advantage that it is inexpensive to purchase the necessary amount of apparatus.

The minimum essentials for earth-resistivity measurements are:

(1) A source of current, direct or alternating, supplied by a battery or generator.

(2) A portable voltmeter or potentiometer for measuring small potential differences accurately.

(3) A portable milliammeter for measuring currents.

(4) Four iron stakes or other forms of electrodes.

(5) An abundant supply of well-insulated wire.

As soon as work in the field begins with such a simple outfit, the following troubles will force themselves sooner or later on the observer.

(1) The effect of *earth currents*.

(2) The effect of *natural currents* due to the electrolysis of local ore-bodies.

(3) The effect of *strays* due to direct currents leaking from dynamos, street railways, and haulage motors in mines.

(4) The effect of *polarization* or galvanic action when iron stakes are driven into the ground and used as electrodes.

(5) The possible effect of *inductance* between the wires leading to the current and to the potential stakes.

(6) The effect of wet ground on the insulation of the long wires leading to the four stakes. This may be called *wire leakage*.

(7) The effect of dampness on the instrument used, particularly from the battery leads to the potential leads. This may be called *instrument leakage*.

(8) In the field the effect of *tilt* of the instrument vitiating small readings.

(9) Skin effect, or the tendency of high frequency alternating current to approach the surface.

Some of these troubles have been eliminated by specially designing the measuring instruments, others are made negligible by increasing the power put into the earth. The question of the most satisfactory arrangement will depend upon terrain conditions. The manner in which each system eliminates the several errors caused by the above disturbing factors will be described in each case. The four methods have actually been compared over the same area, but we shall postpone the discussion of such comparative results until later.

(a) The Porous Pot method

This is the most direct and also the simplest method of measuring earth-resistivity. The arrangement consists of direct measurements simultaneously of the current between two points C_1, C_2, and of the potential between two points P_1, P_2. The values of current and corresponding potential are inserted in the simple formula, p. 106, and the mean resistivity of the ground to a depth about equal to the distance apart of the stakes is found.

Four electrodes, at 100 ft. intervals, are placed in a straight line as already explained. The two current stakes C_1 and C_2 are the usual iron rods or stakes shown in Fig. 41. For the potential electrodes P_1 and P_2, however, non-polarizable porous pots are used, in order to eliminate errors due to polarization effects, which are always present when iron is used. These pots are placed in well-watered holes as already explained on p. 84. The source of current may be three 'B' batteries, of the type used in radio work, connected in series so as to give a potential difference of about 135V. The authors have found that three such batteries, placed snugly in a portmanteau or suit-case, with the two terminals, labelled plus and minus respectively, fastened at each end of the case, may be easily carried about in the field and serve as a useful inexpensive source of current. A small direct-current dynamo might be used but would be more difficult to transport.

The batteries are connected in series with a portable milli-ammeter to the two current stakes C_1 and C_2 (Fig. 49). The porous pots P_1 and P_2 are connected to a potentiometer, and the potential difference E between P_1 and P_2, due to the current from the batteries, is measured. The current I between C_1 and C_2 is recorded at the same time. If there are natural or earth currents present the potentiometer will indicate a reading when no current passes between the current electrodes. The difference between this reading and the reading when the current I is flowing, gives the corrected value of E which is used in the calculations. As a check on the correct reading, the current from the battery is reversed and the readings taken again. The average value of I and E of these two sets of readings may be used to evaluate the resistivity from the equation

$$\rho = 2\pi A E / I.$$

This gives the mean resistivity for the 100 ft. between P_1 and P_2, reaching to an approximate depth of 100 ft. To make a survey over an area, the whole system of pegs and pots is moved forward 100 ft. and the measurements repeated. In this way the average resistivity to a depth of 100 ft. is found along the line. When good conducting material, such as a mass of chalco-pyrite, lies within 100 ft. of the surface, the value of ρ will be less over that region than the normal value, and thus the presence of the good conductor is indicated.

The effect of surface streams, wet clay and swamps is much like that of a mass of conducting ore, and great precautions must be taken to avoid misinterpretation of the indications. Usually surface conditions can be readily allowed for, especially when even a rough topographical map of the region is made or is available. But a layer of damp or wet clay, situated 50 or 100 ft. below the surface, may produce all the effects of ore, and in many places the diamond drill is the only reliable method of revealing this difference.

The resistivity of rock, when measured with direct current, has a higher value than when an alternating current method is employed as the exciting current between C_1 and C_2. Of course direct-current instruments, like the potentiometer and milliammeter already described, cannot be used with alternating current without additional equipment, such as the special form of commutator used by Gish and Rooney in their apparatus. This difference in value will be observed in comparative tests which will be described later.

The important question of the depth of a deposit may often be answered with precision by the earth-resistivity methods, and for this reason alone these schemes are of immense value to the geophysicist and mining engineer. The depth below the surface of a conducting mass at a given point is determined in the following way. The interval of the four electrodes is continuously increased, keeping the centre of the system at the point Z where the depth survey is to be made. Thus starting with an electrode separation of 20 ft. (which will give ρ the average resistivity to a depth of about 20 ft.) the two $C_1 C_2$ electrodes are at distances of 30 ft. from Z. The resistivity measurements are made and ρ is calculated. The electrode separation is then increased to 40 ft., 60 ft., and so on, at intervals the size of which depends upon the accuracy of depth required (Rooney and Gish, 1927, p. 55). The manipulation of the four wires leading to P_1, P_2, C_1 and C_2 demands brief consideration. The current 'leads' may be placed on reels not far from the selected station, and these may be drawn out or wound up by two assistants who can also drive in the pegs or stakes at the previously measured and well-labelled positions. The leads to the pegs P_1 and P_2 should not be coiled because inductance might lead to error. The handling of excessive lengths may be avoided by using wires about 500 ft. long which can be linked by well-insulated connectors. The inductance in the $C_1 C_2$ circuit of course does not introduce any error. The average resistivity ρ for each electrode separation is plotted as ordinate against the electrode separation A as abscissa, and the separation may be thought of as depth.* It has been found by tests over known areas that the average resistivity gradually increases with depth over normal soil, but when a conducting layer is met the value will commence to decrease again. This decrease, or break in the curve of the graph, will continue until the depth passes through the conducting layer, when the resistivity will generally increase

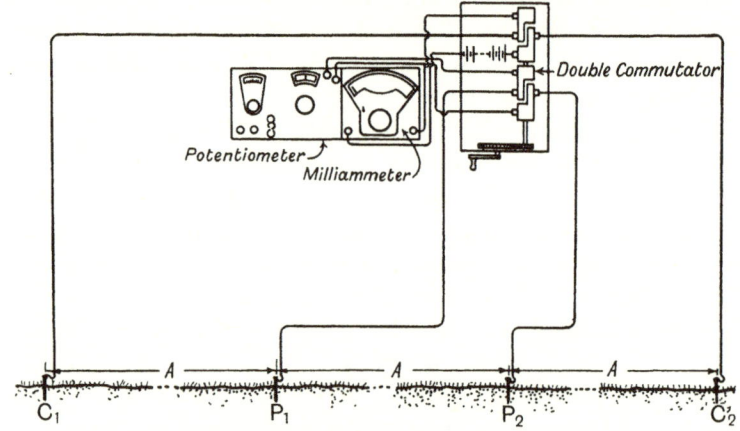

Fig. 50. Diagram of Gish-Rooney apparatus and ground connexions for earth-resistivity measurements.

again with increasing stake separation. From such a graph not only may the approximate depth of the ore be obtained, but also a rough estimate of this thickness may be made.

This scheme for determining *depth*, which has just been described, may be applied to any of the earth-resistivity methods,

* But a thorough mathematical investigation by King (1933) indicates that the change in mean resistivity, due to a good horizontal conductor beneath poorly conducting rocks, should not give the pronounced changes which Rooney found experimentally to yield the result 'stake-interval equals depth'. Hence further work both theoretical and practical requires to be carried out with care and skill. See also Hummel (1931); Tagg (1937); Roman (1941); Ruedy (1945). C. F. Tagg, *Tech. Publ. A.I.M.E.*, No. 755 (1937); I. Roman, *U.S. Geol. Survey Bulletin* No. 927-A (1941); and R. Ruedy, *Can. Journ. Research*, vol. 23 (1945), pp. 57–72.

and it has been proved successful both with the Gish-Rooney method and with the 'Megger'. A survey may also be repeated along the same line over an area, using increased stake separation each time the area is traversed. A series of measurements is thus obtained giving the average resistivity down to different depths at each station. By making surveys along two lines drawn at right angles and obtaining similar results for the two directions, the position, depth and even the strike of a hidden ore-body beneath may be inferred.

(b) The Gish-Rooney earth-resistivity method

Many of the errors in earth-resistivity measurements enumerated on p. 108, may be eliminated by the simple expedient of reversing the current from the battery as already explained in the previous section. The errors 1, 2 and 4 are removed in this way. Gish and Rooney (1923, 1925 a, b) have overcome these difficulties in a satisfactory manner with the apparatus (Fig. 50) they employed. The most important feature is the commutator, which is turned by hand and which imposes a reversed direct current, or type of alternating current, on the ground electrodes C_1, C_2. As will be seen from the figure, the current from two, three or more 'B' batteries passes through the milliammeter, and after it has been established for a brief interval to give time for it to rise nearly to its maximum value, the potential leads are 'made' by the commutator, so as to connect with the potentiometer. The commutator then breaks all contacts, next reverses the current to the ground which passes between C_1 and C_2, but still maintains the former direction in the direct current milliammeter, and after a brief interval the potentiometer indicator is again connected. This cycle of operations is repeated about 30 times a second when the commutator is rotated, and the current which enters the ground may be termed a reversed direct current, rather than an alternating current. The construction, spacing and setting of the commutator is a most important factor. The current from the battery to the commutator is only 'made' when the handle of the apparatus is pressed in against a spring, thus preventing the possibility of

8   E & K

the batteries remaining by chance connected when measurements are not being taken. With the use of such a commutator, a correction factor, which is usually determined experimentally in a region free from serious disturbance, has to be applied to the result. This is due in part to the shape of the wave form of the current communicated to the ground by the instrument. By using such reversed currents, the effects of natural currents, which are unidirectional, as well as all other direct current strays, are usually eliminated. If they do become troublesome,

Fig. 51. Dr Rooney and his earth-resistivity apparatus in the field.

a condenser which allows the alternating current to pass, but not a direct current, may be placed in series with the potentiometer leads.

Leakage from the apparatus is neatly eliminated by enclosing all the apparatus containing the batteries, the commutator and the connexions to milliammeter and potentiometer in a metal box, which is grounded. The potential connexions are also enclosed in flexible earthed conductors. This prevents any leakage of current from the current wires to the potential leads. The method of procedure is exactly similar to that already described in section a (p. 109) except that all four stakes may now be of iron. The four wires from the instrument are connected to the ground electrodes, and the potential of the batteries may readily be tested by using the milliammeter and a suitable resistance in series. The Gish-Rooney method is much more reliable than

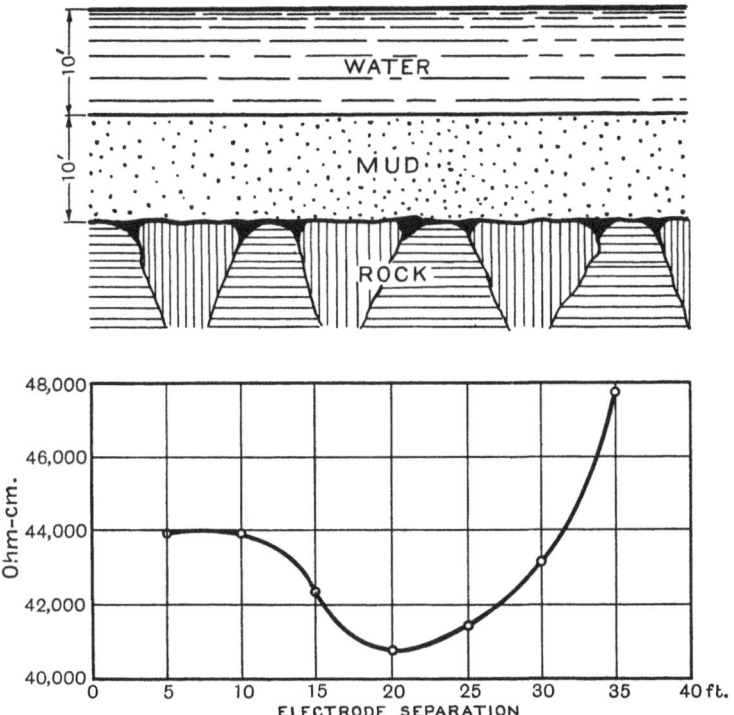

Fig. 52. *Upper:* Profile, Tidal Basin, Washington. *Lower:* Graph showing change of resistivity with variation of electrode separation, indicating depth of rock equal to electrode separation.

other earth-resistivity methods because the stake resistances, which Rooney usually measures, do not enter into the calculations. If the stake resistances are so high as to affect the proper reading of the milliammeter and potentiometer on account of too low a value of the current passing into the ground, then the number of 'B' batteries is increased until the potential difference is sufficient to give reliable readings on both instruments. The whole apparatus is quite compact and portable. A typical photo-

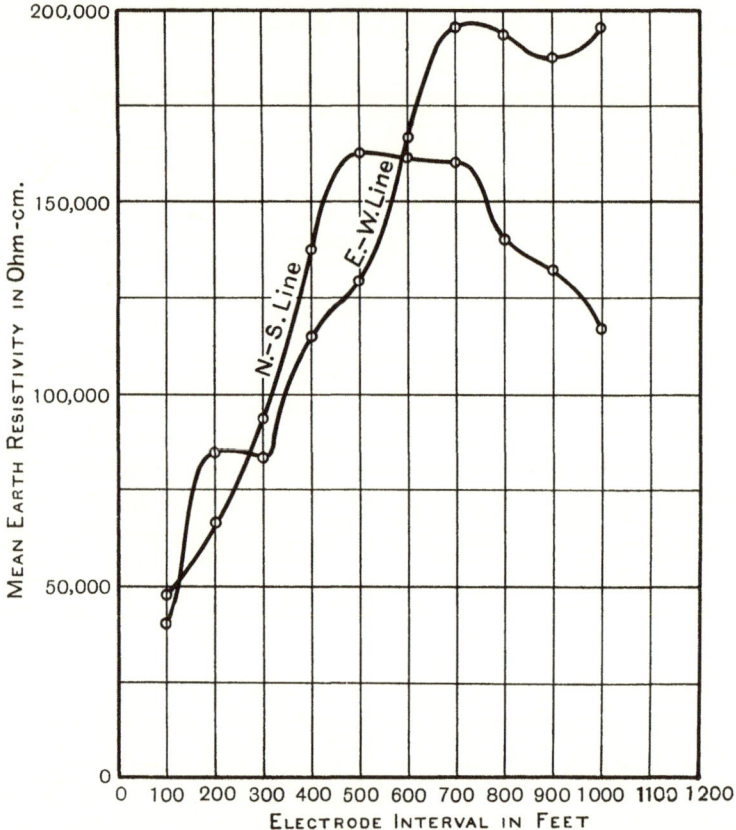

Fig. 53. Diagram giving mean earth-resistivity with increase of stake separation, indicating a conductor at a depth of about 700 ft. (Survey by Dr Rooney.)

graph of Dr Rooney operating his apparatus (Fig. 51) gives some idea of the 'lay-out' in the field.

The accuracy of this method in determining the depth of a conducting layer will be evident from Fig. 52 which gives the results of a survey over the tidal basin at Washington. The method was also applied by Rooney to find the depth of a cap or sheet of magnetite about 20 ft. thick, lying at a depth of approximately 500–700 ft. below the surface. The results of a survey for depth, at a point called 40 W, which was carried out near Mineville, N.Y., are shown in Fig. 53. It will be seen that the resistivity increases up to 700 ft. and then suddenly the slope of the curve changes, as the resistivity slightly decreases. This indicates that a good conducting layer lies about 700 ft. below the surface at this point, a fact which was corroborated by an inspection of the diamond drill cores of the region. The Gish-Rooney method may also be applied for the location of mineral deposits of higher resistivity than the surrounding medium. Clearly the method is also well suited for finding the depths of salt domes—many hundred feet beneath the surface— a field of exploration in which the method may have immense economic value. Interpretation is not always simple. See footnote, p. 112.

(c) The 'Megger' method

While making some comparative tests during the summer of 1928 on the various electrical methods of locating ore, in collaboration with the U.S. Bureau of Mines, some very interesting and quite important results were obtained with a Megger.* Dr F. W. Lee of the Bureau of Mines pointed out that the Megger has much in common with the Gish-Rooney apparatus for measuring earth-resistivity, and that it was at the same time lighter, more compact and easily operated.

Instead of a battery, the instrument contains a small direct-current generator or magneto, turned by hand. On the same shaft there is a current reverser and a potential commutator. There are also the four terminals labelled C_1, C_2, P_1 and P_2 which are joined to the four equally spaced ground electrodes, as in the Gish-Rooney method. A diagram of the internal workings of the

---

* Trade name given to an instrument used for measuring directly insulation, stake and ground resistances, manufactured by Evershed. London. The word is derived from 'megohm'.

apparatus is shown in Fig. 54. It will be seen that the current coil and the potential coil are jointly used to produce a deflexion of the galvanometer needle in such a manner that the ratio E/I is indicated and measured directly in ohms. The frequency of the current reversal is about 50 times a second, and the readings are independent of all direct-current effects, whether earth, natural, stray or galvanic. The above ratio is, however, subject to errors due to induction between the current and potential

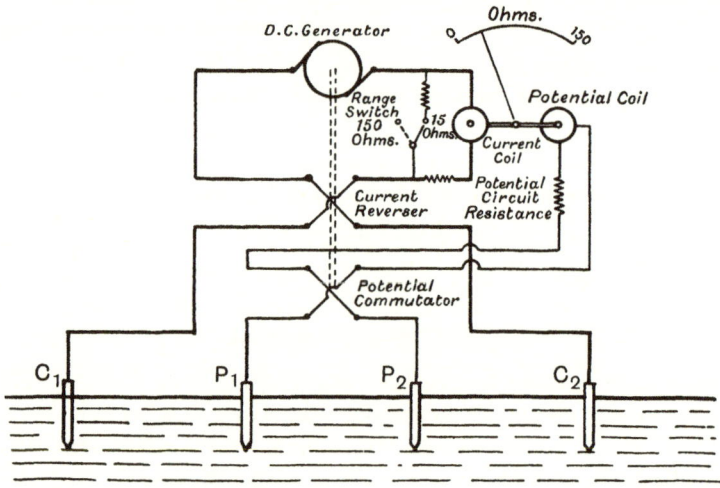

Fig. 54. Diagram showing the Megger, and earth connexions, as used for finding mean resistivities, and inferring depths of conductors under the ground.

lines, and for this reason when the intervals between the stakes become large (1000 ft. or more) troubles from this cause may arise.

The Megger (Drysdale and Jolley, 1924, pp. 385–8), as made for testing 'grounds' and insulation resistance, and for general electrical engineering work, has been improved for geophysical applications. It is known as the Geophysical Megger and differs from the ordinary type of earth-tester in that a resistor AB, Fig. 55, is provided with a sliding contact C, by means of which the operator may balance the potential difference between the terminals P_1, P_2 with the potential drop across CB, as indicated by the zero reading of the galvanometer. The introduction of this potentiometer device eliminates the contact or stake resistances at P_1 and P_2. Other improvements, such as shielded leads and lower range are incorporated in the instrument. The instrument has been used for (a) the determination-of thickness of overburden by the central electrode method, (b) determination of the mean resistivity of the ground to a definite depth

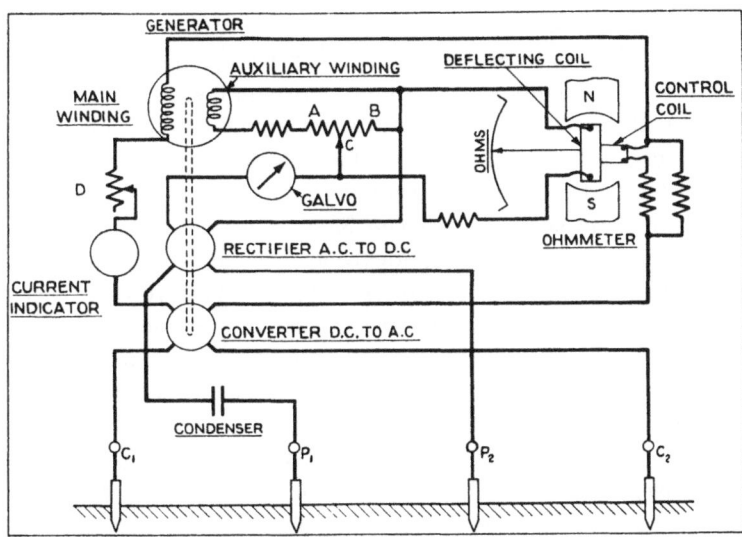

Fig. 55. Geophysical Megger.

by traversing along a given line with constant electrode separations, and (c) the comparison of mean resistivities to the same depth in different directions from a given point. Either point electrodes or line electrodes may be used. Gilchrist, when using the point-electrode method (p. 142), uses four separate electrodes placed in different directions at equal distances from the central electrode and arranges by means of auxiliary resistances that the current through each to the central electrode is the same, thus making the potential bowls more nearly spherical in shape when in homogeneous ground. When using line electrodes instead of points the equipotential surfaces are made more nearly semicylindrical by arranging that equal currents flow to the central line electrode from two distant electrodes, one on each side of the central electrode. If these precautions are not taken, one may, in certain cases, obtain resistivity curves which are not indicative of the actual conditions near the central electrode. Gilchrist found this to be the case in an investigation at the Chickamauga Dam site in Tennessee.

The instrument as obtained from the manufacturers was used by Lee and the authors for locating ore and also for determining its depth. The results were very satisfactory up to stake intervals of 500 ft., or possibly a little more. For small stake separations of about 100 or 200 ft., the results are affected by local surface conditions. A knowledge of the local topography will help to interpret results for such distances.

A large-scale survey with the Megger was made on Barton Hill, N.Y., in an attempt to locate a sheet of conducting magnetite. For this purpose two lines at right angles were surveyed. The results of these surveys are shown in Fig. 56. It will be seen that the conducting ore, as indicated by the sudden decrease in resistivity, was reached at a depth of 600 ft. on the north-south line, and at a depth of 800 ft. on the east-west line. These curves were constructed from the average readings for each stake interval, obtained at a series of equally spaced stations along each line. The resistivity for the 100 ft. stake intervals was of the order of 25,000 ohm-cm. for each and every station. This low value is perhaps due to marshy surface conditions along the whole length. The resistivity increased steadily until at 700 ft. intervals the value reached 125,000 ohm-cm., after which there was a decrease, which we relate to the presence of the conducting magnetite, known from the diamond drill records to lie at about that depth.

The results agree broadly with those obtained over the same area by Rooney, who indeed made his survey after the determination by the Megger just described. It will be noted that his values of mean earth-resistivity are higher than those obtained with the Megger, but that their general character and indications are similar.

Probably the clearest way of depicting the effect of an ore-body, whether cap or lens, is by drawing the equipotential bowls together with the short-circuiting ore-body. Thus in Fig. 57

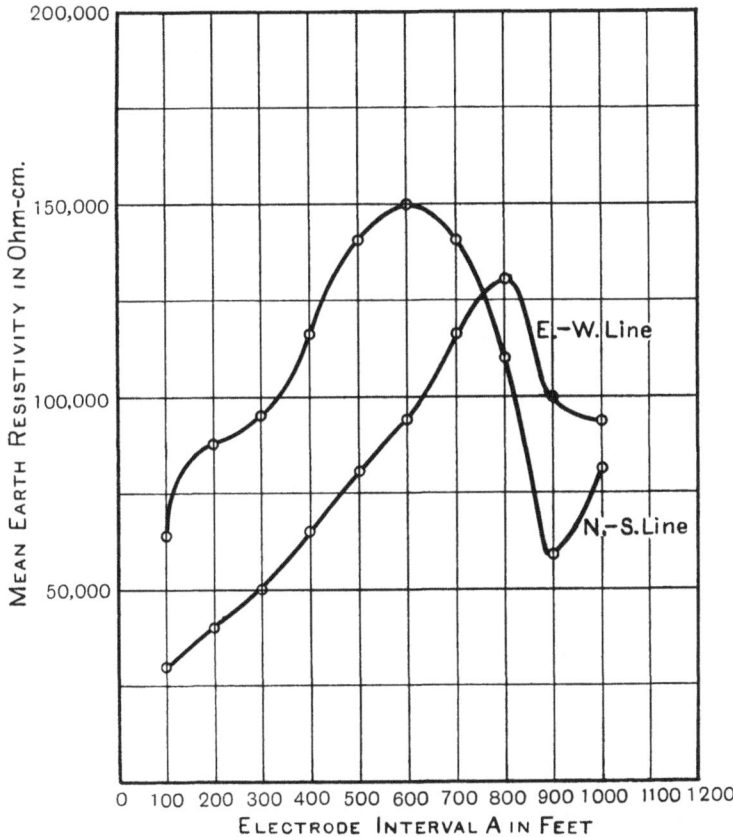

Fig. 56. Diagram showing change of mean resistivity with increasing stake interval. Depth of conductor inferred as 600 ft. on north-south line and 800 ft. on east-west line (Barton Hill, N.Y.). (Survey with Megger.)

there are three diagrams with electrode intervals of 400, 700, 1000 ft. respectively, all drawn to scale in profile together with the ore-body as located with a diamond drill at Barton Hill, N.Y. With the 400 ft. stake interval the effect of the conductor body will be small, but at 700 ft. it is beginning to be pronounced as it enters the right bowl, and is nearly penetrating the left bowl, or equipotential surface.

It is interesting to note that the average depth of the ore is found to be different along the two directions. These average depths in two directions at right angles will give some idea of the general dip of the layer, more accurate determination of

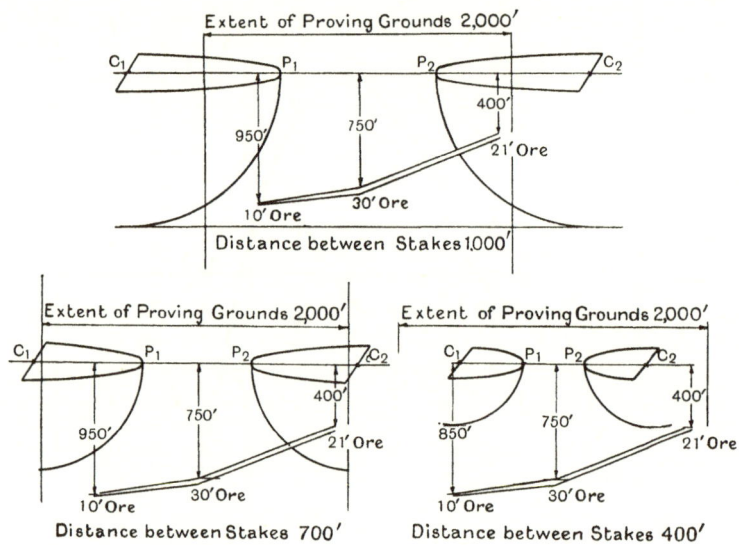

Fig. 57. Three profiles showing equipotentials with stake intervals of 400, 700, 1000 ft., and the magnetite cap as determined by diamond drill at Barton Hill, N.Y.

both dip and strike being possible by repeating the surveys along lines parallel to the two already mentioned. This possibility of locating strike and dip as well as depth of conductors will be discussed further in connexion with another survey.

These results of the application of the Megger to geophysical work were confirmed by the work of Rooney over the same area. In order to test this method further, two lines were run over a point on Fisher Hill, the one in a north-south direction and the other in an east-west direction. The results are shown in Fig. 58, and these are again of particular interest, because they reveal the nature of the strike of the ore deposit. It will be noted that a shallow conductor occurs at a depth of about 20 ft., with a thickness of about 10 ft., and that it runs east and west, because the corresponding effect on the north-south line is small. Such a result again indicates the possibility of finding by this method the *direction* as well as the *depth* of a deposit.

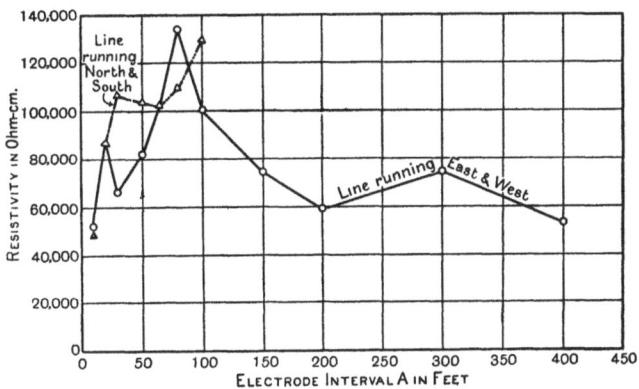

Fig. 58. Diagram showing variation of mean earth-resistivity with increase of stake, or electrode, interval; conductor inferred at depth of 80 ft. (Fisher Hill, N.Y.).

In connexion with the use of the Megger it is of interest to note how the resistance varies with the distances of the potential electrodes P_1 and P_2 from the current electrodes C_1 and C_2, when the distance C_1 to C_2 is kept constant. Fig. 59 shows the results of such a series of experiments over meadowland free from ore deposits and gives an idea of how the resistance between the 'bowls' (see p. 105) varies with their size.

We give one more example of how the Megger may be used in actual field work for determining the location of an ore deposit and for guiding the mining engineer in his choice of preliminary drill-holes. The area shown in the key map of Fig. 60 was surveyed for depths of the deposits at each of the stations shown. Stake intervals from 25 to 600 ft. were used at each of four points, and the results are also shown graphically in Fig. 60.

A cursory glance at these curves indicates the desirability of first drilling at H, where ore may be expected at a depth of 400 ft., and at D, where a conductor should be struck at 300 ft.

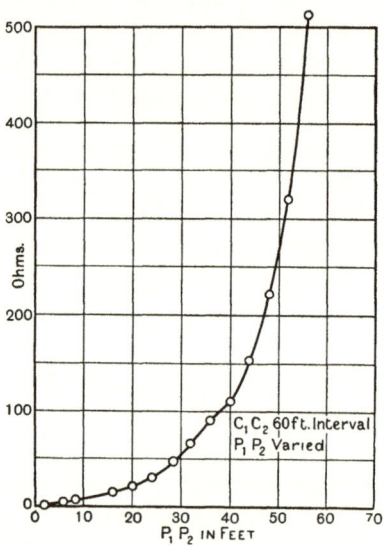

Fig. 59. Diagram showing change of resistance between potential electrodes, when the distance between them is varied.

At all other stations the prospects were unfavourable with the possible exception of station C. The Megger method is a simple one which any mining engineer, or indeed any college graduate in electrical engineering, can carry out without much preliminary training, such as most of the other electrical methods require. The method has not been patented, but great care is needed in obtaining correct interpretation, even under the simplest conditions.

It is of interest to compare the results obtained using the three methods of measuring earth-resistivity over the same piece of ground. For this purpose we chose an area at Fisher Hill, N.Y., and the results of the three surveys for locating the depth of the ore-body are shown in Fig. 61. It must be borne in mind that in this diagram the horizontal line gives stake intervals in feet, but the reader will also remember that he can think of these as *depths* of probing. The vertical line gives the average resistivity calculated from the measurements taken as described above in

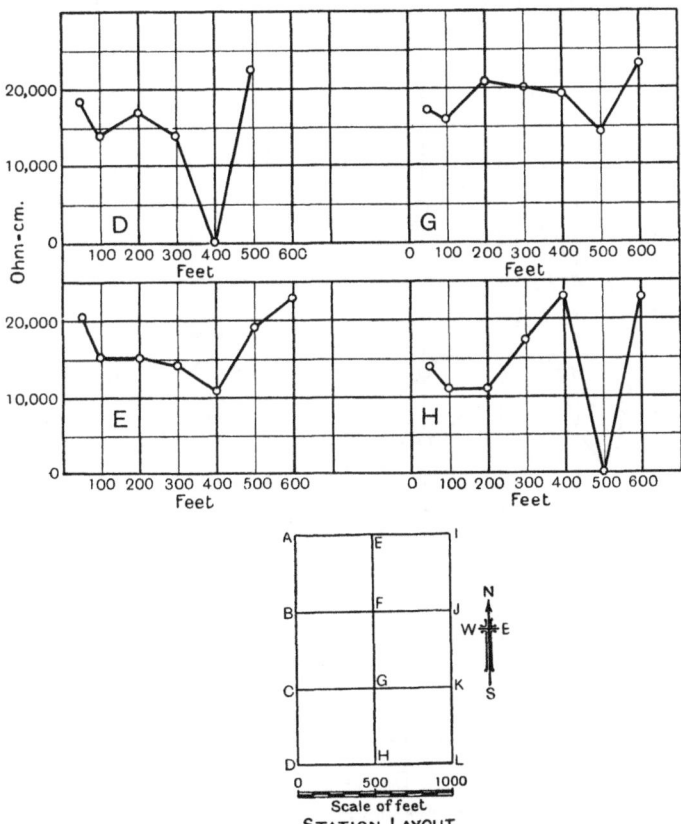

Fig. 60. Key map below, side of each square 500 ft. Results of survey shown above at four stations D, E, G, H. Strong indications at D (400 ft.) and at H (500 ft.).

the case of each method. The same stakes were used for electrodes in each method, except that, for the porous pot method two well-watered holes were made not far from the iron stakes, and porous pots were used as potential electrodes in the first arrangement.

It will be evident at once that the three methods give concordant results, the resistivity rising as the electrode interval increases from 50 to 80 ft., while at 80 ft. all three methods indicate penetration into a conducting layer, as shown by the

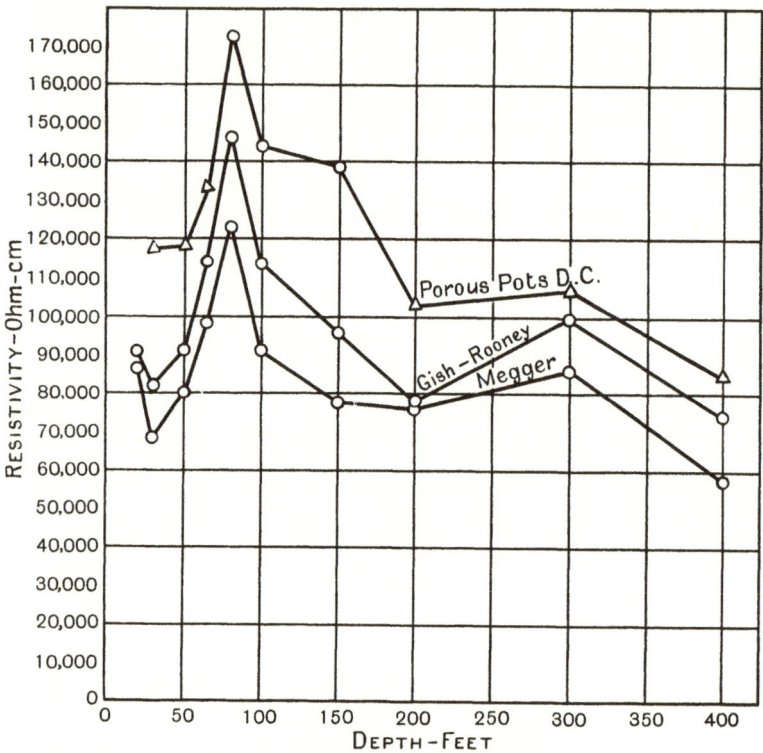

Fig. 61. Comparison surveys at Fisher Hill, N.Y. Porous Pot, Gish-Rooney and Megger methods.

sharp drop in all three curves. This decrease continues up to a depth of 200 ft. and then becomes less marked to 400 ft., the limit of observations made.

The Gish-Rooney method may be taken as our standard for comparison purposes, and the curves show that the Megger read almost uniformly 16 % too low, while the porous pot method gave an average about 18 % too high. In making these measurements no correction factors were used, except the usual one used in calculating the resistivity from the Gish-Rooney readings. One reason why the porous pot method reads higher may be that the direct current resistivity measurements made on specimens of rock have been found to give larger values than when alternating-current methods are used. If proper corrections are made for stake resistances the three curves (Fig. 61) would be approximately coincident.* The results of this test, however, show that any one of these methods may be used for locating an ore-body, a result of importance to the mining engineer who wishes to have a simple method of surveying an area for conducting bodies.

A few words may be written here about a method proposed by Weaver (1928), but only tested by him in the laboratory. The arrangement proposed by Weaver was again a system of four electrodes, but instead of placing them in a row, the two inside potential electrodes P_1 and P_2 were placed on a line at right angles to C_1C_2, passing through the mid-point of C_1C_2 and as far apart as the distance between C_1 and C_2. The current was introduced into the ground through C_1 and C_2, and when P_1 and P_2 were connected through a microammeter, the direction of the flow of current between these two electrodes indicated which was at the higher potential. The system resembles somewhat a Wheatstone's bridge. If P_1 was at the same potential as P_2, then it might mean that the region was perfectly uniform. Any difference in potential indicated dissymmetry in the current distribution caused by inequalities in the conductivity underground. The interpretation presents serious difficulties as quadrantal

---

* R. Dobridge, McGill University, has found that, while the Megger is correctly calibrated for measuring resistances, low values are obtained in work in the field, particularly when the 100:1 ratio is used. The reason is this. The resistance of the voltage coil is about 17,000 ohms for a 1:1 ratio, 16,000 ohms for the 10:1 ratio, but only 1700 ohms for the 100:1 ratio. In the last case the two stake resistances P_1, P_2, amounting to about 1000 ohms, increases the resistance in the voltage circuit from 1700 to 2700 ohms, so that the value of V will be seriously in error and V/I will be too low. This fact is clearly stated in the instructions issued with the instrument. Stake resistances should be measured and adequate corrections made, except when the geophysical Megger is used.

differences are mainly effective. Thus a line of conducting ore symmetrical and perpendicular to $C_1 C_2$ will produce no difference of potential between P_1 and P_2. The method was tried in the field, but we found the determination of the location of the ore-body too perplexing and unnecessarily involved compared with the other methods already described.

(d) The Lee modification of the Gish-Rooney method

Lee (1929) has suggested placing a central potential electrode midway between the two potential electrodes of the Gish-Rooney arrangement, as shown in Fig. 62, in which C_1, C_2 are the current electrodes by means of which the current is intro-duced into the ground, and P_1, P_2 and P_3 are the potential electrodes. The electrode separations (equal to a) $C_1 P_1$, $P_1 P_2$ and $P_2 C_2$ are kept equal as in the Gish-Rooney system, and the electrode P_3 is placed half-way between P_1 and P_2. When a current, I amp., is passed into the ground between C_1 and C_2, which may be either a direct, alternating, or interrupted direct current, the corresponding potential difference V_1 is found between P_1, P_3 and the potential difference V_2 between P_3, P_2 for the same current I. From previous discussion of the Gish-Rooney method, it is evident that the value of V_1/I will measure the mean resistance of the volume A, and the value V_2/I the corre-sponding resistance of B. The left-hand region A will have an average resistivity given approximately by $\rho_1 = 4\pi a V_1/I$ ohm-cm., and the right-hand region B by $\rho_2 = 4\pi a V_2/I$, and it is the relative values of these which are important in exploration. The dividing plane between A and B is the equipotential through P_3, and this will not generally be vertical as in the figure. A survey may be made along a given line, keeping P_3 fixed and increasing the electrode separation a, thus obtaining comparisons of the mean resistivity of the right and left of P_3 for increasing depths. A graph may be drawn using the values of a as abscissae and the corresponding values of ρ as ordinates. If better conducting material lies to the left, then as B encompasses this good con-ducting material, the mean resistivity to the left will be less than that to the right. Hence the left curve will deviate from the right, becoming less at some definite electrode separation. By applying Rooney's rule, the approximate depth to the good conducting material may be found. The results of such a survey made over a pyrrhotite-nickel vein is shown in Fig. 63. The strike of the vein was east-west and the survey was made along a north-south line, the central electrode P_3 being 100 ft. south of the point directly over the vein which was beneath an overburden of about 100 ft. of glacial drift. The ore was thus

$$\sqrt{(100^2 + 100^2)} = 141 \text{ ft. from } P_3.$$

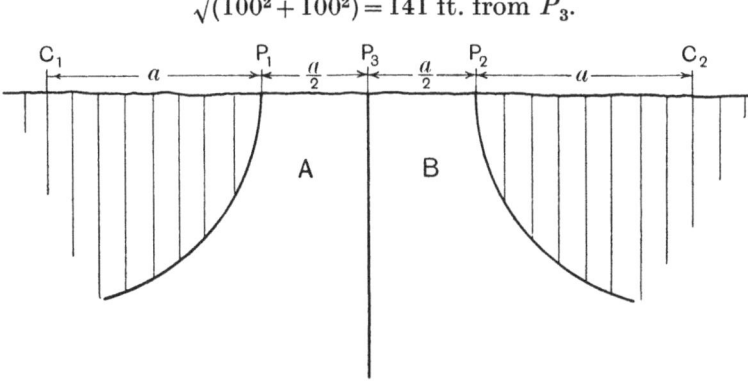

Fig. 62. Diagram of the Lee modification of the Gish-Rooney method of measuring mean resistivities with three potential electrodes.

An inspection of the north and south mean resistivity curves will indicate that they diverge at a point where a is about 140 ft. and that the better conducting material is to the north.

A second survey was made when the central electrode P_3 was placed directly above the centre of the pyrrhotite vein as determined by the magnetometer survey. Lines running north-south and east-west were surveyed so as to compare the mean resistivities in a direction across the strike and parallel to the strike and again to ascertain the thickness of the overburden. The results are shown in Fig. 64. The mean resistivity is about the same along the north-south direction as the east-west direction until we reach an electrode separation of about 100 ft., when the east-west mean restivity becomes less. Hence we conclude that the thickness of the overburden is about 100 ft., a result which is in excellent agreement with the facts obtained from drill core records.

This method may also be used for determining the presence of oil, when the overburden is not too great, and for studying the thickness of tar sands. In the case of the oil, the mean

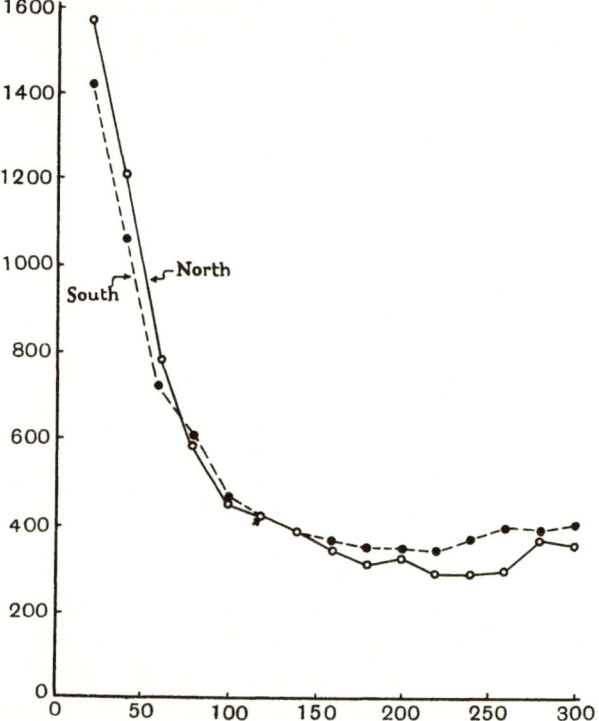

Fig. 63. Mean resistivity curves over a pyrrhotite-nickel dike, by the Lee modification of the Gish-Rooney method.

resistivity will increase on the side towards the oil sands. By surveying over a given point to a depth of 400 ft. and then proceeding to a station on the high-resistivity side, several small oil deposits, averaging one to five barrels a day, have been located in the Bowling Green district of Kentucky by the application of this method.

A region can be traversed by this method moving stations about 500 ft. along a straight line, and exploring to a similar or greater depth; any change of conductivity near or below the surface, whether due to fault, fold, ore-body, oil sands, or other cause will become strikingly conspicuous.

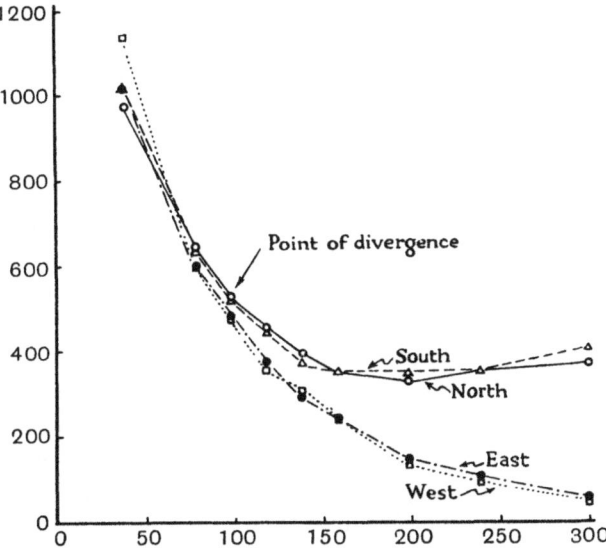

Fig. 64. Mean resistivity curves over a pyrrhotite-nickel dike indicating the strike of the vein.

Interpretation of Earth-resistivity Curves

It was pointed out on p. 112 that Rooney's rule for estimating depth of overburden as being equal to the electrode separation at which a discontinuity in the resistivity curve occurs is an approximation. Many mathematical investigations into this problem have been carried out with the object of finding a simple and yet accurate method of determining the amount of overburden. The simplest case occurs when a layer of homogeneous material of resistivity ρ_1, and thickness h, lies above a homogeneous medium of resistivity ρ_2. The problem is to determine h from the apparent resistivity ρ_a measured on the surface using the Wenner set-up of equally spaced electrodes. Tagg (1934, 1937) has given the theory of this two-layer problem and has worked out a series of curves which are of considerable use in finding accurately the value of h. It may be shown that $\rho_a/\rho_1 = 1 + 4F$, where F is a function representing the sum of an infinite series of terms involving the ratio h/a, where a is the electrode separation, and a power series in k, where $k = \dfrac{\rho_2 - \rho_1}{\rho_2 + \rho_1} = \dfrac{1 - \rho_1/\rho_2}{1 + \rho_1/\rho_2}$. It is evident that k has all possible values between $+1$ and -1, the positive values referring to the cases where the lower stratum has the higher resistivity. A series of theoretical curves are drawn for different values of k using ρ_a/ρ_1 as ordinates and the corresponding values of h/a as abscissae. When ρ_a/ρ_1 is greater than unity, use the values of the conductivities σ_a/σ_1. Such a set of curves are reproduced from Tagg's paper in Fig. 65. The procedure for finding h is as follows. Find first the value of ρ_1 by using small stake-intervals, following the Gish-Rooney method. Then selecting electrode intervals of 50, 100, 150 ft., and so on, determine the apparent resistivity ρ_a for each separation and construct a table of the corresponding values of ρ_a/ρ_1 or σ_a/σ_1. From the theoretical curves find the corresponding values of h/a (and then h) for each value of σ_a/σ_1, for $k = 1 \cdot 0$, $0 \cdot 9$, ..., to $0 \cdot 1$. The resulting table will then show that for some particular value of k, the values of h as determined for the various electrode separations will be approximately the same. The average value of these h's will then give the depth of overburden. The following results obtained by Tagg will make the procedure clear. The value of ρ_1 was found to be 6703 ohm-cm. and from the apparent resistivity at each stake interval, the following table may be constructed.

Electrode interval in ft.	Apparent resistivity in ohm-in.	Ratio σ_a/σ_1
150	8,960	0·748
200	10,740	0·625
250	12,320	0·544
300	13,860	0·483
350	15,220	0·441
400	16,480	0·407

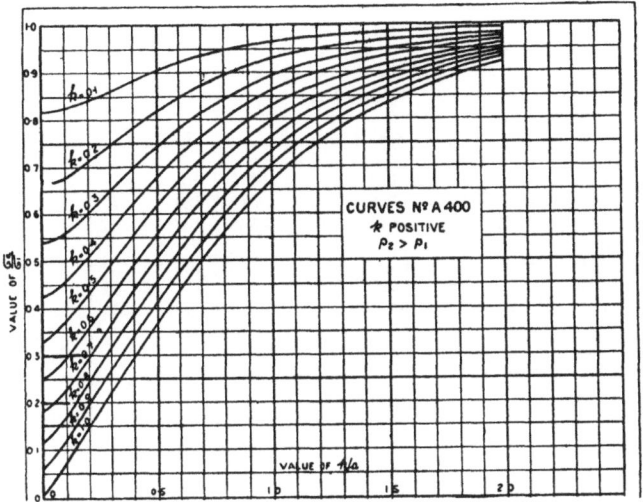

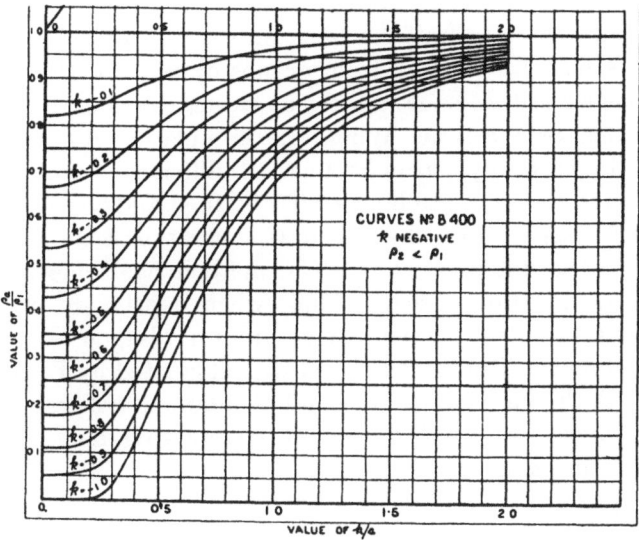

Fig. 65. Tagg's curves giving the relation of ρ_α/ρ_1 and σ_α/σ_1 with k.

From the curves given in Fig. 65 the following table is obtained.

Value of k	150 ft. $\sigma_a/\sigma_1=0{\cdot}748$		200 ft. $\sigma_a/\sigma_1=0{\cdot}625$		250 ft. $\sigma_a/\sigma_1=0{\cdot}544$		300 ft. $\sigma_a/\sigma_1=0{\cdot}483$		350 ft. $\sigma_a/\sigma_1=0{\cdot}441$		400 ft. $\sigma_a/\sigma_1=0{\cdot}407$	
	h/a	h	h/a	h	h/a	h	h/a	h	h/a	h	h/a	h
1·0	1·19	179	0·915	183	0·770	193	0·675	202	0·61	214	0·560	224
0·9	1·12	168	0·850	170	0·705	176	0·610	183	0·545	191	0·500	200
0·8	1·045	157	0·775	155	0·640	160	0·545	163	0·485	170	0·435	174
0·7	0·96	144	0·700	140	0·565	141	0·478	143	0·41	144	0·36	144
0·6	0·87	130·5	0·620	124	0·485	121	0·390	117	0·325	114	0·28	112
0·5	0·785	118	0·525	105	0·39	97·5	0·295	74	0·226	78	0·17	68
0·4	0·66	99	0·42	84	0·27	67·5	0·16	40	0·06	21		
0·3	0·525	79	0·26	52	0·03	7·5						
0·2	0·315	47										
0·1												

From this table it is evident that the values of h agree fairly closely for $k = 0{\cdot}7$. The mean value is 142·6 ft., which will be the depth of overburden.

[The three-layer resistivity problem has been treated in a similar manner by Wetzel and McMurray (1937).]

Vertical depths to uniform horizontal strata of different resistivities may be deduced from graphs drawn when the sums of the mean resistivities are plotted as ordinates against electrode separations as abscissae. When the cumulative resistance is plotted in this way, straight lines of different slopes corresponding to the separate layers are obtained and the points of intersection as read on the electrode separation scale give the depths to the different layers. The method has some mathematical justification (Ruedy, 1945) when depths are not too great and the strata have fairly uniform resistivities. It is simpler than the Tagg method and that suggested by I. Roman (1941). Both the Rooney rule and this cumulative resistance method serve as useful guides in interpreting depths from resistivity measurements.

The apparent resistivities deduced from the Megger surveys shown in Fig. 56 have been plotted as cumulative resistance graphs in Fig. 66. An inspection of the plots in this latter figure indicates discontinuities at depths (electrode separations) of approximately 320, 575 and 785 ft. The results are in fair agreement with those deduced by Rooney's rule, except the discontinuity at about 300 ft. was not noted in Fig. 56. Also the break at about 600 ft. observed on the N–S curve in the Rooney graph is not evident in the cumulative resistivity plot on that line but is quite evident on the E–W line. The advantage of using all available methods of interpretation will be obvious.

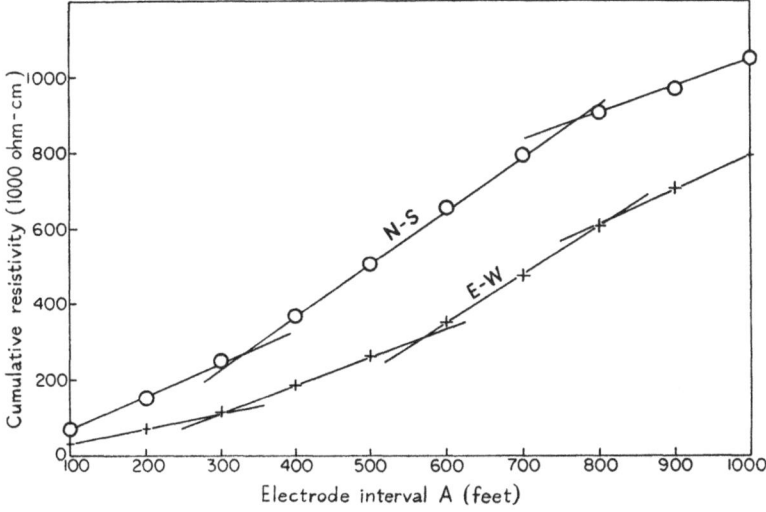

Fig. 66. Change of cumulative resistance with electrode separation.

The cumulative resistance method of interpreting the results of resistivity measurements has been very successful in determining overburden in highway and dam-site engineering (Moore, 1952). The depths met with in such work vary from a few feet to 200 ft. and more. The accuracy obtained is quite sufficient and very useful for such construction purposes.

(e) The Single-method Probe

An alternative method to that of Gish and Rooney is the exploration of the potential field around one only of two electrodes. For this purpose one electrode consisting of a stake is placed

over the exact spot under which it is proposed to probe electrically rather than with the more costly method of the diamond
drill. This scheme may be specially commended to geologists,
engineers and mining men who are anxious to find the depth of
any discontinuity underground. Two electrodes are used; one,
as explained, at the station of exploration, the other anywhere
in the earth, but a long way from the first. If the depth probe
is for a few hundred feet, the second or 'far' electrode should
be at least four or five times that distance away. A long conducting insulated wire must connect the far electrode to one or
more 'B' batteries, each 45V., arranged in series. A short wire
will connect the other pole of the battery to the near or 'home'
electrode.

The potentials near the home electrode are explored along any
one straight line, on the earth's surface, which passes through the
electrode. The equipotentials in homogeneous ground will be very
nearly hemispheres, with the home electrode as centre. This is
true because the distant electrode is too far away to distort them.

It is easy to think of these hemispheres as *bowls*, or perhaps
it may help to regard them as resembling successive layers in an
onion which has been cut into halves.

In Fig. 67 the centre C is the home station or electrode,
B is the battery, A is an ammeter, while the far electrode is
500–1000 ft. away!

The rims of the bowls are on the surface, and we may suppose
that the lowest drawn in the diagram is penetrating into a *better*
conducting material than the higher stratum. The last two bowls
will thereby be partially *short circuited* and the potential between
them will drop, or be decreased below normal. The figure will
also be deformed thereby in a perfectly known manner, but we
need not worry about that at present.

If now we take a potentiometer and two electrodes and
measure the potential differences between D and E, then between
E and F, and so on, step by step, we shall find an unwonted low
value for the potential between F and G, because our (imaginary)
bowls are partly short-circuited by the good conducting layer.

If the lower layer is a *worse* conductor than the upper strata,

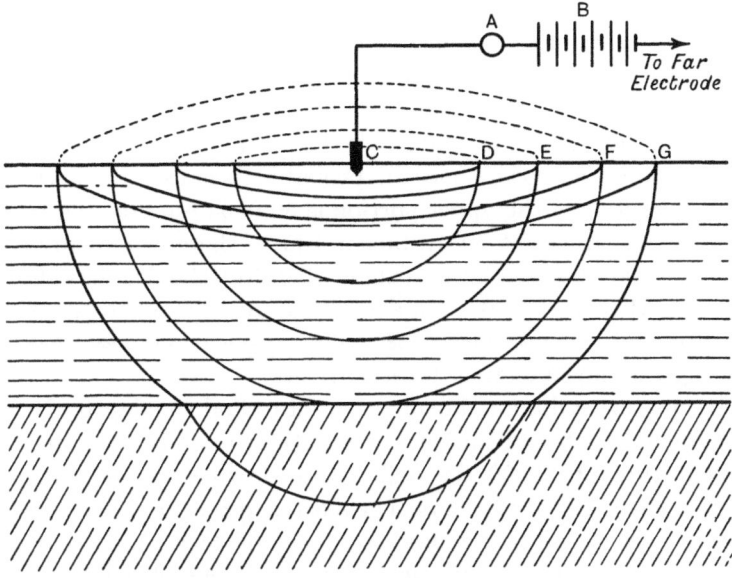

Fig. 67. The single-electrode probe in profile showing the equipotentials
with the lowest drawn entering a better conductor.

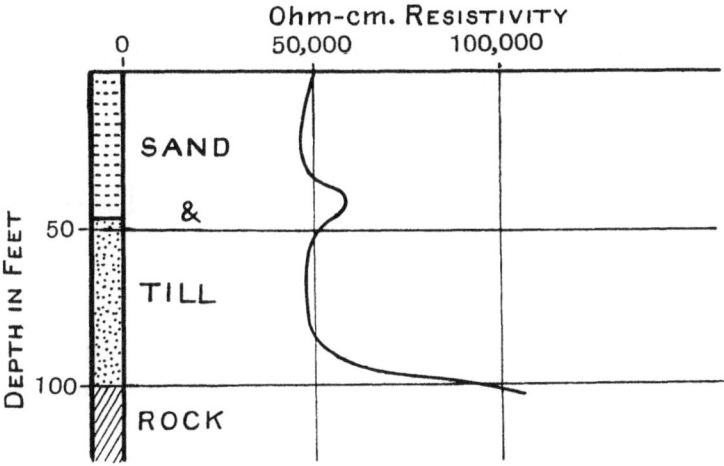

Fig. 68. Depth resistivity curve determining depth of rock for a dam
site, using single-electrode probe.

then the potential between F and G will be larger because the bowls are partially *insulated* as compared with the better conducting overburden. The result may be indicated by a diagram (Fig. 68).

The upper 100 ft. are sands, clays and till, which conduct far better than the rock beneath, probably because of moisture. The resistivity of the upper portion is 50,000 ohm-cm. and there is a slight increase at 50 ft., but at 100 ft. there is a steady rise of resistivity to much larger values, therefore the depth is 100 ft. down to the rock. This is an actual example from the paper by Crosby and Leonardon (1928). The application was that of finding the depth of rock for a dam site.

In one case a diamond drill was used to find the depth of rock under sand, clay and till. After a month or more of work and an expenditure of \$9000 the drill broke down without reaching rock, on account of boulders. The electric probe gave the correct depth to rock in half a day at a cost of \$65. It is scarcely necessary to add that this method can only detect underground changes which involve variation in *electrical* conductivity.

The necessary mathematical calculations are simple. The current I that enters at C (the electrode or stake) must pass from shell to shell. Let r be the radius of any shell and let dr be the small distance to the next shell, then if ρ be the resistivity, a and b the radii of two shells, the resistance between them is

$$R = \int_b^a \frac{\rho\,dr}{2\pi r^2} = \frac{\rho}{2\pi}\left(\frac{1}{b} - \frac{1}{a}\right),$$

but $R = V/I$, where V is the difference of potential between the shells of radii a and b, hence

$$\rho = \frac{2\pi ab}{a-b}\frac{V}{I}.$$

It is sometimes convenient to take measurements with $b = \frac{1}{2}a$, when $\rho = 2\pi a V/I$, which has the appearance of Wenner's formula, but not the substance.*

* Here again, King's investigations indicate that the geometrical picture and the mathematical result here given are alike too crude to fit the actual circumstances.

Now plot ρ against depth, and a variation in ρ will mean better or worse conductors as the case may be, and the depth of this change is noted. Although direct-current methods are here described, the use of alternating currents or of the Megger is clearly possible.

Leonardon has used this method of Schlumberger to map the bedrock of dolomite and magnesian limestone under boulder clay near Morrisburg on the River St Lawrence. By electrical exploration over 101 stations in a region of 100 acres he prepared a contour map of the bedrock (Fig. 69) superposed on the contour map of the surface elevation. Checks were made at four stations by actual drilling with satisfactory results. This investigation was made with a view to building a dam across the river. The total cost of the electrical survey is quoted as $4500, or about £900.

One more example will suffice to illustrate the possibilities of the single-electrode probe method. Professor T. Alty* applied this method to determine the thickness of overburden above rock-salt deposits in England. The geological structure of the field was well known from numerous bore-hole records and consists of a surface covering of drift below which four or five hundred feet of marl overlie a layer of almost pure rock salt, extending several hundred feet below.

Alty placed his distant electrode 3500 ft. away, used non-polarizable copper sulphate porous pot electrodes 100 ft. apart connected to a Cambridge 'unipivot' potentiometer for measuring the potential differences, and an ammeter in the circuit supplying the current to the two electrodes from a 300V. battery of accumulators. The current was reversed and measurements repeated at each station to eliminate the effects of the natural earth currents, the average values of I and V being used in the calculations of the mean resistivity. The values of ρ were found for different values of a, the interval $a-b$ being kept 100 ft.

In Fig. 70 the results of plotting ρ against a are given and it is evident that as soon as a becomes equal to the depth of the top of the rock salt, which possesses a high resistivity, found by

* Privately communicated to the authors.

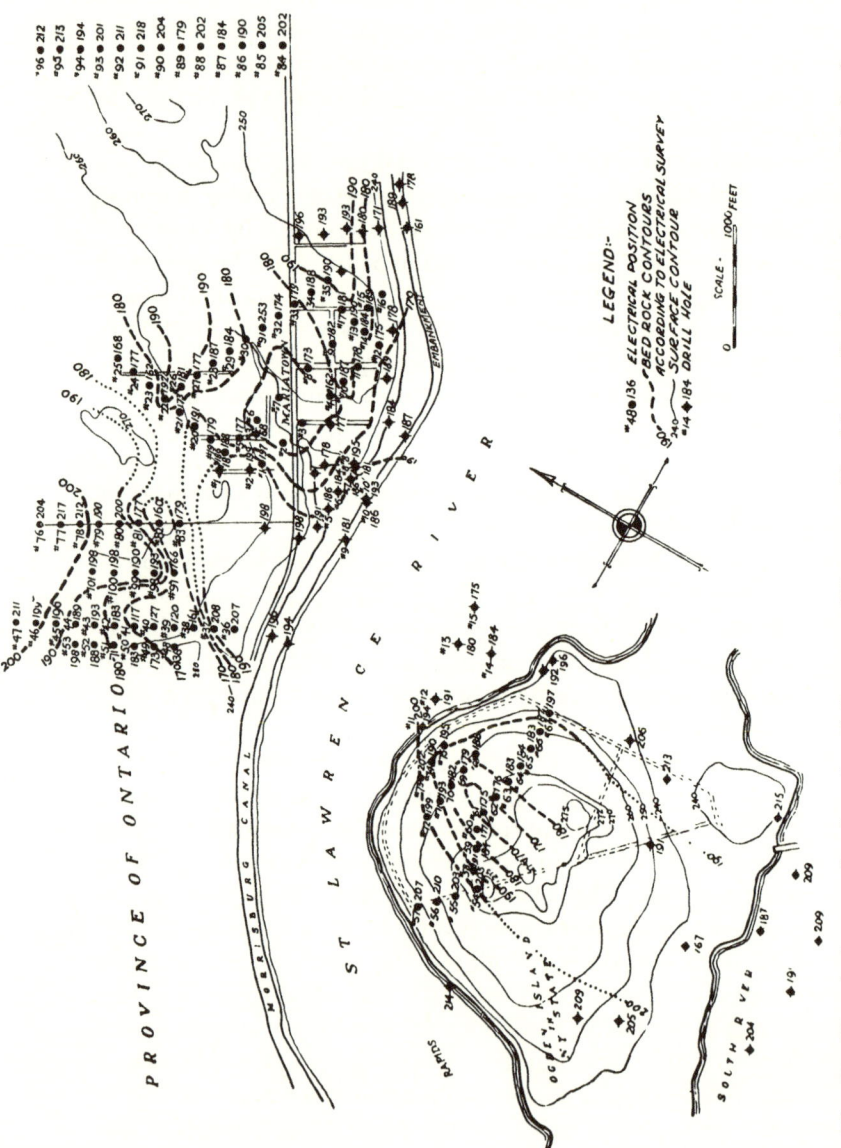

Fig. 69. Electrical survey for dam site, St Lawrence River, near Morrisburg. Surface contours, and bedrock contours are shown. Difference gives depth of glacial drift.

actual laboratory tests to be about 2×10^6 ohm-cm., there is a sudden rise in the value of the mean resistivity. The depth found in this way checked remarkably well with the actual amount of overburden in the places where the results could be compared with the drill-hole records.

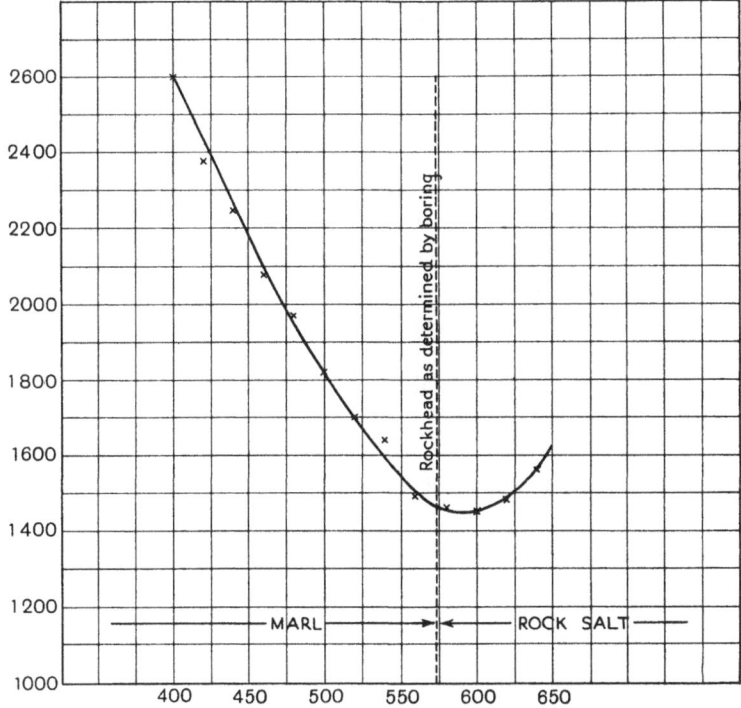

Fig. 70. Resistivity curve, single-electrode probe, detecting rock salt (high resistance) under marl. Cheshire, England (Alty).

Alty was able also to detect the top of the conical cavity formed in the rock salt by the water which is forced down the bore-hole to dissolve the salt and then pumped to the surface in the form of brine. Bore-holes reach nearly to the bottom of the layer of salt and, as the salt dissolves, the conical cavity grows upwards. The cavity filled with brine is a good conductor, hence when a survey by the single-probe method is made over the surface, the poorly conducting dry salt layer is first located by the rise in resistivity and, as greater depths are reached, the presence of the better conducting brine in the cavity is detected by the sudden decrease in mean resistivity. This diminution occurred in one survey when a became 860 ft. which he associated with the depth to the conducting cavity.

Dr L. Gilchrist* prefers to use four electrodes, each placed north, south, east and west, four or five hundred feet from the central electrode. The four electrodes are joined to the same pole of the battery, the ground resistance of each being adjusted to be equal either by driving more iron stakes in the ground and watering to reduce the resistance or by inserting auxiliary resistance in other cases until the currents from each of the four distant electrodes to the central one are equal. In this way a more symmetrical distribution of current about the central electrode is obtained and the potential bowls will be more nearly spherical. Also with this arrangement, surveys may be carried to greater distances from the central electrode with the same electrode separation. When only one single far electrode is used, surveys should not be made to distances greater than one-fifth of the electrode separation. Gilchrist considers the extra labour in running out and adjusting the resistances of the four electrodes justified by the results he obtains.

A striking example, due to Gilchrist, is shown in Fig. 71 which gives the mean resistivity for different depths over an area near James Bay, N. Ontario, in which lignite occurred below an overburden of muskeg and boulder clay, with a current electrode separation of 800 ft. It will be noted that the resistivity at first increases, but that there is a marked decrease as soon as the bowls enter the better conducting lignite. The results were so encouraging that the Ontario Government sent parties to the region in the autumn of 1930 to carry out further geophysical investigations.

Gilchrist (1950) has also used a somewhat modified lay-out for surveying deep below the surface from a drill hole (Fig. 72). Line electrodes, several hundred feet long, are laid out sym-

* Privately communicated to the authors.

metrically about the current electrode C_1 at the bottom of the drill hole, about a half mile away on the surface from the drill hole. Each of the line electrodes in turn serves as the other

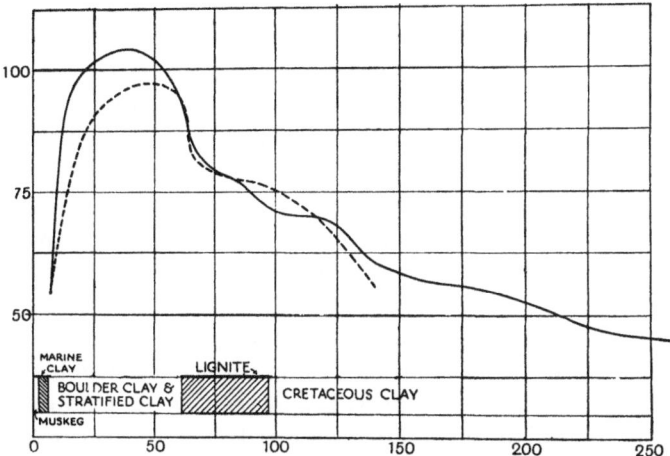

Fig. 71. Resistivity survey over lignite deposit under boulder clay, James Bay, N. Ontario (dotted curve). Megger survey (full curve). Potentiometer and milliammeter survey.

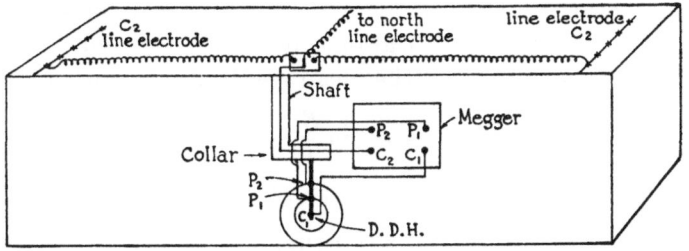

C_1, C_2 - Current electrodes P_1, P_2 - Potential electrodes
Collar is 1125 ft. below surface

Fig. 72. Gilchrist drill-hole survey method.

current electrode, C_2, their contact resistances having been adjusted to be equal. The two potential electrodes, P_1 and P_2, make contact with the side of the drill hole and are connected by insulated wires to the potential terminals of the geophysical Megger. If the distance of these electrodes from C_1 be a and b

respectively, then the mean resistivity ρ of the material between the two potential surfaces, is $\rho = \dfrac{4\pi ab}{b-a} R$, where R is the Megger reading in ohms. The interval $b - a$ is kept small, about 50 ft., the values of ρ measured for different distances of b from C_1, when each of the four line electrodes are used as C_2. If there is any change in the mean resistivity in the four directions, the curves will indicate this fact and also at what depth such changes occur. Furthermore, by placing C_1 and P_1 down one drill hole and C_2, P_2 down another, he has obtained information about the mean resistivity of the medium between the two drill holes which has been helpful in deciding where the next hole should be drilled (Gilchrist *et al.* 1950, pp. 3–12).

Lee (1936) has designed a convenient modification of the apparatus used for measuring ground resistivity which he has called the geoscope. Its merit lies in the fact that no calculations are required in determining the resistivity from the readings. The instrument reads directly the resistivity, and the apparatus is applicable to any of the standard methods, such as those using the Wenner formula with equally spaced electrodes or the Lee modification. The theory of the instrument is briefly this. If a is the electrode separation, then with the four equally spaced electrodes (p. 106) the mean resistivity is ρ, where

$$\rho = 2\pi a \frac{E}{I},$$

which becomes when a is in feet, E in millivolts and I in milliamperes
$$= 191aE/I = 191Ea/I.$$

The current introduced into the ground through the current electrodes is made numerically equal to the electrode separation a, or to some convenient fraction of a, such as $\frac{1}{1000}$ of a. Thus, if a is 400 ft. I is made 0·400 mA. When I is made numerically equal to a, the resistivity $\rho = 191E$. If the potentiometer circuit be so arranged that the indication of potential difference on the potentiometer of the instrument E_m really means $1\cdot91E$ then
$$\rho = 100E_m \text{ ohm-cm.},$$

and the instrument becomes direct reading except for the multi-plication by the factor 100. Thus, if the reading were 121, the resistivity would be 12,100 ohm-cm.

A simple circuit diagram of how this adjustment is made in the instrument for use, applying the Gish-Rooney method (p. 113), is shown in Fig. 73. The current electrodes C_1, C_2 are

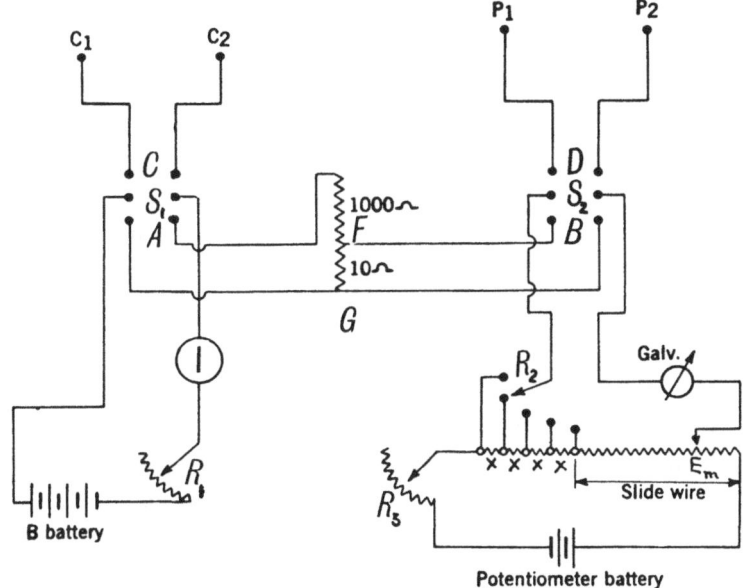

Fig. 73. Circuit diagram of Lee's geoscope.

connected to the two terminals C of a double-pole double-throw switch S_1. The current from a 'B' battery passes through a variable rheostat R_1 and milliammeter I to S_1. The other poles A of the switch are connected by resistances 1000 ohms and 10 ohms in series. The two ends of the 10-ohm coil are connected to the poles B of another double-throw switch S_2. The other poles D are joined to the potential electrodes P_1, P_2. The switch S_2 is connected to a potentiometer consisting of a slide wire, marked off into 100 equal divisions and four equal coils X, each having a resistance equivalent to that of the slide wire. The potentio-meter battery is joined in series with an adjustable rheostat R_3.

The switch R_2 connects one pole of S_2 to the slide wire and the other pole of the switch is joined through the galvanometer to the movable contact of the potentiometer.

To adjust the potentiometer wire, the switch S_1 is connected to A and the current I adjusted to 10 mA. by means of R_1. This produces a potential drop across FG equal to 100 mV. S_2 is then connected to B, the potentiometer switch set at 1 and E_m at 91, so the potentiometer reading is 191. The current through the potentiometer is then adjusted by means of the rheostat R_3 until the galvanometer indicates a balance. When this is the case, the actual potential difference of 100 mV across FG is balanced by 191 divisions on the potentiometer. When S_2 is connected to D, the instrument will multiply any potential difference across P_1, P_2 by the factor 1·91 and so indicate resistivity directly. To operate the instrument S_1 is connected to C, the current I adjusted to be numerically equal to the electrode separation in feet and the potential, read on the potentiometer, will give mean resistivity directly in ohm-cm.

OTHER APPLICATIONS

The resistivity methods described in this chapter have found useful application in fields other than that of locating mineral deposits and ore-bodies. They have been used with great success to the determination of overburden in connexion with road building, already referred to (Shepard, 1935; Moore, 1952), for the location of cavities in limestone formations at dam sites and in railway construction (Stipe and Kelly, 1937), and in finding ground-water resources (Swartz, 1937; Paver, 1945). They all have applications in special cases, and it is left to the trained judgement of the geophysicist, aided by the knowledge of the geologist and the mining engineer, to decide which method may be used with the greatest chance of success and with the least expense. There are, however, other methods, also electrical, which may be more suited to the particular location which is to be surveyed. These will be described in the next chapter.

PROBLEMS

1. In a survey over an area in which sulphide ore was expected to exist under shallow overburden, the self-potential method was used. One non-polarizable pot was kept stationary at the zero point and the differences in potential between this point and the moving pot at points 50 ft. apart along a line are listed in the following table. The figures give the potential of the moving pot relative to the stationary one. Where would you recommend the best region to drill for locating the possible ore-body?

Station	Potential difference in mV.	Station	Potential difference in mV.	Station	Potential difference in mV.	Station	Potential difference in mV.
0	0	8	− 400	16	− 900	24	− 50
1	0	9	− 170	17	− 800	25	0
2	− 10	10	− 180	18	− 900	26	0
3	− 45	11	− 145	19	− 900	27	+ 8
4	− 445	12	− 190	20	− 1000	28	0
5	− 430	13	− 200	21	− 1000	29	− 2
6	− 200	14	− 280	22	− 500	30	0
7	− 200	15	− 400	23	− 200		

2. The following readings were taken at Barton Hill using the Gish-Rooney method of measuring the mean apparent resistivity for various electrode separations. The average surface resistivity was 47,450 ohm-cm. Using Tagg's curves for a two-layer problem find the approximate depth of overburden in feet and the resistivity of the lower layer.

Electrode separation in ft.	Apparent mean resistivity in ohm-cm.
200	66,500
300	94,000
400	113,300
500	129,400
600	166,400
700	195,200
800	193,100
900	187,500
1000	195,200

3. In an investigation to determine the depth of overburden to a conducting layer of brine at Malagash, Nova Scotia, the following readings were taken with a Megger with the four equally spaced electrodes at the intervals given. The centre of the system was kept

constant and the surface layer was found to have a resistivity of 2900 ohm-cm. Determine the probable depth of overburden and the resistivity of the brine layer.

Electrode separation in ft.	Apparent resistivity in ohm-cm.	Electrode separation in ft.	Apparent resistivity in ohm-cm.
40	2850	220	1290
60	2710	240	1130
80	2530	260	990
100	2355	280	870
120	2170	300	780
140	1980	320	710
160	1800	340	670
180	1630	360	650
200	1455	380	645

4. The following readings were taken with a potentiometer between two non-polarizable porous pot electrodes 50 ft. apart along a line in British Columbia, the region being covered with overburden. The potential of the forward electrode with reference to the hinder one is given in each case. Determine the most probable region for the location of oxidizing ore.

Station		Potential difference in mV.	Station		Potential difference in mV.
1.	0–50	− 1	19.	900–950	−10
2.	50–100	− 1	20.	950–1000	−19
3.	100–150	− 3	21.	1000–1050	− 3
4.	150–200	− 2	22.	1050–1100	−14
5.	200–250	− 2	23.	1100–1150	− 6
6.	250–300	− 4	24.	1150–1200	− 3
7.	300–350	−11	25.	1200–1250	+ 4
8.	350–400	+10	26.	1250–1300	− 2
9.	400–450	− 5	27.	1300–1350	+ 5
10.	450–500	−22	28.	1350–1400	+14
11.	500–550	+15	29.	1400–1450	+17
12.	550–600	−13	30.	1450–1500	+13
13.	600–650	+12	31.	1500–1550	+ 9
14.	650–700	+ 5	32.	1550–1600	+17
15.	700–750	−15	33.	1600–1650	+ 4
16.	750–800	− 5	34.	1650–1700	+ 8
17.	800–850	− 8	35.	1700–1750	− 4
18.	850–900	−15			−

5. To check the estimate of overburden above a small pyrrhotite vein in the Sudbury basin, readings of the apparent resistivity were taken with a Megger at the following intervals. The surface

resistivity was measured as 96.8×10^4 ohm-cm. Calculate the over-burden and the resistivity of the vein material.

Electrode separations in ft.	Apparent resistivity in 10^4 ohm-cm.
80	77·0
120	57·9
160	38·2
200	24·8
240	17·2
280	18·2

6. In an investigation in Kentucky, it was necessary to estimate the thickness of a layer of limey sandstone deposited above a layer of limestone. The four equally spaced electrode method was adopted using a Megger instrument. The resistivity of the sandstone was found by using small electrode intervals to be 1,567,000 ohm-cm. From the following readings estimate the thickness of the sandstone layer. Calculate also the mean resistivity of the limestone.

Electrode interval in ft.	Mean resistivity in ohm-cm.
40	1,212,000
60	782,000
80	590,000
100	451,000
120	424,000
140	370,000
160	344,000
180	324,000

7. The following readings were taken using the four equally spaced electrode method with the object of determining the depth of glacial drift above bed-rock. Using the cumulative resistance method, find the approximate depth of overburden.

Electrode interval in ft.	Apparent resistivity in ohm-cm.	Electrode interval in ft.	Apparent resistivity in ohm-cm.
50	2828	350	1675
100	2600	400	1545
150	2380	450	1465
200	2190	500	1460
250	1980	550	1500
300	1815	600	1620

8. In an investigation to find the depth to different strata in the Sudbury basin, the following readings were taken of the mean resistivities for different stake intervals, using the Wenner layout.

What are the depths as determined by the cumulative resistance method?

Electrode interval in ft.	Mean resistivity in ohm-cm.	Electrode interval in ft.	Mean resistivity in ohm-cm.
40	8650	240	662
80	2310	300	1020
120	1020	360	1720
160	588	420	3000
200	743	480	4900

9. A geophysical survey, using the self-potential method, was made over some claims in Colorado. The non-polarizable electrodes were 50 ft. apart and the potential of the forward pot was measured in terms of the hinder one. From the readings along one line, where would be the most probable location of the ore?

Station	Potential difference in mV.	Station	Potential difference in mV.
1. 0–50	− 8	15. 700–750	− 9
2. 50–100	− 5	16. 750–800	+ 6
3. 100–150	− 6	17. 800–850	+14
4. 150–200	− 7	18. 850–900	+10
5. 200–250	−13	19. 900–950	− 5
6. 250–300	−33	20. 950–1000	+ 6
7. 300–350	−29	21. 1000–1050	+ 5
8. 350–400	− 5	22. 1050–1100	+ 6
9. 400–450	0	23. 1100–1150	+ 4
10. 450–500	+ 1	24. 1150–1200	+ 4
11. 500–550	0	25. 1200–1250	− 2
12. 550–600	0	26. 1250–1300	− 1
13. 600–650	+ 1	27. 1300–1350	+ 1
14. 650–700	0	28. 1350–1400	+ 1

10. To determine the depth of Quaternary gravel a resistivity survey was made in the Laochunmiao oil field. From the following results of such a survey, find the depths of the Quaternary and Tertiary beds.

Electrode interval in ft.	Mean resistivity in kilo-ohm cm.	Electrode interval in ft.	Mean resistivity in kilo-ohm cm.
45	49·6	300	99
120	125	360	96
150	160	420	100
180	180	480	105
240	190	540	110

11. In constructing a new highway, the depth of overburden was estimated by the cumulative resistance method. From the following readings, estimate the excavation necessary to reach bed-rock.

Electrode interval in ft.	Mean resistivity in ohm-cm.	Electrode interval in ft.	Mean resistivity in ohm-cm.
5	7810	30	5120
10	5600	35	5980
15	4980	40	7600
20	4710	45	7980
25	4600	50	7220

CHAPTER IV

ELECTROMAGNETIC METHODS

INTRODUCTION

Electromagnetic methods have been widely used for locating ore-bodies. They provide a means of discovering conducting masses beneath overburden under conditions in which the resistivity methods might prove difficult. They are especially suitable over dry or desert areas, frozen lakes and snow-covered land. They have been employed to a large extent in exploration work in Sweden, Germany, the United States and Canada, as well as in many other countries. Most of the arrangements are operated under patents granted to the different inventors or companies, but the actual principles underlying the schemes are all well known to the physicist, and in many cases the application is the only novel feature. We have already described the magnetic and electrical methods, and now in this chapter we shall deal with those devices which employ an artificial electromagnetic field. Such fields may be produced by either (1) a horizontal loop or a system of conductors, or (2) a vertical loop. An alternating current, usually but not always of audible frequency, is circulated round such coils of wire, producing in the neighbourhood electromagnetic waves. These waves spread out in space both above and below ground and may stimulate currents by induction, according to the well-known laws of Faraday and Maxwell, in conducting masses of ore beneath the surface. As a result of this excitation the ore-body itself will re-radiate and alter the field above the earth. There results distortion, either in (1) direction, (2) intensity, or (3) quality of the electromagnetic field produced by the primary loop. This modification of the primary field is detected and interpreted according to the particular method adopted for locating the hidden conductors.

The apparent simplicity of the electromagnetic methods, however, is complicated by the fact that the detecting coil receives the radiations from all the different conductors in its vicinity, and in many cases the signals from the exciting loop as well. The coil detects only the resultant of these various fields. The resultant electromagnetic field has in the neighbourhood of excited ore-bodies peculiar properties which are expressed by stating that it is *elliptically polarized*.

ELLIPTICAL POLARIZATION

There is frequent reference in the literature of applied geophysics to the fact that in general the electromagnetic field in the neighbourhood of excited ore-bodies is *elliptically polarized*. The idea is familiar to physicists, but a brief statement of the situation may not be out of place.

If there are two circuits, and in one of them a current be varied in any manner, started, stopped, increased, decreased, or alternated, then in the other closed circuit there will be an electromotive force giving rise in general to an *opposing* current.

So when a loop horizontal or vertical, is used in prospecting and excited with an alternating current all the conducting bodies in its neighbourhood are stimulated in consequence, and a multitude of opposing currents arise. These are out of phase then with the primary current.

Although in an electromagnetic field there is at every point, and at every instant, both magnetic and electric intensity, which may each be represented by a straight line or vector, yet it is more convenient in description to deal with one of them, and we will, therefore, select the magnetic vectors.

If magnetic vectors at a point are all of the same phase and of the same frequency then their resultant can be found by the ordinary well-known parallelogram law of vectors.

If we deal with two vectors of the same frequency which are at right angles to one another, and the one a quarter phase (90°) behind the other, then the resultant vector *revolves* with its extremity on an ellipse. This is the well-known case of two harmonic motions at right angles which give always an elliptic motion, including as special cases a circle and a straight line.

Briefly, in mathematical language, we have

$$x = a \sin 2\pi ft,$$
$$y = b \cos 2\pi ft,$$

where f is the frequency, t is the time, a, b the amplitudes.
Then

$$\frac{x^2}{a_2} + \frac{y^2}{b^2} = 1,$$

the equation of an ellipse. So also for any two vectors in one plane

$$x = a \sin 2\pi ft,$$
$$y = b \sin (2\pi ft + \alpha),$$

where α is a phase difference, the elimination of $2\pi ft$ leads at once to an ellipse whose equation is

$$\frac{x^2}{a^2} - \frac{2xy}{ab} \cos \alpha + \frac{y^2}{b^2} = \sin^2 \alpha.$$

But in our practical case of an excited loop and of excited ore-bodies and conductors there are vectors in various directions and various phases at every point of the field. What is their resultant?

The answer is unambiguous. It is a single vector whose extremity describes a *plane ellipse*. Such a field is called elliptically polarized.

The proof is not difficult.

Any vector can be replaced by three components of the same phase along three selected lines or axes at right angles. Thus any one vector can be replaced by

$$X = a \sin (\theta + \alpha),$$
$$Y = b \sin (\theta + \alpha),$$
$$Z = c \sin (\theta + \alpha).$$

Where θ is written in place of $2\pi ft$, α is a phase angle, and a, b, c are constants.

Thus for any number of vectors we have

$$X = \Sigma a \sin (\theta + \alpha),$$
$$Y = \Sigma b \sin (\theta + \alpha),$$
$$Z = \Sigma c \sin (\theta + \alpha),$$

where Σ denotes addition, and these are equivalent to

$$\left.\begin{array}{l} X = A_1 \sin \theta + B_1 \cos \theta, \\ Y = A_2 \sin \theta + B_2 \cos \theta, \\ Z = A_3 \sin \theta + B_3 \cos \theta. \end{array}\right\}$$

The three sine terms give rise to a single vector, and the three cosine terms give rise to another single vector, which is 90° out of phase with the former. Hence the result is that the final vector moves in a plane, and that its extremity follows an ellipse; there is elliptic polarization.

The practical result of this is important. It is not possible by a single position of a coil to determine a direction which will by its minimum current (and audibility) indicate an ore-body. The more complex difficulties of this subject are summarized in an article by Ambronn (1928b).

A detecting coil will only give complete silence when its plane is in the same plane as the ellipse of polarization for the magnetic vector. It is possible in many instances to take a suitable coil, free to move in azimuth and dip, and actually to place it in the plane containing the ellipse. We will call this the SILENCE PLANE.

On rotation through other positions it may be possible to get a minimum, either broad or sharp. But there is one, and only one, plane of complete silence. When found, the coil can be placed at right angles to the major axis, usually near or in a vertical plane, and then the flux through the coil can be measured with amplifier, rectifier (copper oxide) and micro-ammeter. Similarly, the coil can be placed at right angles to the minor axis and the flux again measured.

An exploration of the ellipse of polarization is not so difficult as its mathematical description, but neither one nor the other will lead directly to the discovery of ore-bodies, and it is, therefore, necessary to discuss some of the more reliable methods.

Before doing so, however, it is necessary to point out that in the above work we have assumed a single frequency f, but in practice it is difficult to get a pure sine wave free from all harmonics.

Suppose, for example, that one coil is balanced against another so that frequency f is eliminated, then, owing to capacitance or inductance, there will not be a balance for a different frequency f'.

Or again, in one scheme* a horizontal coil is used and a vertical coil is also used at the same place. These extremities are connected to a condenser and a balance can be obtained by variation of inductance or of capacitance, or by varying the number of turns in one coil. In this way the fundamental frequency can be balanced out, but its harmonics will *not* balance out.

It is well known that in all but the more modern gramophones the bass notes, as of the violoncello, or of a man's voice, were entirely absent, and yet the ear and the brain were able to infer their presence from experience. The same is true of the ordinary telephone transmitter and receiver. So, when prospecting with feeble reception under amplification, it is most difficult to secure elimination or balance with the main 500-cycle frequency when numerous harmonics are present. There are various filters which will choke these either at the generator or at the receiver, preferably the former. Hence the conclusion:

It is important to use a generator of as pure note as possible, and to eliminate by filters any harmonics which may be present.

In this way a survey will be made more swiftly, and its accuracy will be enhanced.

The other point is well made in the words of Ambronn:

It is only by taking into account the elliptic polarization of the electromagnetic field of thee arth currents that it is possible to carry out a scientifically accurate investigation of the subsoil by alternating current.

RESONANCE AND THE IMPORTANCE OF TUNING

When a direct, uniform and continuous current is passed through a coil the relation between electromotive force, current and resistance is given by Ohm's law.

$$E = RI,$$

* The Bieler-Watson method, see p. 170.

but when an alternating current is used the corresponding relation is
$$E = ZI,$$
where Z is called the impedance. The current lags behind the electromotive force by a phase angle ϕ whose cosine is R/Z. The impedance Z is a blend of three factors, the resistance R, the reactance due to inductance $X_L = 2\pi f L$, and the reactance due

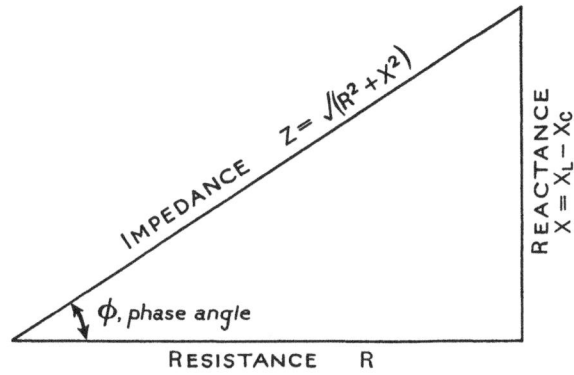

Fig. 74. The relation between impedance, reactance, resistance and phase angle. For resonance tuning make X equal to zero, so that Z equals R.

to capacitance $X_C = \dfrac{1}{2\pi f C}$, where f is the frequency, L the self-inductance in henries, and C the capacitance in farads. If we write $X = X_L - X_C$, then a right-angled triangle shows the connexion between resistance, reactance, impedance and phase angle. It is desirable to make Z as small as possible, where $Z^2 = R^2 + \left(2\pi f L - \dfrac{1}{2\pi f C}\right)^2$, and this can be done by making $2\pi f L = \dfrac{1}{2\pi f C}$, and then $Z = R$, or the impendance has the lowest possible value equal to the resistance alone. Hence in all measurements with coils, whether using headphones, or rectifier and galvanometer, it is desirable to introduce a condenser with adjustable capacity so that the relation holds

$$2\pi \text{ times the frequency} = \frac{1}{\sqrt{(LC)}}.$$

Example

A five-turn loop, 50 ft. in diameter, was excited with 1·52 amp. of 500-cycle frequency. A keen observer, with a coil of 841 turns held horizontally and connected with good headphones, walked away from the centre of the loop to 363 ft., when the signals became just inaudible. He then introduced the condenser, and rapidly tested for a maximum effect by altering the dials. He now walked to a further total distance of 944 ft. from the centre of the loop before the signals again became just inaudible. This was done without the use of the amplifier. With amplification and a larger current in the loop, there is no doubt that signals could be read about a mile away. The main point, however, is the fact that the distance reached was nearly trebled by resonance reception. The inverse cube law of distance holds, so that in this case the effect was increased twenty-five-fold. It is often useful to employ a dial type of condenser, and sometimes a vernier condenser for fine adjustment. It is quite unnecessary to carry these for constant use in the field, however, because a small condenser of suitable capacitance such as those used in radio work can be permanently attached to the receiving coil so as to secure resonance, and the headphones can be attached to terminals which lead to the coil and to the condenser in parallel.

HORIZONTAL LOOP METHODS

The location of conductors by electromagnetic methods depends largely on two factors. In the first place the conducting ore-body must be excited by some form of radiating loop, and secondly the electric currents induced in the hidden conductor must be detected on the surface with special types of coils and registering instruments. The question naturally arises, What is the best form of loop to use, and how may the signals from the ore-body be received to the greatest advantage? This question has been answered in different ways, many geophysicists possessing their own patented methods, a fact which has stimulated new and further developments, but has obscured to some extent the relative merits of the various electromagnetic devices.

The electromagnetic methods fall naturally into two distinct classes, depending upon whether the exciting loop is placed in a horizontal plane on the ground, or is supported vertically in specially designed frames and holders. In each class we have different arrangements for receiving the disturbances caused by the presence of the ore-body and for determining from what direction the secondary electromagnetic waves come. The receiving apparatus may be designed so as to measure either direction, intensity or phase of the radiation from the conductor, and the methods differ merely in the way in which either one or two of these qualities are determined. The interpretation then follows from the observations made at known points over the area surveyed. We shall commence with those methods using a horizontal loop for excitation of the ore-body by describing a few representative arrangements of this class.

(a) Parallel-wire method

We have already described, on pp. 89–94 of Chapter III, the method of laying long bare copper wire electrodes on the ground and determining equipotentials with the help of exploring 'prodders'. When alternating current is used there will be an electromagnetic field produced in the area between and near the wires. This field will be the result (i) of the currents in the wires, (ii) of the currents through the ground between the two electrodes, and (iii) of any currents induced in the ore-bodies which may be present. If the region is free from conductors, then at a few hundred feet from the electrodes the electromagnetic effects will be mainly due to currents flowing through the ground between the copper wires.

The resultant electromagnetic field may be detected by a coil of several hundred turns of wire wound on a wooden frame. In order to determine the direction of the maximum of the resultant magnetic field, the coil is mounted in an outer frame in such a way that it may be tilted at any desired angle. To provide for setting in any azimuth, the frame is supported on a transit tripod, fitted with levelling device and scale to read the azimuth. In Fig. 75 is shown such a search coil, 2 × 3 ft. in area, wound with

400 turns of no. 26 double silk-covered copper wire. The ends of the wire are connected to suitable clips or terminals and a telephone receiver is connected either directly to the coil or, when necessary, through a three-stage audio-frequency amplifier placed in the circuit. The coil is tuned to the frequency of the exciting loop by connecting a condenser of appropriate capacity across its ends. The parallel wires are excited by alternating

Fig. 75. A search coil in the field. (Caribou Mountain, Colorado.)

current of 500-cycle frequency, so that an audible note may be heard in the headphones. For this purpose a specially designed portable generator driven by a gasoline engine is useful. The generator should be capable of giving from 5 to 10 amp. at 110 of 220 V. It has already been pointed out that in a region which is excited by an alternating magnetic field from a loop, the presence of conducting ore-bodies will so disturb the resulting field, that very little, if anything, can be deduced from direction measurements alone. We have to form a mental picture of a magnetic vector with its extremity racing round in an ellipse with the frequency employed. In consequence of this ellipticity of the resultant magnetic field, the minimum obtained by listening with the headphones to the current induced in the coil, which is rotated through any azimuth or dip, becomes very broad, and. no actual silence point is usually determined except in what has been termed the plane of silence, which contains the polarization ellipse. The more nearly the ellipticity of the ellipse approaches a circularly polarized field, the broader will the minimum become. With the size of loops and distances between electrodes used in practice, the major axis, due to the field produced by the loop, is usually much greater than the minor axis of the ellipse.

The method of procedure is to set the exploring coil on its tripod at each point in turn of the previously blocked-out area, and determine the direction of the maximum magnetic vector of the resultant electromagnetic field. To find the plane of the polarization ellipse, the coil is set vertical and the ends attached to the input side of a good amplifier, the output terminals being connected to a pair of good headphones which the observer places on his ears. It is often convenient for the observer to wear a cap, like that used by aviators, to protect the ears from any external noises such as wind or the hum of the generator that is used to supply the alternating current to the electrodes. External noise prevents the observer from making accurate settings of the coil. The coil is turned to a position of minimum intensity and the azimuth noted. If the coil is now rotated through 90° and then tilted, another minimum will usually be found. By slight adjustment of both tilt and azimuth the silence position may now be found, and the plane of the coil will then be in the plane of polarization of the magnetic vector. The azimuth and tilt are noted and similar observations are repeated at various points over the area to be surveyed.

From these readings the *presence*, but not the position of a disturbing body may be found, especially if it lies near the surface. This method, however, is not reliable. Sometimes the coil is kept vertical and that plane of minimum sound is found which gives the direction of the horizontal component of the maximum magnetic vector. If the coil be now turned through a right angle, and if the intensity of the induced current is measured with a rectifier and microammeter, then the reading will be proportional to the horizontal component of the resulting field. When the horizontal component at the various stations over the area is found, it is possible to locate the edges of the conductor beneath the surface, because the horizontal component will fluctuate (either increase or decrease) from its normal value when passing over the boundaries of the ore-body. The horizontal component of the field due to the currents in the wire electrodes will decrease as the distance from the electrode increases, and at 300 or 400 ft. it would generally be very small, and then the field due to the electrodes will be practically vertical. It is the disturbing conductor which thus causes the change in the horizontal field. Any variation in the current used in the exciting wires or loop will also affect readings in the field, and hence precautions must be taken to keep the exciting current constant. Of course the vertical component of the magnetic field may also be determined, but experience has shown that mere direction is not sufficient for ore detection. Though this parallel wire method has been used yet the loop method now to be described is perhaps to be preferred.

(b) Large horizontal loop method

Instead of using two bare parallel wires, a large circular or rectangular loop of insulated wire is laid on the ground and a current of from 5 to 10 amp., supplied by a 500-cycle alternating-current generator, is passed through the loop. This creates the electromagnetic field that is used to excite the ore-body. The resultant electromagnetic field, due to the current in the loop, will be approximately vertical at distances greater than a few hundred feet from the loop. If the region enclosed by the loop contains a good conducting body, then the electromagnetic field of the loop, which may be represented by $H = H_0 \sin 2\pi ft$, will induce a current in the conductor which will be given by $I = I_0 \cos 2\pi ft$, which is 90° out of phase with the current in the primary loop. The loop and the ore-body resemble in many respects an induction coil or transformer, except that in the case of the ore-body the current in the ore is governed nearly always by the resistance, the self-induction of the ore-body being small. The magnetic field created by the secondary current will thus be out of phase with that due to the exciting loop.

To survey a region using a large horizontal loop, three methods may be followed. In the first the strength of the vertical field at one point is compared directly with the vertical field at another station. Instead of vertical fields, horizontal field intensities may be compared. In the second method, equipotentials are located electromagnetically and plotted. The third scheme is to find the silence plane as described on p. 161 and then determine the size of the axes of the polarization ellipse by measuring the strength of the current in the coil when it measures (i) the horizontal component in the silence plane, and (ii) the vertical component in the silence plane. The phase relation between the conjugate diameters of the polarization ellipse may also be deduced. We shall first describe the comparison method.

(i) *The comparison method using large horizontal loops*

The comparison method has been used with success by Lundberg and Sundberg in Europe and in America (Lundberg, 1928; Sundberg *et al.* 1925). To make such a survey the authors used two similar coils, each 6 sq.ft. in area and containing 400 turns of insulated wire. These coils are joined in opposition with an amplifier and headphones in the circuit. When the coils are placed, the one directly over the other, there is absolute silence in the headphones. When one coil is moved 100 ft. along a given line and both coils are placed horizontally on the ground, then if there is any disturbing influence such as a conductor nearer the one coil than the other, the intensity of the electromotive force induced in the two coils will be different. As a result the note of the exciting current in the loop will be heard. To avoid any capacity effects it has been found from experience in the field that the amplifier should be inserted in the middle of the double-flex wire which connects the two coils together. By tilting one of the coils the sound in the phones may be brought to a minimum and the angle of tilt—which is the angle the plane of the coil makes with the horizontal—is then read. The secant of the angle of tilt will give the relative strength of the vertical field, at the station where the coil is tilted, to the field at the other coil which is kept horizontal and is considered as unity. The tilting of the coil brings in the effects of any horizontal component due to the ore-body, and for this reason we are introducing an error which will, in some cases, be serious. In other cases, however, the horizontal field is not sufficiently intense to vitiate the method. Proceeding along the line in this way, the field at each station is compared with that at the previous station, and from the readings a general relation of the strengths of the vertical components all along the line may be calculated.

Surveys made in this way have the advantage that any change in the intensity of the current in the exciting loop will not affect the comparison readings. It is, therefore, free from errors due to fluctuations in speed of the petrol engine used to drive the generator. In field work it may happen that the plane of elliptical polarization changes so much both in its direction and dimensions from the one station to the next, that the currents induced in the two equal exploring coils are considerably out-of-phase. The result is a broadening of the minimum to such an extent that reliable readings are difficult to secure. This difficulty may perhaps be overcome to a large extent by using a capacitance bridge, but such extra equipment encumbers the progress of a survey (see Hedstrom, 1937). When such difficulties arise it is often simpler to adopt some other method.

When the survey is completed along one line, it may be repeated along another line parallel to the first, by which process another set of figures are obtained giving the relative strengths of the vertical fields at the stations along this run. By making a set of observations in a direction at right angles to the lines already surveyed, the different series of field strengths may be tied together, and in this way the value of the vertical field at each station over the area is found in terms of any one reading which may be regarded as the standard.

When such a survey is made, the next question is, What should be the variation in the vertical and horizontal components of the field when passing over a conductor? To answer this pertinent question we must remember that the vertical field is the resultant of the field due to the current in the loop and the current in the conducting ore-body. If the ore-body lies in the central portion of the loop, we may for practical purposes suppose the field due to the loop constant and vertical. The horizontal component of the magnetic field, caused by the secondary current flowing in the ore-body, will increase to a maximum as we cross the boundary of the ore and then decrease to a minimum when over the centre of the ore-body. The horizontal component of the field will again increase to a maximum when passing from a region over the conductor. When an exploring coil, kept vertical so as to receive only the horizontal component of the magnetic field caused by the secondary current in the ore, is carried along a line crossing the ore-body, the intensity of the induced current will thus reach a maximum at the boundary of the body, that is, when passing either on to it or off from it. Hence, when the horizontal components are measured over an area, the presence and boundaries of the conductor may be found by the sudden rise and fall in the strength of the induced current in the exploring coil, as indicated in Fig. 76. The sound in the headphone receivers will be enhanced on each side of the conducting body when the horizontal component is investigated.

The vertical component of the magnetic field, caused by the secondary current of the ore-body, will give a minimum sound over the boundaries and a maximum over the centre. This vertical component will, however, be in a direction just opposite to that due to the exciting loop, and it must be remembered that the detecting coil measures the *resultant* field. To obtain a mental picture of what we should expect, we see that the variable vector due to the ore-body must be subtracted from the constant (if near the centre of the large loop) vertical component due to the current in the loop. The result is that the vertical field of the loop will be weakened as we pass over a conductor, as shown in Fig. 76.

In the foregoing discussion we have supposed that the two currents, the primary in the loop and the secondary in the ore-body, are just 90° out-of-phase. This, of course, is not generally the case, for when several conductors are scattered about and are excited by the current in the loop, both the vertical and horizontal components of the magnetic fields due to the currents in the separate ore-bodies will not be in phase either with the

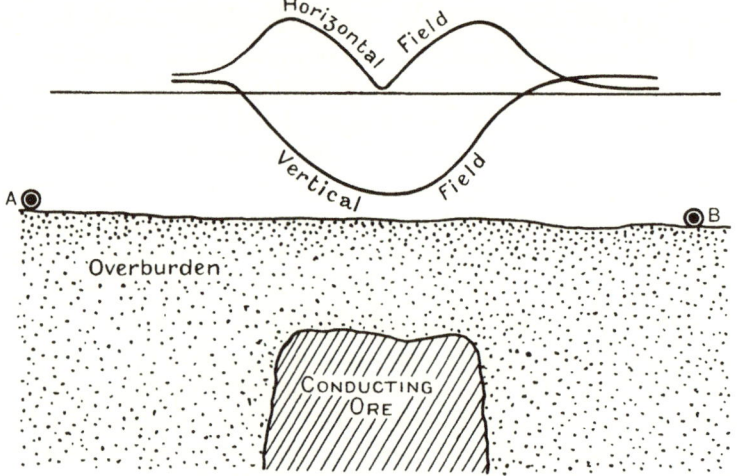

Fig. 76. A profile illustrating the change in the horizontal and vertical components of the magnetic field over an excited ore-body, using alternating currents in a loop A, B.

primary field or with each other. Even in the case of a single ore-body the difference in phase between the field from the ore and that from the loop may be of use in estimating the resistivity of the ore, because the change in phase will be governed to some extent by the resistance of the ore-body. Some geophysicists advocate the measurement of phase rather than intensity in making a survey and deduce the presence of conductors from the shift in phase, as will be described later.

Instead of working inside a large horizontal loop, using the methods already described, in many cases it is advantageous to work outside the loop. When this is done, the energizing loop is usually made smaller and circular in form, with a diameter of about 200 ft. The field due to the coil will again be nearly vertical when surveys are made at distances of more than about 300 ft. from the centre of the loop. It is now the variations in the horizontal components which are most easily detected as the result of the presence of a disturbing conductor. The same methods may be followed outside the loop as were described for the surveys inside, and both methods of surveys have been adopted in the field.

Surveys have been made by Lundberg in Sweden, in Northern Manitoba and elsewhere (Sundberg *et al.* 1925; Lundberg, 1928), using the comparison method described in this section, measuring both the horizontal and vertical components of the magnetic field. The Swedish deposits were particularly favourable for electrical methods, but when the ore is deep, say 600 ft. below the surface, the success of electromagnetic methods of any type was found by the writers to be uncertain.

(ii) *The 'equipotential' method using large horizontal loop*

Instead of using the Sundberg method, as applied by Lundberg in Sweden and elsewhere, equipotential lines or contours may be determined over an area using two similar coils for making the exploration. The procedure is to keep one coil stationary (in a horizontal plane) at a given point. The observer then walks away carrying the other coil (supported horizontally in a convenient manner around his waist) and wearing the headphones. He is able to find points about 100 ft. apart at which the sound in the headphones is either zero or a minimum. When this is the case both coils will be on the same 'equipotential' line. The different points found on this equipotential line are staked out, and it will close within a few feet if the observations have been taken by a trained observer. If now one coil is kept on the equipotential line and turned through an angle of about 25° from the horizontal, then the other coil is carried in a direction at right angles to the equipotential line which has been found, until the sound is again zero or a minimum. The new point is used for determining a second 'equipotential' line, and the strength of the vertical magnetic field along this new equi-potential line will be about nine-tenths of that along the first.

When the area has been mapped out in this way, the positions of the stakes are surveyed and their locations transferred to a chart. It will then be found that when over the boundary of a good conducting ore-body the field will be less. Such a survey can be made rapidly and, if the troubles arising from elliptical polarization of the field are not too great, the conducting body may be found. In Fig. 77 the results of a survey carried out in the manner described, over an area on Fisher Hill, N.Y., are given. The ore-body was located at A and its presence was con-firmed in this case by other methods.

The accuracy with which such electromagnetic surveys can be made depends a great deal on the current in the exciting loop being of one definite frequency free from harmonics. When balancing out the sound in the headphones, if other harmonics are present, the listener becomes confused and finds it difficult to determine when the fundamental note is at a minimum. This point is mentioned here again, for the authors' experience with these methods has shown that when the exciting current is rich in harmonics, as, for example when a type of buzzer was used, the methods failed entirely (Eve and Keys, 1928).

(iii) *Intensities and phase methods inside a 'horizontal' loop*

The field method of measuring intensities and phase has been developed and applied to Sundberg (Sundberg and Nordstrom, 1928). To apply this method, the strike and dip of the polariza-tion ellipse of plane of silence in an exploring coil are found, as already explained on p. 161. Instead of determining the ampli-tudes along the principal axes of this ellipse, Sundberg measures the vertical and horizontal components of the field along the strike of the polarization ellipse. If the vertical and horizontal components are found at a series of stations along any line, then the ore will generally be found beneath the place where the horizontal component is a maximum and where the vertical component passes through a point of inflexion.

Experiments have been made both in the laboratory and in

the field on the determination of depth, using the horizontal
loop and measuring the horizontal field at the centre, and the
change in phase in the secondary current caused by the presence
of a conducting layer at different depths below the surface.
Often oil is associated with layers of salt water, and in these
cases the detection of the conducting salt solution may lead to
the discovery of oil. The depth of the conducting layer is determined by comparing the strength of the horizontal component

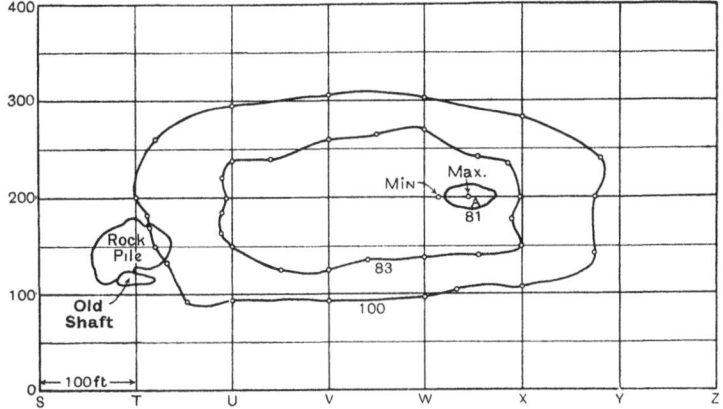

Fig. 77. The results of an 'equipotential' electromagnetic survey
with a loop (700 × 400 ft.) over an area on Fisher Hill, N.Y.

of the magnetic field along a line which passes over a region in
the central part of the large loop. For this purpose the current
in the loop must be kept constant or, if this is not possible, the
values of the horizontal components must be compared with
the field produced by the current in the wire. From the shape
of the curve drawn from the readings taken at the different
points, the depth of the conducting layer may be deduced. The
lower the values of the horizontal field strength the deeper will
the layer of conducting material be below the surface.

When the phase relation between the secondary and primary
fields due to the conducting layer and the loop respectively are
measured, it is found that the rate of change of phase may also
give an independent indication of depth. The phase difference

is less in the centre than on either side (Friedl, 1927). Both interpretations are, however, based upon the results obtained in laboratory practice and from working over deposits of known depth.

THE BIELER-WATSON METHOD

The apparatus consists of two coils, the one vertical, the other horizontal, mounted together at right angles on a light upright staff resting on the ground, and held or steadied with one or both hands. The ends of both coils (Fig. 78) are connected to the opposite plates or faces of a condenser of suitable capacity, so that the vertical coil is tuned to the frequency (500 ∼) of the large horizontal *loop* excited by an alternating-current generator. The amplifier and headphones are in series with the horizontal coil (Bieler, 1928; Watson, 1931).

If a station is selected within the large loop and if the apparatus is held quite upright there, then in the absence of conductors or of ore-bodies there will be no excitation in the vertical coil, and therefore no turns of the horizontal coil will be required to secure a balance. It must be understood that this is a null method, and that complete silence is obtained in the headphones when the number of turns in the *horizontal coil* is suitably adjusted. The word 'loop' refers to the insulated wire laid on the ground, generally forming a rectangle a few thousand feet long and wide. The word 'coil' refers to the detecting apparatus, and the horizontal and vertical coils are each rectangles, 1 or 2 ft. long and broad.

Next let us suppose that there is one ore-body in the vicinity of the apparatus. The alternating current in the main loop will induce an alternating e.m.f. round the conducting body.

The resulting current around the conducting body underground will induce an e.m.f. in the vertical coil if the lines of magnetic force produced by this current thread this coil, but not if they lie in the plane of the coil. The e.m.f. across the condenser will be 90° of phase behind the current in the vertical coil. But, since this coil is tuned to the particular frequency used, the flow of this current is governed only by the resistance of the vertical coil and is therefore in phase with the e.m.f. generated in the coil by the field threading it. Hence, the e.m.f. across the condenser will be 90° of phase behind the e.m.f. induced in the vertical coil. Now, in practice, the field inducing an e.m.f. in

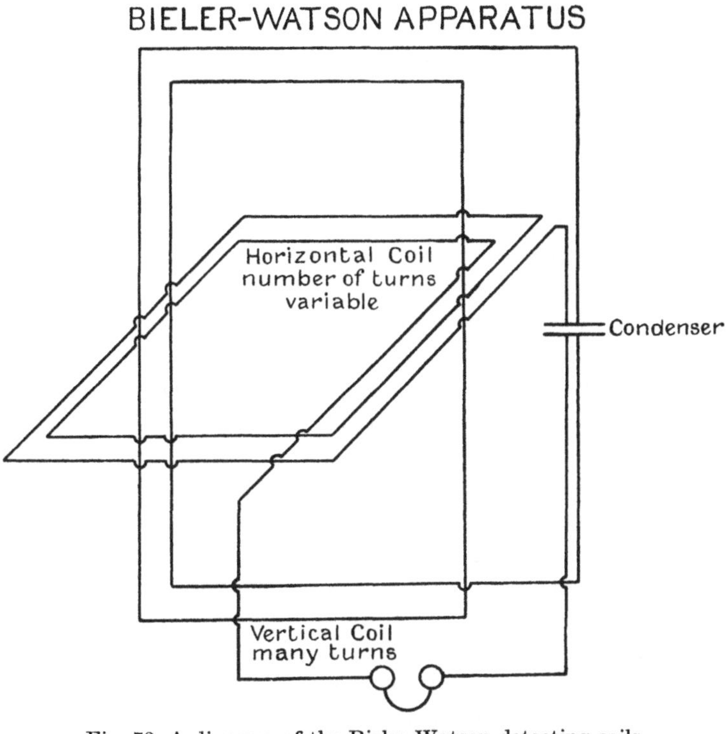

BIELER-WATSON APPARATUS

Fig. 78. A diagram of the Bieler-Watson detecting coils, with condenser and headphones.

the vertical coil differs in phase by 90° from that inducing an e.m.f. in the horizontal coil, so that the e.m.f. across the condenser will either be in phase or 180° out-of-phase with the e.m.f. of the horizontal coil. This latter condition is always maintained through the use of a reversing key in the horizontal coil circuit.

There are then two essential adjustments:

(1) A reversing key for the horizontal coil relative to the condenser.

(2) A step-by-step switch which can throw in more or less turns of the horizontal coil, as required, until complete silence is possible when the staff is nearly or quite upright. A slight swinging of the staff and coils by the hand facilitates the adjustment.

The steps are 0, 1, ..., 9 turns with one switch, and 0, 10, ..., 90 turns with a second switch.

Another viewpoint may be helpful.

To show that two fields exist which are 90° out-of-phase, we turn back to p. 155 and consider the plane of silence which contains the ellipse of polarization. We realize that the axis major and axis minor vectors are 90° out-of-phase, and it is the two fields represented by these vectors that are compared. If we place the vertical coil perpendicular to the axis minor, and the horizontal coil perpendicular to the axis major, then, by adjustment of the turns in the horizontal coil, we can get complete silence and this position will be 'sharp', as in radio direction finding under good conditions. It may, of course, first be necessary to use the reversing key, which will indicate whether the coils are in phase or in opposition and balanced. In the actual apparatus when the correct position is found the key turns *towards* the ore-body. In practice it is found more convenient not to hunt by rotation for perpendicularity to the axis minor, but to measure the number of turns necessary for a balance when the vertical coil faces north and south; and also to measure the number of turns when it faces east and west. These numbers are plotted to scale, both along the north-south and the east-west lines on the map at the point representing the place of observation. The resultant vector is a measure of the minor axis as compared with the major axis. 'No conducting body; no minor axis! Near a large conducting body; a large minor axis!'

The instrument can detect the magnitude of this ratio, minor axis to major axis as low as one in a thousand. It seems important to secure a frequence in the primary loop that is free from harmonics, or it will follow that when a balance is obtained for the fundamental note, the harmonics prevail to a most troublesome extent, and a good balance is difficult even with practice. If there are a number of ore-bodies there will be a resultant horizontal vector due to all these bodies, but, on approaching any one of these ore-bodies, there will be a large vector pointing towards it. This method has been tested in

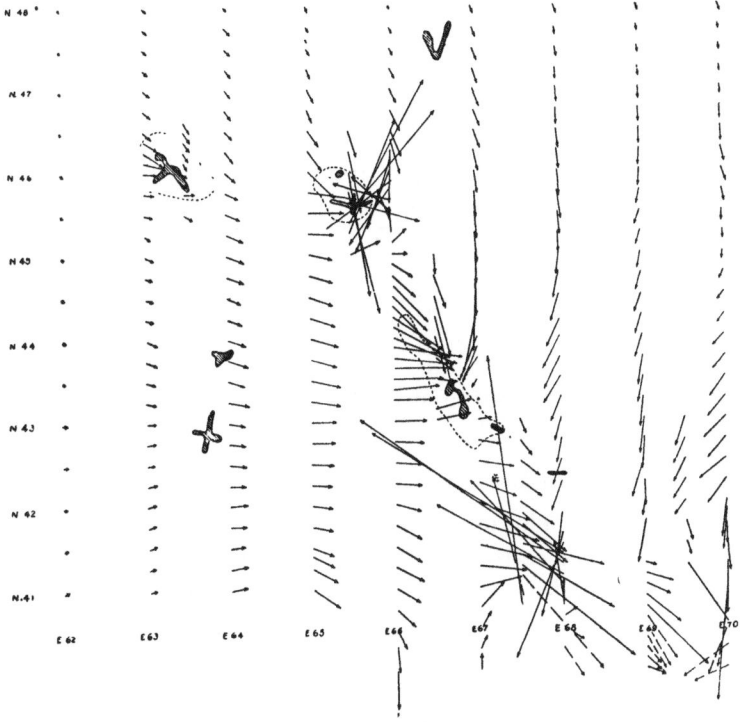

Fig. 79. A survey made at the Amulet Mine, Quebec,
by Bieler and Watson, using their method.

Northern Quebec (Figs. 79, 80) and in the Sudbury region by the inventors and it has promising characteristics. The trouble in Northern Quebec is the presence of wet clay bodies near the surface which not only have a screening effect but also simulate the behaviour of veins or lenses of ore. This trouble is common to several methods and makes Northern Quebec a difficult region for electrical prospecting.

On hilly ground when the loop is not horizontal the primary magnetic vector will not be vertical, and it is imprudent to work near the boundary of the loop (say within 100 ft.). It is, of course, possible to work outside as well as inside the loop, but the primary field is weaker outside.

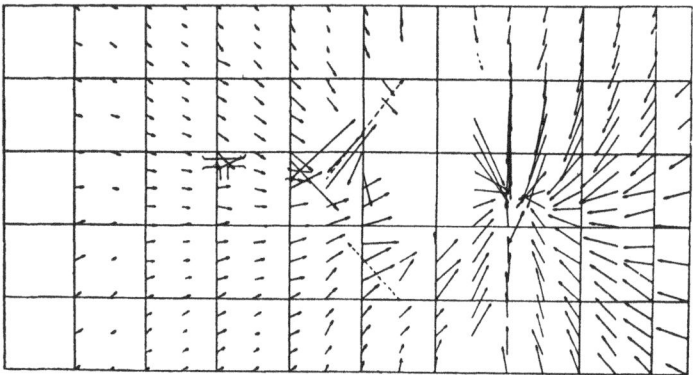

Fig. 80. An electrical survey at the Aldermac Mine, Quebec Province, by Bieler and Watson, using their method.

VERTICAL LOOP METHODS

Instead of employing horizontal wires or a horizontal loop for creating the electromagnetic field which excites the ore-body or other conductors, several methods have been developed in which a vertical loop is used. In such arrangements a vertical loop of ten or fifteen turns of insulated wire is constructed in a rectangular or triangular form with sides about 20 ft. long. The actual dimensions are not of prime importance, but the loop must be made in a collapsible form, easy to transport and set up in rough and wooded country. The loop may be supported either on a low tripod or from the top of a pole, so that it may be kept vertical and rotated in any azimuth. We have used a triangular loop, each side 20 ft., of eight turns of no. 14 well-insulated copper wire, the vertex of the triangle being supported on a swivel at the top of a 20 ft. pole held vertical by four guy ropes which were attached to pegs in the ground in much the same fashion as when a tent pole is kept vertical (see Fig. 81). The base of the triangular loop was stretched over a 20 ft. piece of deal made with grooves in the ends to hold the eight turns of wire. The prospecting firms naturally use much more refined devices, but the principle is the same.

The loop is supplied with a powerful electrical current of constant frequency which may be changed from about 500 cycles

Fig. 81. A triangular vertical loop excited with
alternating current.

per second to 80,000 cycles, depending upon the method used. In the case of the 'Radiore' process, the outfit resembles a modern radio broadcasting station, in which the wave-length may be varied accurately in steps of 5000 cycles as desired. The constancy and purity of the frequency is again an important feature.

The electromagnetic field proceeds from the loop, excites the ore, and the combined effects are determined at different stations by means of a search coil similar to that described on p. 159.

The detection is essentially a direction one, and the principles involved resemble very closely the location at sea of a ship by means of directional wireless. In the case of ore detection, the observer has the direct radiation from the loop as well as the signal from the conducting ore to contend with, and it is the direction of the latter disturbance that he is usually interested in.

A very important factor enters into all such electromagnetic methods of searching for conductors, and that is the ability of the radiation to penetrate through rock and soil, for it is necessary for the electromagnetic waves to penetrate from the loop through the rock to the ore-body and then induce currents of sufficient strength to re-radiate back waves that must travel up to the surface and still be strong enough to be detected in the search coil. It is known that radio waves will not penetrate through more than about 60 ft. of sea water. The ordinary over-burden has not the conductivity of sea water, but some experiments carried out in a railway tunnel at Montreal on two different occasions have proved definitely that waves of 40 m. cannot be detected 1500 ft. inside the mouth. Radiation of longer wave-length, such as the waves used as the carrier in broadcasting, could be detected right through the three-mile tunnel, but as wires and rails were present the experiments could not settle whether the radiation came through the rock, the air, or along the rails and wires. Some experiments carried out in a mine at Caribou, Colarado (Eve, Keys and Denny, 1927; Eve and Keys, 1928; Eve, Steel et al. 1929), seemed to indicate that broadcasting waves would penetrate through 500 ft. of rock. The results, however, were not conclusive, and further tests on the point were carried out in the Mammoth Cave, Kentucky (see p. 189).

(a) The Mason, Slichter and Hay method

This method has been described by Mason (1927). The method is essentially one of direction and is therefore worthy of careful consideration. A vertical loop is set up and excited with alternating current of 60,000 cycles (5000 m. wave-length), and a line is surveyed perpendicular to the face of the loop. Stations are marked off on this line every 100 ft. from the loop for a distance of about 2000 ft. A coil is used for detecting the field, and this coil may be made in circular form and attached to an ordinary transit base as shown in Fig. 82. This instrument, of course, improves the accuracy with which angles may be read,

Fig. 82. A circular detecting loop attached to a transit.

but our experience has been that the larger rectangular coil shown in Fig. 75 gave more satisfactory results. When the smaller circular coil is used, the number of turns of wire is increased. A good three-valve amplifier must be placed in the circuit in order that accurate settings may be made at distances over 500 ft. from the loop. It is usual to use headphones and find the position of silence or minimum, but a rectifier or thermocouple with microammeter may be employed.

It has already been explained that the alternating current in the loop will induce a current in a conducting ore-body which in turn will thus produce a secondary field. The strength of the secondary current in the ore will depend upon the strength of the magnetic field at the ore produced by the loop and upon the rate of change of this field, that is, the frequency of the current in the loop. The higher the frequency the greater will the current in the ore-body be, but, owing to the greater absorption of the higher frequencies and other complications, there is an upper practical limit to the frequency which may be used.

The detecting coil is set up at each station in turn and two readings are taken. The one reading is called the strike and is found by keeping the plane of the coil vertical but turning into that azimuth at which the sound in the headphones is a minimum. The plane of the coil will then be parallel to the horizontal component of the maximum magnetic field. The deviation of the plane of the small coil from the vertical plane which is at right angles to the face of the loop is a measure of the strike. These angles will not be great, amounting to only a few degrees when the conductor is deep down. The second reading is called the dip, which is a measure of the angle through which the coil, when initially in a horizontal plane, must be turned to obtain minimum signal strength. This is easily measured by setting the plane of the coil horizontal and then tilting it until the minimum sound is heard and measuring the angle of tilt. The values of the strike and dip are found at each station along the line.

Over homogeneous and level ground free from conductors both the strike and the dip will be zero all along the line. If, however, the line passes near a vein or ore deposit which acts as a conductor, the resulting field will be disturbed and there will be variations in both readings. The strike will first increase in one direction as the vein is approached, and return to zero at the point nearest the deposit. As the coil is taken further from the loop along the line, the readings will gradually increase in the opposite direction, reach a maximum, and then decrease again to zero when the conductor is left behind. The dip, on the other hand, will increase to a maximum near the centre of the conductor and then decrease to zero as the observations are taken further beyond the ore-body. In Fig. 83 curves are drawn showing the type of readings that one might expect when passing near but not directly over an ore deposit. The solid curve gives the readings of strike and the broken curve the readings of dip.

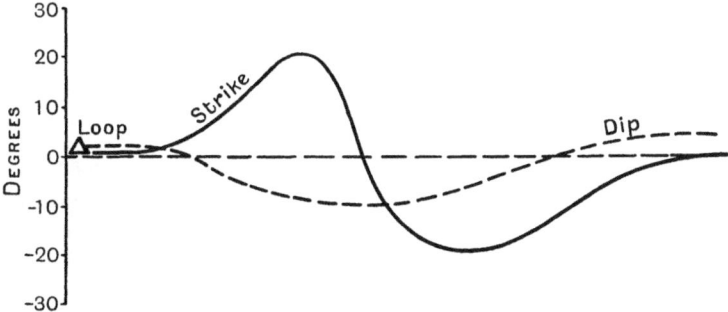

Fig. 83. A diagram illustrating the change in strike and dip of a search coil in passing over a conducting ore-body.

In rough and thickly wooded country the method would not be so readily carried out. The change in altitude of detecting coil with reference to the loop would alter the readings found, even in the absence of conductors, so that an observer should be careful in interpreting results obtained when the maximum readings are about 20°. It is clear that changes in the level of the ground will produce corresponding changes in the dip.

(b) The Radiore process

The Radiore method is essentially one of direction finding in which a conducting ore-body is excited by an alternating current in a vertical loop, and the position of the electrical axis of the conductor is determined by finding, with a suitable coil, the direction of the secondary electromagnetic waves which the ore has re-radiated. Here again the theoretical success of the method depends upon the penetrability of radio waves through the less conducting rocks and other materials which surround the ore.

The method is applicable only to *conductors*, which may, moreover, be disseminated, but the method will not locate such nonconducting sulphides as stibnite and sphalerite. Large masses of (relatively) poorly conducting ores will give about the same indications as small amounts of better conducting material. The conductivity of the gangue in which the ore occurs affects the ease with which the ore may be located. Hence, ore lying under damp clay may not be found if the conducting clay gives the main response.

An alternating electromagnetic field is used to excite the conductor, and consequently it will be those minerals which have a low resistance, as measured with alternating current, which will be most readily found. This also accounts for the possibility of exciting disseminated ore. The separate parts of the conductor with insulated gaps intervening behave like a group of condensers, the capacity of which we may call C. The electrical resistance of such a disseminated body, as measured with direct current, would be extraordinarily high. The effective resistance of the disseminated ore as measured with alternating current of frequency f may be low and is given by the capacitative reactance

$$X = \frac{1}{2\pi f C}.$$

It is evident from this that the higher frequencies will give a larger current through such disseminated ore. But, when the frequency is too high, induction effects are introduced and the absorption of the electromagnetic waves in the rock becomes great, so that there results an optimum range of frequencies for use in such a method.

The apparatus consists of two parts. The sending or energizing source is a vertical pancake loop of about twenty turns of wire wound on an insulated frame that may be about 20 ft. square. This frame is supported on a tripod and may be turned in any azimuth. The current is supplied by batteries or by a small 150 W. hand-cranked alternator of 600 cycles to a small broadcasting equipment. The loop serves as the antenna for radiating wireless or radio waves whose frequency may be varied in steps of 5000 cycles all the way from 30,000 to 50,000 cycles. The equipment will also permit of broadcasting lower frequencies of 50–3000 cycles, should these be required. These radio carrier waves pass out through the air and into the ground from the loop A as shown in Fig. 84. Suppose that CC is a conductor lying within 500 ft. of the surface, the electromagnetic radiation from the loop will stimulate or excite currents of the same

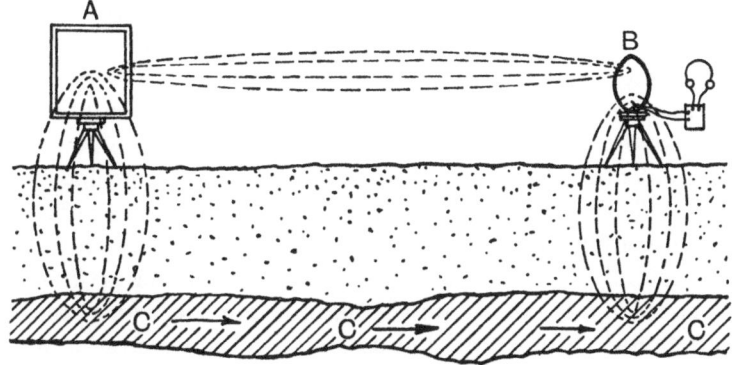

Fig. 84. Diagram showing a large coil, A, excited with radio frequency, which stimulates a long conducting ore-body, CC, giving rise to secondary induced currents. Detector coil B is excited by both primary and secondary fields.

frequency in the conductor CC, and these secondary currents will flow back and forth along it much as wired wireless will travel along telegraph wires. The currents in the ore will themselves cause a radiation of the same frequency, and so, at points on the surface above, we shall have a resultant electromagnetic field at any instant due to the direct radiation from the loop and to the re-radiation from the ore-body. It is the radiation from the ore-body which we wish to detect, but the presence of the direct radiation makes this somewhat difficult.

To detect the radiation, an exploring coil B (Fig. 84), of the type already described on p. 177, may be used. For practical convenience the Radiore Corporation use two coils arranged like a 'pair of spectacles' supported on a tripod in the centre, the effect being just the same as with a single coil. This detecting coil is tuned to the same frequency as the emitting loop and is supplied with amplifier and headphone receiver. The loop is first set up at a given point and then, at a distance of a few hundred feet away, the detecting coil is placed on its tripod. Both the loop and coil are vertical, and the loop is turned so that the loop and coil are in the same vertical plane.

When the loop is energized with current of a given frequency the coil will receive radiation directly from the loop. If no ore-body is present, when the coil is turned until its plane is horizontal, no current will be induced in the coil. When, however, an ore-body is present, then the radiation from it will induce a current in the coil even when horizontal. As a result of the presence of a conducting ore-body, there will be two magnetic vectors at any point which in general will not be in phase and which will be of different intensities. The resultant vector will therefore be elliptically polarized. In such cases direction finding becomes difficult, for the minima found when the coil is tilted are not sharp; in fact, we obtain the 'figure eight' diagrams familiar to students of direction finding, but, as shown in Fig. 85A, with a broad minimum. The minimum may be made sharp, however, by bringing the two waves into phase or nearly so, when the shape of the polar curve will be like that shown in Fig. 85B. In these diagrams the strength of the induced e.m.f. in the coil or signal strength in the headphones is represented by the length of the vector and the direction of the vector indicates the direction of the plane of the coil. The elimination of phase difference between the direct radiation and that from the ore-body may be brought about by changing the frequency of the transmitter.

To illustrate the principle underlying the method in which the direction finding is applied, suppose the loop is set up over one end of the ore deposit and observations are made with the coil at a number of equi-spaced points on a straight line which crosses above the ore, then we may readily draw vector diagrams of the result to be expected. In Fig. 86 let O represent the electric *axis* of the ore-body CC shown in Fig. 84. The line EF of Fig. 86 is at right angles to CC of Fig. 84, and the coil B is moved to the various points on this line EF as indicated. The

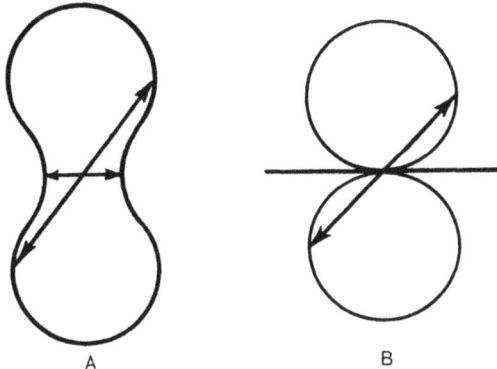

Fig. 85. Direction-finding diagram. A, minimum broad; B, minimum sharp.

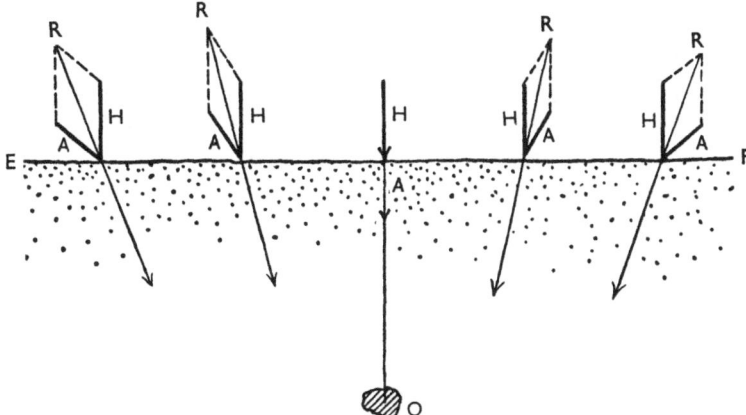

Fig. 86. Diagram in which upright lines H indicate the effects of the exciting coil (primary); the inclined lines A show the effects of the secondary or induced current due to the ore-body O, which is perpendicular to the paper. The resultants of H and A (parallelogram law) point toward but below the ore-body O.

electrical axis of an ore-body is the point in the conductor at which the current may be supposed to be concentrated. At each point on the surface we shall have a constant vector due to the radiation from the loop which we may designate by H. If the plane of the coil were kept vertical at each station, then the strength of the maximum current induced in the coil, in the absence of an ore-body, would be represented by H. Similarly, if only the effects of the radiation from the ore-body were received, the direction of the plane of the coil for maximum current would be represented by A. The direction of A will change from station to station and its magnitude may vary a little also, especially if O is near the surface. If these two fields are in phase, then we may obtain the direction of the resultant by combining the two vectors at each point by the parallelogram law and obtain the resultants R as indicated in Fig. 86. These R's reversed will point to a position a little below O.

To find the direction of R, the coil is turned at each station to the position of minimum sound in the headphones, the frequency of the transmitting loop altered (and the coil tuned to the same frequency) until the minimum is sharp. Such change in frequency may not always be necessary in practice. When the silence position is found the direction of R will be along the perpendicular to the plane of the coil. The direction of the resultant magnetic vector R of the field found in this way at each station is plotted, and the position and depth of the electrical axis of the conductor deduced from the diagram as indicated.

We have supposed in the foregoing description that the direction of the conductor has been found and the loop set up over it. This is not always so simple to accomplish. A preliminary survey may be made and indications found where the ore-body lies and the loop then set up over this point.

The interpretation of direction is often much more complicated than in the simple case we have described. For example, we have supposed that there is no refraction of the electromagnetic waves in going from the ground into the air above. This is not the case. Refraction may take place and the waves in the air become so distorted that the actual reading would indicate the ore-body as being above the ground ! In such cases the wave-front may be determined from a number of readings at different points and the approximate position of the conductor found. In Fig. 87 a theoretical case of distortion of the wave-front is shown, as found by the direction-finding coil, when the direct radiation from the loop is eliminated. When, however, such directions are combined with the horizontal field due to the direct waves from the energizing loop, the resultant field will, in most cases, indicate a point below the surface. The interpretation of such readings so as to give the approximate depth of the conductor is largely the result of practical experience in the field.

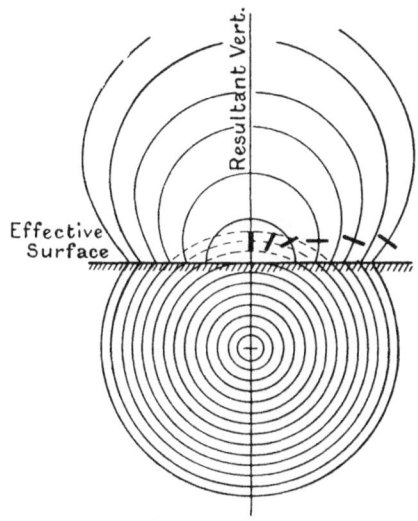

Fig. 87. Profile showing the refraction of electromagnetic waves passing from uniform ground into the air above, indicating the positions of the plane of the search coil as set for maximum signal reception, at right angles to the magnetic vectors.

The efficiency of many of the electromagnetic methods is dependent on the mutual inductance between the excited loop above ground and the conducting ore-body at a distance D below the earth. As the loop is equivalent to a large thin magnet or dipole the effect at a point on the ore-body varies inversely as the *cube* of D. The efficiency of the method is no less dependent upon the mutual inductance between the stimulated ore-body and the coil at a distance D' above it. Here again the effect varies inversely as the cube of D'. Hence for the double journey detection is dependent upon the product $\dfrac{1}{D^3 D'^3}$, and it is only due to the great sensitivity of the telephone that ore-bodies can be detected at a depth of a few hundred feet. If both distances are doubled the reduction of strength of the received note in the phones is lowered to one sixty-fourth part of its original value.

There is another factor, however, to be considered inasmuch as the intervening earth is always a partial conductor, and it will therefore to some extent absorb the alternating electro-magnetic disturbances, both on their downward and their upward journeys, to an amount depending upon the frequency employed as well as upon the conductivity of the earth.

For effective work it is desirable to have large current and area, many turns but low impedance in the loop, and also large area, many turns and low impedance in the receiving coils. The low impedance with many turns can, for the higher frequencies, only be achieved by tuning with capacitance (see p. 156).

While the penetration into the earth and back to the coil is less effective both for high frequencies and for low resistance, yet the actual stimulation of the ore-body is greater for the higher frequencies. Hence there may be an optimum frequency for given conditions of depth and conductivity. For example, if the resistance of an overburden is 10^4 ohm-cm., and if the frequency be 10,000 cycles, the intensity of the field may be reduced at a depth of 100 ft., by absorption, to about 52 % of the surface value. If the frequency is increased fivefold, to 50,000 cycles, the intensity would be reduced to 26 %, and thus the induced e.m.f. in the ore-body becomes 2·5 times as great as before.*

The Radiore method has been used extensively in the United States and in Canada, and the company holding the patents and operating the apparatus are sufficiently confident of the validity of their results to guarantee that they will not overlook any con-

* The Radiore method is, however, usually operated with 500-cycle frequency.

siderable conductor within 400 ft. of the surface. Their maps consequently show a large number of veins and ore-bodies, and it requires the skilled training of geologists and mining engineers to determine what are the best indications to develop.

The problem of absorption by the earth over the short distances used in geophysical exploration is not very different from that of the penetration of wireless (radio) waves into the earth. With the rapid development of radio broadcasting following the conclusion of the First World War, and the use of electromagnetic induction methods in geophysical prospecting, the penetration of rocks by such waves was of considerable importance. The results of some of our investigations in this connexion are of interest.

PENETRATION OF ELECTROMAGNETIC WAVES AND INDUCTION INTO THE EARTH

Electromagnetic effects and radiations may penetrate into the earth, sometimes to considerable depths, to an extent depending both upon the character of the electromagnetic disturbance, and upon the nature of the material penetrated.

The depth at which these effects or waves can be detected or measured must involve the further question of the sensitiveness of the detecting instrument, so that it is more rational to state the depth at which an effect of given magnitude at the surface is reduced to a stated fraction or percentage.

If the source on the surface is a point source, then beneath the earth there will be a spreading of the radiation involving the inverse square of the distance in homogeneous material.

Good conductors of electricity, such as silver and copper, tend to absorb electric disturbances and act as electric screens. Highly permeable substances, such as permalloy, are good absorbers of magnetic effects. Many thin, alternating layers of copper and permalloy make an effective screen against all electromagnetic effects.

Ordinary country rock free from conducting or magnetic materials should, therefore, be penetrable by electromagnetic effects to great depths, and it is on this account that electrical detection of conducting ore-bodies is possible, certainly to depths of many hundred feet.

This is true both of direct current, and also of alternating current, provided the change of frequency is low, when the resistance of the material is the main factor. As the frequency increases, however, the penetration falls off rapidly. Thus ordinary light, which has a frequency of several million million cycles a second, is a purely electromagnetic effect which certainly cannot penetrate rock or earth.

In the case of a coil or conductor, carrying alternating current, and placed beneath the earth, there is absorption by the over-burden, but it is necessary to remember that the excitation of the earth beneath the coil is also a factor in any consideration of measurement.

During the First World War a French submarine received signals by a well-insulated coil outside the hull, connected to an amplifier and phones within the boat, up to depths not exceeding 60 ft. Similar results have since been obtained by the U.S. Naval Research Department. Now the resistivity of sea water is about 100 ohm-cm., of clear river water about 7000, and of rocks about 10^4–10^5 ohm-cm., so that sea water conducts one hundred to a thousand times better than limestone or granite. Hence the penetration into or through rock should be much greater than through sea water.

Again an English newspaper stated several years ago that wireless broadcasting was easily received in the Severn Tunnel except during the passage under the saline water of the estuary.

These considerations led to an investigation in the Mount Royal Tunnel, Montreal, which is $3\frac{1}{2}$ miles long, and has a maximum overburden of 300 ft. In June 1926, it was found that wireless waves of 411 and of 1300 m. could be detected throughout the whole tunnel, but waves of 40 m. wholly vanished at 1500 ft. within the tunnel, although there were at that point only 48 ft. of rock above the tunnel. In the summer of 1927 experiments were made at the Caribou Mine, Colorado, and broadcasting (267 m.) from Denver, 50 miles away, was received clearly at a depth of 220 ft., and with some difficulty 550 ft.

underground. Further investigations (Eve, Steel *et al.* 1929) were made in the Mount Royal Tunnel, in April 1928, and the signal strengths were measured by Steel, and others, of 43, 55, 411, 1400, 17,000 m. wave-length. Again the two shorter waves did not penetrate far into the tunnel, while all the longer waves were detected from end to end falling off exponentially towards the centre. Here again there was uncertainty as to whether the signals came through the rocks, through the openings or along numerous rails, cables and pipes.

In order to remove all further doubt Lee selected the Mammoth Cave, Kentucky, which was entirely free from conductors, which had about 300 ft. of overburden, mostly limestone, and was so long and circuitous that transmission through the mouth was definitely proved ineffective.

Using a 300 ft. horizontal antenna in River Hall, morse signals were received from six different long-wave stations, while speech and music were made audible with a loud speaker, 300 ft. underground, received from Cincinnati (700 kc., 429 m.) 200 miles away, from Louisville (820 kc., 366 m.) 90 miles away, and from Nashville (650 kc., 461 m.) about 100 miles away. These signals unquestionably penetrated 300 ft. of rock.

Finally a ten-turn horizontal loop resting on the earth 100 ft. in diameter, with an alternating current of 500 cycles, and 2·3 amp., was detected with coil and headphones, without any tuning or amplifier, in the cave, at a position where the signals must have penetrated through 900 ft. of rock.

So far these results deal with detection, but it was most desirable that definite measurements should be made. These involved the problem of measuring, with some accuracy, alternating currents of the order of 10^{-8} amp. The currents could of course be magnified to any desired extent with amplifiers, but exact measurement in such cases presents much difficulty in the laboratory, and practical experience in the field showed that measurements with amplifiers might be most erratic. Lee suggested the use of copper-oxide rectifiers and a direct-current galvanometer. A horizontal coil in the cave received by induction an alternating current from the loop above ground; the coil was tuned with suitable capacitance, and the output was led to the rectifier, and thence to a Leeds and Northrup wall galvanometer clamped to a firm table. It was necessary to use a thermostat, and to screen the rectifier from light to avoid photoelectric effects.

Calibration was effected by a small auxiliary coil at a few feet distant. If I is the current in the loop, M the mutual inductance between loop and coil; if i is the current in the auxiliary loop, m the mutual inductance between the auxiliary and receiving coil; then if the deflexion of the direct current galvanometer is the same in both cases, it is not hard to show that $MI = mi$, provided there is no absorption of the electromagnetic effects by the ground. We found for 125 ft. of sand and limestone that MI was less than mi by about 5 %, and this was a measure of the absorption, a result not out of keeping with the theoretical results to be explained shortly. Work with higher frequencies, 10–20 kilocycles, was also carried out, but a large crop of difficulties arose (see *Studies of Geophysical Methods*, 1931, 1932).

THEORY OF THE ABSORPTION OF ELECTROMAGNETIC WAVES

If Maxwell's equations are written down and solved for a horizontal plane wave descending vertically into the earth an absorption term will appear, for either electric or magnetic intensity of the form e^{-Az}, where e is the usual Napierian base, z is the depth below the surface and A is equal to $2\pi f\kappa/c$, where f is the frequency, c is 3×10^{10}, and

$$\kappa^2 = \frac{\mu}{2} \left\{ \sqrt{\left(\epsilon^2 + \frac{4\sigma^2}{f^2} \right)} - \epsilon \right\}.$$

Here we have μ the permeability equal to 1,

$$\epsilon = (\text{dielectric constant})/4\pi = 0 \cdot 8 \text{ perhaps},$$

while σ is the conductivity in electrostatic units $= 9 \times 10^{20}/10^9$ times the resistivity in ohm-cm. For the frequencies under discussion ϵ can be ignored and $\kappa^2 = \sigma/f$, so that the absorption factor reduces to

$$\exp\left[-\frac{2\pi}{c} \sqrt{(\sigma f)} z \right]$$

or equal to

$$\exp\left[-2\pi \sqrt{\frac{f}{\rho}} z \right],$$

provided ρ the resistivity is expressed in electromagnetic units.

This absorption term seems to be at least an approximation to the correct value even for the case of induction between a loop above ground and a receiving coil directly beneath it underground, at least for 500 cycles. There are other methods of attack such as that due to Zenneck, summarized by Fleming (1919) and further discussed by Eve (1930). The most complete solution has been given by Sommerfeld (1905), but it is difficult to apply this theory to the short distances involved in geophysical prospecting.

In any case, we may rest assured that the absorption is approximately exponential, decreasing with depth, and large for high frequencies and conductivities.

A numerical example will make the matter plainer. For a frequency of 500 cycles and for 124 ft. of rock the percentages of absorption for different frequencies are theoretically as follows:

Resistivity in ohm-cm.	Absorption (%)
10^4	16
10^5	5
10^6	$\frac{1}{1000}$

On the other hand, suppose that the resistivity is fixed at 10^5 ohm-cm. and the frequency is varied, we have then:

Frequency	Percentage absorption (124 ft.)	Thickness to half value in feet
500	5	1600
21,400	32	250
43,000	39	170

It is convenient to get away from exponentials, and, as in radioactivity, to write down the thickness which will reduce the full surface effect to half value as in the last column just quoted.

Dellinger (1919), using retarded potentials, gives the strength of the field H at a large distance R from the aerial in the form

$$H = \frac{h2\pi f}{CR} I_0 \cos 2\pi f \left(t - \frac{R}{C}\right) + \frac{h}{R^2} I_0 \sin 2\pi f \left(t - \frac{R}{C}\right),$$

where f = the frequency, t = the time, h = the height of the antenna, I_0 = the maximum current in the antenna, C = the velocity of the waves.

It will be seen that the equation for H is composed of two terms. The first which varies directly as the frequency and inversely as the distance has been called the induction term, while the second or radiation term, exactly out-of-phase with the first, varies inversely as the square of the distance and its intensity is independent of the frequency. These ideas refer to the waves travelling outwards from the antenna along the ground, and do not refer to those reflected or refracted from the Kennelly-Heaviside and Appleton conducting layers which may be about 60 and 120 miles above the earth's surface.

Peters and Bardeen (1932, pp. 103–22) have deduced from theoretical considerations that the optimum frequency for greatest penetration is given by the relation $h\sqrt{(f/\rho)} = 10$, h being the depth in metres, f the optimum frequency in cycles per second and ρ the resistivity in ohm-cm. Haycock, Madsen and Hurst (1949), and Wait (1952) have recently made a further study of this subject.

It will be noted by the reader how much work remains to be done to bring practical measurement and sound theory into good agreement.

PARALLEL WIRES

An interesting investigation was carried out at Cross Keys, N.J., by Bowen and Gilkeson (1930).

A single conductor, 34 ft. above the ground and 8500 ft. long, was grounded at both ends and stimulated by an alternating-current generator in series, so that the return current was through the earth. A parallel insulated wire, 500 ft. long, was also grounded at both ends and an alternating-current potentiometer in series with this wire indicated the potential between the grounds. The distances between the parallel wires were varied from 21 to 1000 ft. The frequencies were 60 cycles, and also various values from 100 to 1000 cycles per second.

From a knowledge of the current in the long wire and of the potential difference between the ends of the short wire, it was possible to deduce the resistance and the impedance of the earth, and of the phase angle between them.

The resistivity was found to vary somewhat with the frequency, and was as follows:

Frequency (cycles)	Resistivity (ohm-cm.)
60	2400
200	2700
500	3700
1000	5000

These values seem unusually low, but the investigators find that the values obtained agree well with the hypothesis of an upper layer 130 ft. thick, with a resistivity of 400,000 ohm-cm., beneath which there is a good conducting layer of 2000 ohm-cm.

This condition might well be verified or checked with the Gish-Rooney or some equivalent method. It appears that most of the current is carried by the lower layer. In any case the investigation indicates the possibility of detection, by this parallel-wire method, of geological discontinuities and of good conducting ore-bodies underground.

OTHER ALTERNATING-CURRENT METHODS

Broughton Edge and Laby (1931) and others (Lundberg *et al.* 1931) now use an alternating-current potentiometer, measuring both potential and phase differences. Lundberg recommends surveys both forwards and backwards over a conductor. The ordinates of the graphs are algebraically added together, and a distinction may be observed between conductors near the earth's surface and those deeper in the ground.

Lundberg uses a line electrode, well pegged to the ground, as exciter, in place of the point electrode. He then runs lines at right angles to the electrode, using three probes with 50 ft. intervals, and balancing to zero with headphones the current to the middle probe by suitable resistances in his *ratio compensator* ('Racom' method). The hemispherical bowls around the central electrode must therefore be replaced by coaxial half cylinders round the electrode. Remember to replace the formula (p. 102)

$$R = \frac{\rho}{2\pi} \left(\frac{1}{b} - \frac{1}{a} \right) \quad \text{by} \quad R = \frac{\rho}{\pi L} \log_e \frac{b}{a},$$

where L is the length of the line electrode.

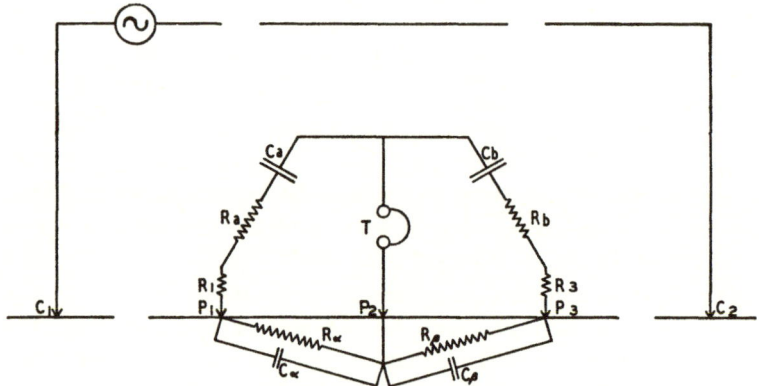

Fig. 88. Ratiometer method.

The alternating-current ratiometer method of Broughton Edge and Lundberg is essentially a variable ratio-arm alternating-current bridge with provision for phase adjustment. Alternating current of 500 cycles per second or less is introduced into the ground between two stakes or line electrodes C_1, C_2, Fig. 88, which may be a mile or more apart. In the central portion of the intervening region three potential electrodes (usually equally spaced) P_1, P_2 and P_3 are introduced into the ground and the potential difference between P_2 and P_3 compared in magnitude and in phase with that between P_1 and P_2. A fourth potential electrode P_4 is then inserted and the comparison made between

P_3 and P_4 with that between P_2 and P_3. Since these potential differences are proportional to the effective impedances between the equipotentials associated with each point, the effective impedances between any two sets of points may be compared by proceeding in this manner. In general, the impedances tend to be higher at the edges of a good conductor and lower over the conductor than the average value in barren ground.

The principle of the Broughton Edge method may be deduced from considering the schematic diagram, Fig. 88. Let the contact resistances of P_1 and P_3 be R_1 and R_3 respectively. Then using resistances and capacities in each arm of the bridge as shown, by suitable adjustments a condition of silence in the phones may be obtained, or when a galvanometer with copper oxide rectifier is used the current may be made zero. If the resistances and capacities introduced into the two arms are R_a, C_a and R_b and C_b respectively, we shall have

$$V_2/V_1 = Z_2/Z_1 \quad \text{or} \quad V_2 = Z_2/Z_1 . V_1,$$

where Z_1 and Z_2 are the impedances in the two arms. Now

$$Z_1 = R_1 + R_a + jX_a \quad \text{and} \quad Z_2 = R_3 + R_b + jX_b,$$

where

$$X_a = -\frac{1}{2\pi f C_a} \quad \text{and} \quad X_b = -\frac{1}{2\pi f C_b}$$

are the reactances in the two arms respectively. It also follows that the phase angles are

$$\tan \theta_1 = X_a/(R_1 + R_a) \quad \text{and} \quad \tan \theta_2 = X_b/(R_3 + R_b).$$

The values of X_a and X_b may be determined from the values of C_a and C_b in the two arms of the ratiometer when the frequency of the alternating current is known. It is evident, however, that neither the phase angle nor the modulus of either impedance can be found since R_1 and R_3 are unknown. To overcome this difficulty, Broughton Edge uses large values of impedances in the arms (over 50,000 ohms) so that the values of R_1 and R_3 may be neglected in comparison. Under these conditions

$$V_2 = \sqrt{\frac{R_b^2 + X_b^2}{R_a^2 + X_a^2}} V_1,$$

and $(\theta_2 - \theta_1)$ may then also be found.

Lundberg works with a somewhat similar ratio-arm bridge but uses inductances in place of capacitances and, by making double balances, practically eliminates the effect of contact resistances at the potential electrodes.

Dr H. G. I. Watson at McGill has made a modification of Broughton Edge's method by which the contact resistances at the potential stakes are eliminated. The principle of this method (which involves a double balance) is briefly as follows. Suppose R_α, R_β and C_α, C_β, Fig. 89, are the effective resistances and capacities of the ground between the equipotentials through P_1, P_2 and P_2, P_3 respectively. Then for a null effect in the phones, theory shows that we shall have

$$\frac{R_\beta}{R_\alpha} = \frac{C_a}{C_b} \left[1 - (2\pi f)^2 C_b R_\beta \{ C_\beta (R_b + R_3) - C_\alpha (R_a + R_1) \} . \right]$$

When this adjustment is made, a second set of ratio arms, in series with the first, is now connected in the circuit as shown in Fig. 89 by connecting the contact H to G instead of simultaneously to both E and F. The setting in the arms $P_1 A$ and $P_3 B$ are not altered, but adjustments are made of the resistances and capacities in the arms AC and BD until silence is again obtained in the phones T. We shall then have a similar expression for the ratio R_β / R_α with the additional terms added for R_x, R_y, C_x and C_y. By subtracting numerators and denominators we eliminate the unknowns, R_1 and R_2, and finally obtain

$$\frac{R_\beta}{R_\alpha} = \frac{C_x}{C_y} \left[1 - (2\pi f)^2 C_y R_\beta \{ C_\beta R_y - C_\alpha R_x \} \right].$$

If one of R_x and R_y (say R_x) is made $= 0$, C_y kept at the order 10^{-9}, $(2\pi f)^2$ of the order 10^7 or less, the second term in the brackets tends to be small compared with unity since the other quantities of this term are usually small. Thus we obtain $R_\beta / R_\alpha \doteqdot C_x / C_y$ and the phase change $C_\alpha R_\alpha - C_\beta R_\beta \doteqdot C_y R_y$.* With this arrangement the potential ratios V_1 / V_2 will be approximately proportional to C_y / C_x and the phase change will be given by $C_y R_y$. Thus, by making two settings at each station, the readings

---

* Actually $\cot^{-1} (2\pi f C_\alpha R_\alpha) - \cot^{-1} (2\pi f C_\beta R_\beta) = \cot^{-1} (2\pi f C_\gamma R_\gamma)$.

of the second set are sufficient to give the potential ratios and phase change. Fig. 90 shows such an apparatus built for experimental purposes, and in Fig. 91 a typical curve obtained with the apparatus along a line traversing a vein of galena below an overburden of more than 40 ft. is given.

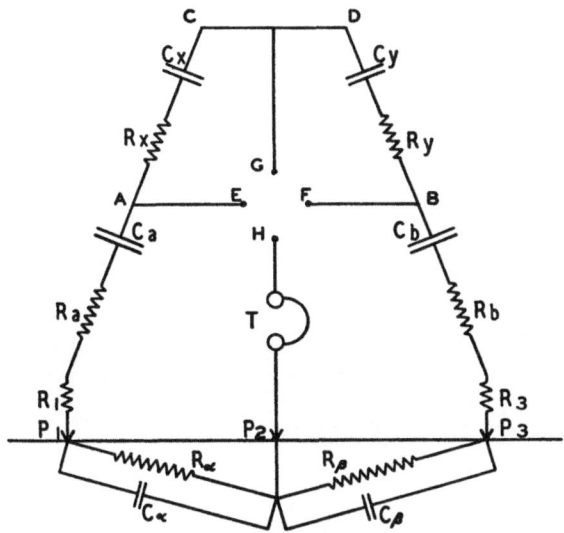

Fig. 89. The double balance ratiometer.

AERIAL ELECTRICAL PROSPECTING

Electromagnetic principles of geophysical prospecting have been applied to aerial surveying from an aircraft for locating ore and oil formations. Such surveys can be carried out rapidly, providing regional information as to faults, shear zones and conducting bodies, which would be favourable areas for further investigations on the ground. The apparatus used by Lundberg (1950) consists of a coil of many turns of insulated wire, wound round the fuselage of an Ansco twin-engined aircraft (Fig. 92), through which an alternating current is passed. The frequency is kept constant at 1000 cycles and the current at about an ampere, supplied by a special generator carried on the plane.

The current in the coil produces an alternating magnetic field, similar to the loop on the ground surveys, which stimulates currents in any conductors over which the aircraft passes. The secondary magnetic field from the ground beneath is picked up by a small coil attached below the starboard nacelle, as seen in

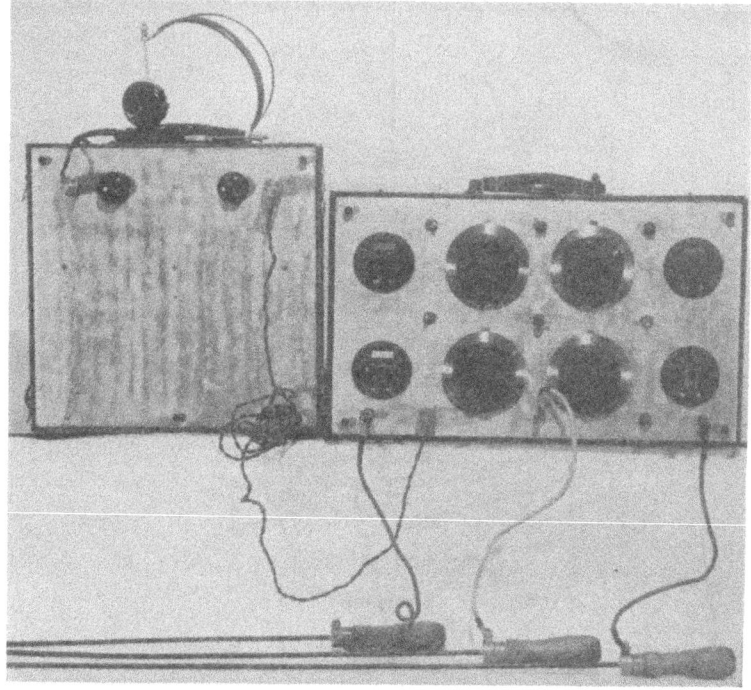

Fig. 90. The double bridge apparatus with amplifiers and telephones. The three electrodes are shown properly connected to the apparatus.

Fig. 92. The circuit arrangements are such that the secondary current in the pick-up coil is balanced out when the aircraft is over an area free from conductors. The signals induced from the ground are recorded on moving films by light delicate pens, the sensitivity of which is high and may be varied. The phase changes are also recorded. The result of a test flight with this equipment over a nickel-bearing sulphide vein in the Sudbury

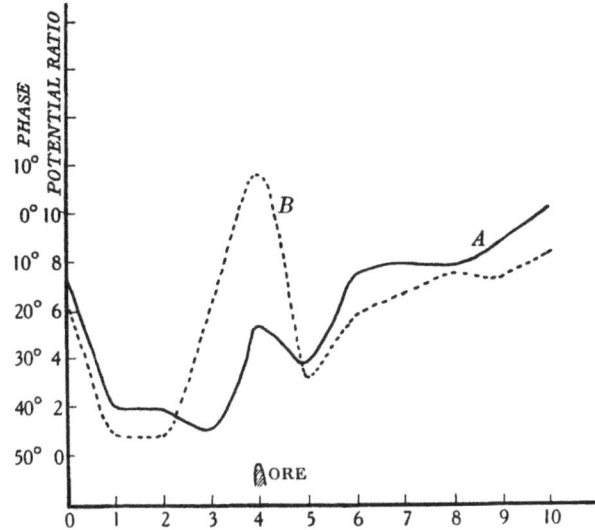

Fig. 91. *A*, potential ratio curve; *B*, phase curve, over a vein of galena with the modified Broughton Edge method.

Fig. 92. Aircraft fitted for airborne electrical survey

area is shown in Fig. 93. The method is of particular value in covering wooded areas, swamps and lakes, when rapid regional surveying is desirable.

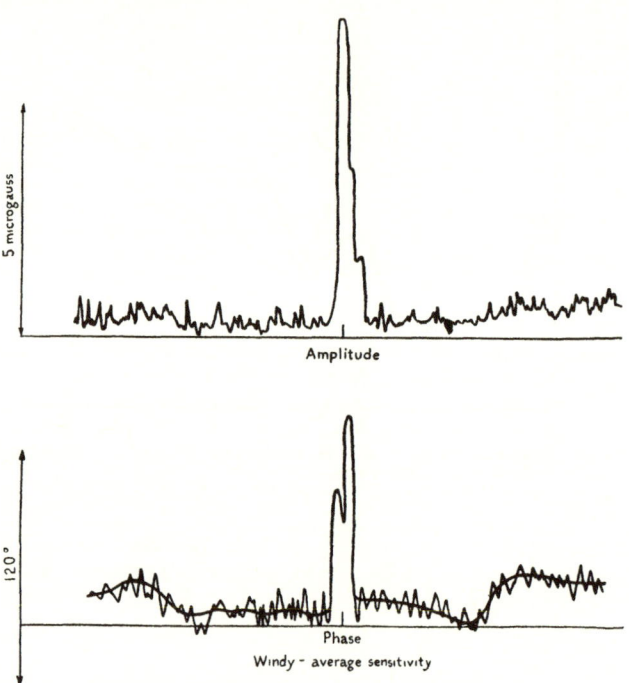

Fig. 93. Record of flight with airborne detector (Sudbury, Ontario).

PROBLEMS

1. A coil has a resistance of 200 ohms, a self-inductance of 50 mH. and is connected to a condenser of capacity $10 \mu F$. A current of 0·20 mA. at 500 cycles flows in the circuit. Find the reactance and impedance in the circuit. What is the phase angle? What e.m.f. is induced?

2. In the above problem, what extra capacity would you need to make the circuit tuned to the 500-cycle current? Assuming the same induced e.m.f., what would the current be?

3. The following readings were taken with the Bieler-Watson apparatus over some claims in the Desmeloises township, P.Q. The surveys were made along lines running north south, at stations 25 ft. apart, the first reading being at the south ends of the lines, which were 50 ft. apart. Determine the location of the conductor.

Station	Line *A*		Line *B*		Line *C*		Line *D*	
0	5E	26N	12E	36N	16E	36N	11E	30N
1	9E	45N	10E	59N	11E	47N	8E	32N
2	2W	88N	19E	100N	5E	60N	4W	32N
3	0	170N	12E	170N	12W	70N	7W	28N
4	53W	100N	2W	88N	20W	61N	4W	24N
5	34W	43N	31W	69N	29W	30N	5W	28N
6	8E	37S	22W	3N	27W	13N	13W	16N
7	9E	47S	15W	15N	24W	3S	13W	10N
8	22E	22S	—	—	—	—	14W	2N
9	3E	54S	4W	27S	20W	14S	12W	5S
10	20E	50S	—	—	—	—	9W	6S
11	17E	39S	1E	37S	4W	19S	7W	9S
12	18E	36S	—	—	—	—	5W	12S
13	17E	30S	5E	28S	4E	21S	7W	12S

4. Determine the position of the conducting vein and approximate depth of overburden from the following readings of dip of the exploring coil in the radiore method. The stations were on a north-south line 25 ft. apart and the exciting coil was 1200 ft. west of the line, directly over the vein.

Station	Dip in degrees	Station	Dip in degrees
0	18·2 S	7	2·0 N
1	17·9 S	8	13·0 N
2	16·8 S	9	17·7 N
3	15·7 S	10	19·5 N
4	13·0 S	11	19·9 N
5	8·8 S	12	20·0 N
6	2·0 S		

5. A survey was made along a north-south line using the Mason-Slichter-Hay method. The exciting loop was at the south end of the line and the values of the dip and strike were measured at stations 100 ft. apart. Determine the position of the conductor. (The angles are in degrees.)

Station	Dip	Strike	Station	Dip	Strike
0	0	0	11	1·5	0·9
1	0	0	12	2·5	1·5
2	0	0	13	5·0	6·8
3	0	− 0 5	14	2·5	12·0
4	0	0	15	0	17·6
5	0	− 0·6	16	− 20·0	15·0
6	0	0	16½	− 25·0	6·0
7	0·7	0	17	− 20·0	− 1·7
8	0·7	0	18	− 12·0	− 10·0
9	0·8	0·4	19	− 4·0	− 10·0
10	1·5	0·8	20	0	− 4·6

CHAPTER V

GRAVITATIONAL METHODS

GRAVITATION

Man actually sees the stars move around him daily while they retain to each other relative positions which are nearly fixed. Sun, moon and planets also appear to revolve round the earth daily, and to follow complex paths relative to the stars.

The Greeks found that the simplest explanation was a central sun around which the earth and other planets revolve. The more complex picture of a central earth was advocated by Ptolemy and was retained for a thousand years. Copernicus revived the heliocentric picture, and the exact measurements of Tycho Brahe enabled Kepler to determine the elliptic motion of the planets with the sun in one focus, and to establish the simple law that for all planets the squares of the periods of revolution varied as the cubes of the larger axes of their orbits. Newton conceived that any particle of matter m attracted any other particle m' at a distance r from it with a force measurable, in suitable units, as mm'/r^2. He further proved mathematically that a large uniform sphere had the same gravitational effect, outside it, as if its whole mass were concentrated at the centre. He thus found that the motion of the moon round the earth and of the planets round the sun conformed to his simple law. He extended his theory to embrace the motion of comets and the movement of tides. His theory of gravitation appears to hold for the whole universe. A slight but complex correction of Einstein rests on a profound natural philosophic basis and completes the rational picture, at least for the present.

The Newtonian laws and principles are used in the search for ore and oil in the modern gravitational methods of applied geophysics.

A small mass on a light thread constitutes a simple pendulum, and a small swing to and fro takes a time t sec. which is equal to $2\pi \sqrt{(l/g)}$, where l is the length of the thread and g is the acceleration due to gravity. The mass of m g. will be pulled by the earth at its surface, with a force of w dynes, where $w = mg$ and g has a value of about 980 cm. sec.$^{-2}$, by which it is meant that a freely falling body would have its velocity accelerated at a rate of 980 cm. per sec. in every second of time. The variations of g with altitude and with latitude are given in such tables as Kaye and Laby's, and need not be quoted here.

Now the simple pendulum is mainly ideal, for no mass is small enough, and no string light enough, to give a very correct value for g. Hence a reversible pendulum may be used, such as Kater's, consisting of a rod with two knife-edge supports at such places that they are not equidistant from the centre of gravity, but yet so that the times of oscillation about the suspension points are nearly equal. This instrument is not used, however, in large- or small-scale gravitational surveys, and it need not receive attention here.

The determination of the value of gravity as measured by g is connected with the question of isostasy. At considerable depths, a few miles underground, the constituents of the earth are at a pressure where they have at least viscous yield, so that there must be ultimately a balance of the superimposed masses. Large regions of mountainous country such as the Himalayas, the Rocky Mountains, or the Andes, have to balance against the depths of the Pacific or Indian Oceans. The relatively light sea water must have beneath it sufficiently dense material to compensate for the mass of the mountain ranges, and in long ages such a balance is perhaps achieved. The whole surface of the earth is, as it were, floating on a viscous and yielding magma, and it is even possible that large motions slowly occur *sideways* upon it, as in Wegener's fascinating hypothesis.

In large-scale gravitational surveys, a free pendulum is swung in a partial vacuum, and the slow change is measured at which it gains or loses time as compared with an adjacent electrically driven pendulum of nearly equal rate. This is done by noting the number of 'eclipses' which occur, say in two days, as determined by two small attached mirrors, one on each, reflecting light from a lamp, to eye or scale. Now the period of the electrically driven pendulum is recorded on a chronograph, whose rate is checked by radio time signals, which in turn are controlled from an observatory with star transit observations. Ultimately the time of the swing of the pendulum is referred to the rate of rotation of the earth. There are regions, sometimes small, sometimes many miles in extent, where gravity is markedly in excess or in defect of its normal value. These anomalies may be local, and the cause near the surface, or wide in area and deep seated. Their further investigation by torsion-balance methods is likely to throw light on geological and geophysical problems of importance.

Cavendish used his torsion balance to determine the value of the constant of gravitation and the mean density of the earth. He was the first to 'weigh the earth', that is, to determine its total mass.

In c.g.s. of absolute units it is customary to measure in dynes the force F due to the attraction between two small masses m and m' g. at a distance r cm. from one another, so that

$$F = \gamma \, \frac{mm'}{r^2},$$

where γ is the constant of gravitation, not to be confused with g. A mass of m g. is pulled by the earth with a force of w dynes, so that $w = mg$.

Rightly or wrongly, the physicist here diverges from the engineer in his viewpoint, and regards the 'pull of the earth'* as a force w dynes, and the quantity of material pulled as m g. The latter is invariable owing to the 'conservation of mass', under static non-radiative conditions, whereas the weight varies with depth, altitude, latitude and the presence or absence of dense materials in the neighbourhood concerned. The force of gravity on 1g. mass, expressed in dynes, will consequently depend upon the local value of g, measured in cm. per sec.[2]. The variation in this force is expressed in units called a *gal*, which corresponds to 1 cm. per sec.[2]. The unit derives its name from the famous sixteenth-century natural philosopher, Galilei Galileo.

* This may be a convenient fiction! See Eddington (1928).

but in practice it is found to be more expedient to use the unit *milligal.* (a thousandth of a gal.) or *microgal.* (a millionth of a gal.) in expressing gravitational anomalies.

THE CAVENDISH TORSION BALANCE

Cavendish suspended horizontally a light rod by a thin fibre. At the ends of the rod (Fig. 94) were two small heavy balls. Under normal conditions such a rod would be 'astatic', and point in any direction, but for the couple due to the torsion of the thread, which gives a zero position. Two large lead balls were then moved slowly up to the rod until they were near the small heavy balls, and by their gravitational attraction at right angles to the rod tended to turn them until the couple was balanced by the torque of the thread. By oscillating the system (rod and small balls) which had a calculable moment of inertia, it was possible to find the coefficient of torsion of the thread, and hence deduce the constant of gravitation.

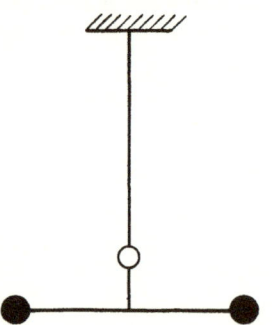

Fig. 94. Cavendish type of torsion balance.

It is then clear that, if a Cavendish torsion balance is in the neighbourhood of a single mountain, the rod will tend to turn round and point straight at the mountain, and that a direction at right angles will be one of unstable equilibrium. The ball nearer the mountain is always pulled harder, because it is closer, than one more remote. And the same will be true of a large dense body under the earth; a Cavendish torsion balance will tend to point at it. Both balls will always be subject to attraction, so that a couple will usually result. The position of stable equilibrium can in simple cases be determined from the principle that potential energy tends to transform into kinetic.

It is not always realized that it would be possible to conduct a geophysical survey using a Cavendish type of torsion balance alone, and that underground geological features such as faults, anticlines or synclines might thus be detected or inferred. Not that such a scheme is recommended, for today it would be like fighting with one arm tied behind the back.

Fordham states that Marcel Brillouin used a Cavendish balance in Spain, and that his was probably the first attempt to apply torsion-balance methods to mining. He also worked out the theory of the gradiometer.

THE EÖTVÖS TORSION BALANCE

The great advance made by von Eötvös was in the use of two balanced masses at unequal levels. The mass m is about 60 cm. lower than the upper mass m', and the two are held by a horizontal light aluminium rod about 40 cm. in length (Fig. 95). The masses, each about 25 g., are cylinders made of platinum or gold, or gold-plated brass, and the whole system, with a mirror, is suspended by a wire of platinum iridium, about 4×10^{-3} cm. in diameter and 50 cm. long, whose torsion coefficient must be determined by the oscillation method, adding subsidiary known masses if required. A first inspection of the instrument might suggest that its use is to measure the change of gravitation with height, or $\partial g/\partial z$, but this is precisely not the case. On the contrary, the direction and magnitude of the vector—*horizontal change of g per unit length*—is one of the objects attained, by determining $\partial g/\partial x$ and $\partial g/\partial y$ in two directions at right angles.

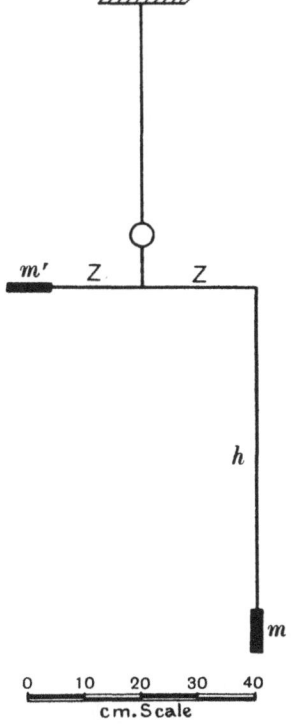

Fig. 95. Eötvös type of torsion balance.

Consider a heavy mass below the earth and on the reader's side of Fig. 95; this mass attracts the lower mass m with a larger force than it attracts the higher and more distant mass m', and there will thus be a couple which turns the balance round, until checked by the increasing torque or couple due to the twist of the suspension wire.

The extent to which the balance will turn must depend on the mass at either end of the rod m, on the length of the rod $2l$, and on its azimuth α, on the difference of levels of the masses h, and the torsion coefficient of the suspension, τ. Indeed, the sensitiveness of the instrument is measured by mhl/τ, and Eötvös achieved a value of about 77×10^3 for this quantity. E. Süss, Budapest, by further improvements, has raised this figure to 115×10^3. Of course the more sensitive the instrument the longer it takes to come to rest, and the more subject it becomes to air currents due to temperature differences, which can, however, be reduced by the use of a hut shelter and of a triple aluminium casing around the instrument (Fig. 96). So fragile an apparatus has to be carried over rough country, and therefore it is necessary to have clamps which will close gently and firmly on the rod and on the masses and allow them to be moved from place to place without damage.

The observations of the displacements of the rod due to variations of gravity can be made visually with a small electric lamp, mirror and scale, but today they are mainly made photographically on small plates, which can be developed and fixed in the hut directly they are removed from the instrument. The path of light is guided by prisms and lenses which require ingenuity as well as mechanical and optical skill in their arrangement, manufacture and use in the field.

In any case, a distance D is involved from mirror to scale, or from mirror to plate, and a deflexion n must be measured from a zero reading n_0. So that $n - n_0$ must depend on D, α, l, h, m, τ, also on the characteristics of the gravitational field at the place of observation, which indeed are what we want to determine, quite free from the constants of the particular instrument that we happen to employ.

It has already been pointed out that the direction and magnitude of a vector, namely, the horizontal gradient of gravity, is one of the main characteristics that are determined. This can

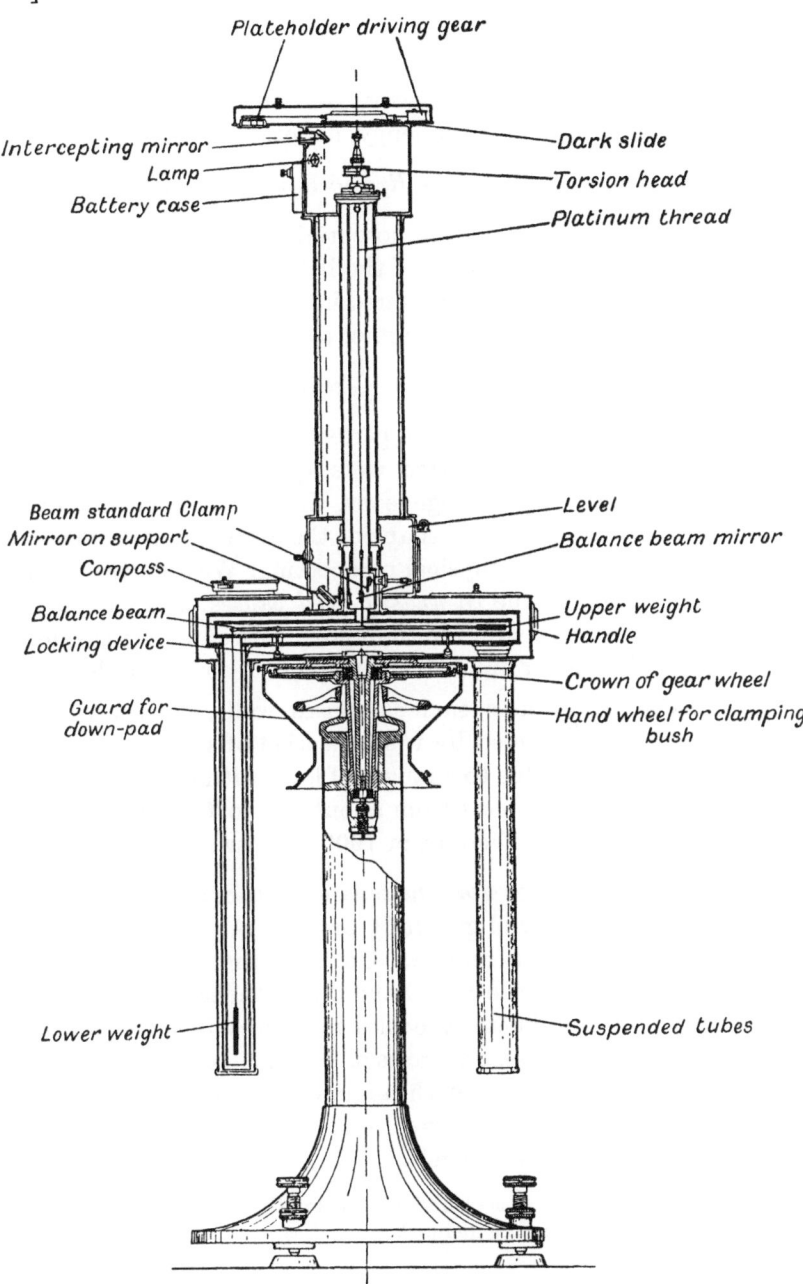

Fig. 96. Eötvös torsion balance.

be found by the simple device of having one mass lower than the other mass on the balance. But there is another, equally important, but rather more evasive quantity involved, known as the horizontal directive tendency, denoted by H.D.T.,* whose magnitude is $g\left(\dfrac{1}{R_2} - \dfrac{1}{R_1}\right)$, where R_1 and R_2 are the principal radii of curvature of the level, or equipotential, surface (gravitational of course) at the central point of the balance. As g is measured in cm.sec.$^{-2}$, and R_1, R_2 are measured in cm., the above quantity has no linear dimensions and it cannot correctly be termed either a force or a couple. Certainly $\dfrac{1}{R_1} - \dfrac{1}{R_2}$ reveals how much the equipotential surface diverges from a perfect sphere, which would be its natural shape if the earth were quite round, free from rotation, and perfectly homogeneous. It must be remembered, however, that the spheroidal shape of the earth and its rotation do not interest us in geophysical prospecting. We seek the local lack of homogeneity due to *underground* causes, and therefore corrections must be made for the above as well as for surface inequalities. The latter are conveniently made under four heads:

(1) Special local causes, such as houses, walls, ditches, which can sometimes be avoided by wise selection of stations.

(2) Terrain effects due to changes of level within 100 ft. or so.

(3) Topographical effects from about 100 to 1000 ft.

(4) Cartographical effects from 1000 ft. outwards.

It is clear that masses *near* the balance, either above or below its centre of gravity, will generally produce a larger disturbing error than those more remote. On the other hand, distant masses by their very size may have a preponderating effect.

As to the terrain effects, a local level survey is made so that the levels are well known on circles of 1·5, 3, 5, 10, 20, ..., 100 m. radius in eight directions, with angular intervals of 45°, from the centre of the station. Corrections are made for the several sectors according to their levels.

* The word 'curvature' is frequently used in the United States rather than 'horizontal directive tendency'.

Similar methods of correction apply to topographical effects when the levels need not be known with the same accuracy or detail, while the larger scale cartographical corrections may be usually deduced from charts or maps of the district. It will be noted that in all cases these corrections involve the densities of the attracting masses, and therefore average densities must be determined from an examination of specimens of soil and rock for each mass under consideration, which may vary from a local mound to a distant mountain or river valley.

The area chosen for each station must be cleared and levelled over a circle about 5 m. in radius, nor should the slope of the ground near the station exceed one in eight.

On this area a light-proof ventilated hut (4 × 4 × 8 ft.) must be placed on a firm foundation, and inside it the portable Eötvös balance must be erected and orientated so that its angle with the true north is known either from accurate maps, or from astronomical observation, or is deduced from the compass on the instrument. For the *direction*, as well as the magnitude, of both the horizontal gravity gradient and of the H.D.T. have to be determined for each station selected, and to be plotted accurately on the chart as a first step to the interpretation of the underground characteristics. It is clear that a single reading cannot give sufficient data, and in fact five readings are necessary, while it is convenient to take a sixth or check reading. After the instrument has been orientated and levelled an hour or more must elapse before a reading is taken, during which time a steady position is attained by the balance. The apparatus is then rotated through 60°, and after another interval a second reading is made, and so on, up to 300°, when the photographic plate may be developed. Eötvös used a double balance, as in Fig. 96, but with two beams side by side. Schweydar adopted a Z-shaped balance (Fig. 97) where the upper mass is firmly upheld by a light aluminium rod, while the lower is similarly suspended. It is of course possible to have a double Z-balance combining these arrangements.

By this method the number of necessary positions is reduced from six to three, and steps of 120° rotation every hour will prove sufficient. Usually a clock is employed which at the end of an hour will temporarily light the small electric lamp, so that a record is made on the photographic plate, after which the clock slowly turns the apparatus through 120°, and leaves it there for another hour, and then again lights the lamp, and so forth, to a total of about twelve observations.

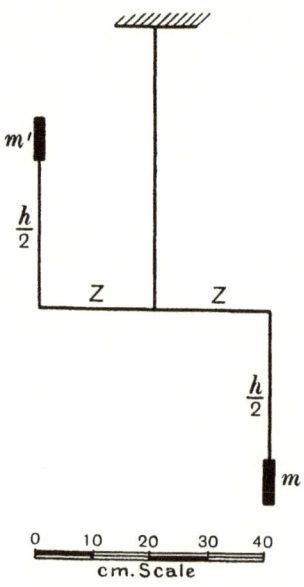

Fig. 97. Schweydar's Z-shaped modification of the torsion balance.

This arrangement leaves the region free from all disturbance, relieves the operator of his troublesome rounds, and permits him to sleep at night. At the end of the time he develops and fixes the plate, and then finds on it a number of dark spots, six to twelve giving the positions of the two balances, while others record the temperatures, for which corrections must be made. The plate is then measured with a glass or mica graticule, after which computations and corrections are worked out and the results duly plotted. The time of one observation was 2 hr. originally, but it was reduced by Schweydar to 40 or 50 min.

DENSITY

Now the densities of some of the constituents of the earth may be grouped as follows, stated in g. per cm.^{3}.

Oil	0·87–0·91
Water	1·00
Sea water	1·01–1·05
Loose sand and earth	1·3
Coal, soft	1·2–1·5
Coal, hard	1·4–1·8
Clay	1·8–2·6
Sandstone	2·3
Rock salt	2·15
Gypsum	2·3
Serpentine	2·6
Gneiss	2·6
Limestone	2·7
Porphyry	2·7
Granite	2·75
Diabase	2·9
Zincblende	4·06
Pyrrhotite	4·8
Pyrite	5·0
Magnetite	5·2
Chalcopyrite	5·7

It must constantly be borne in mind that detection is made, not of densities, but of the difference of densities of one region as compared with another.

The vertical anomaly, Δg_z, in the gravitational force on 1 g. mass at P (Fig. 98), due to a sphere of radius a cm. whose centre is z cm. below the surface, may be calculated from Newton's law. If the difference in density between that of the sphere and surrounding rock is $\Delta\sigma$, then

$$\Delta g_z = \gamma \tfrac{4}{3}\pi a^3 \frac{\Delta\sigma}{R^2} \cos\theta$$

$$= \gamma \Delta\sigma \tfrac{4}{3}\pi a^3 \frac{z}{(x^2+z^2)^{\frac{3}{2}}},$$

where $\gamma = 6\cdot66 \times 10^{-8}$ c.g.s. units and Δg_z is in gals.

If, instead of a sphere, the deposit is in the form of a cylinder of infinite length and radius a, then

$$\Delta g_z = 2\pi\gamma\Delta\sigma \, \frac{a^2 z}{x^2 + z^2}.$$

$\Delta\sigma$ is this case being the difference in density per unit length.

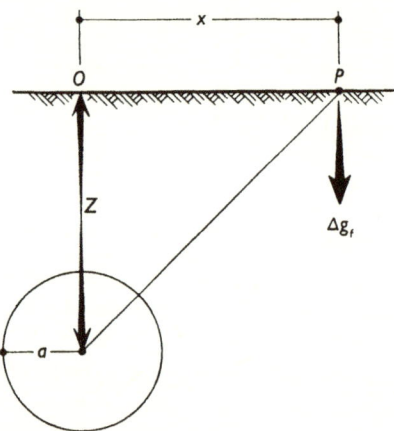

Fig. 98. Gravitational anomaly due to a buried sphere.

Consider a sphere of rock salt, density 2·15, in surrounding strata of density 2·6, then the difference of density is 0·45. Suppose the sphere is 300 m. (990 ft.) in diameter and that the *top* of it is 300 m. below ground. The vertical loss of attraction, immediately above it on the ground, would be $0\!\cdot\!14 \times 10^{-3}$ dyne per g., and this is too small to be measured by the pendulum method, the sensitivity of which amounts to 1×10^{-3} dyne per g. mass, or about one part in a million of the value of g.

On the other hand, the horizontal gradient of gravity would, at the most, and at a distance of a radius and a half from the point on the ground immediately over the centre of the sphere, amount to $2\!\cdot\!7 \times 10^{-9}$ dyne per g. per cm. So that the sphere *could* be detected by the torsion balance, the sensitivity of which is about 10^{-9} dyne per g. per cm., or one part in a million million. It is clear that the Eötvös balance ranks high among the more sensitive types of apparatus. So also a London underground tube, 15 or 16 ft. (5 m.) in diameter and 250 ft. (75 m.) deep, would give a vertical decrease of gravity—0.0063×10^{-3} dyne per g., not detectable by pendulum; but a maximum horizontal gravity gradient $= 6 \times 10^{-9}$ dyne per g. per cm., quite readily detectable by torsion balance.

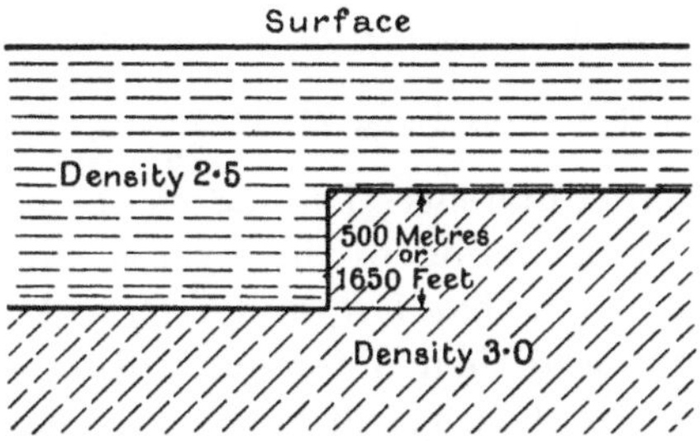

Fig. 99. An underground ledge for the discussion of its effects on a torsion balance.

We will consider one more rather instructive case where there is an underground ledge, somewhat ideal in character, and clearly explained by Fig. 99. If a pendulum is swung first over one region and then over the other there will be a difference in vertical pull equal to 0.010 dyne per g., which could readily be detected, and this is independent of any reasonable depth. As for the horizontal gradient, that would be a maximum over the fault and would be

Depth of ledge	Dyne per g. per cm.
10 m.	261×10^{-9}
1 km.	27×10^{-9}
10 km.	3×10^{-9}

so that the fault could theoretically be just detected at a depth of 18 miles!

Before giving the mathematical theory of the torsion balance a few remarks on the H.D.T., or horizontal directive tendency, or 'curvature', will not be out of place. On the top of a long underground ridge of heavy rock, which is convex upward and so constitutes a hidden anticline, the torsion balance (Cavendish or Eötvös) tends to set itself parallel and along the ridge, and lines can be drawn whose magnitude and direction represent the H.D.T. vector $g\left(\dfrac{1}{R_2} - \dfrac{1}{R_1}\right)$, where R_1, R_2 are the principal radii of curvatures of the *level* or *equipotential* surface at the point under consideration, as explained in the subsequent theory.

Here the *radius* of curvature R_1 is greater than R_2, and the *curvature* $1/R_1$ is less than $1/R_2$, while $\dfrac{1}{R_2} - \dfrac{1}{R_1}$ might be termed truly the curvature difference. The radius of the earth is about $6\cdot4 \times 10^8$ cm., and the radius of the level surface will approximate to that value, so that $1/R_1$ and $1/R_2$ will generally be of the order 10^{-9} cm. and their difference will be less than either of them, so that 10^{-12} c.g.s. will be a convenient unit for the difference of curvature $\dfrac{1}{R_2} - \dfrac{1}{R_1}$.

But in practice $g\left(\dfrac{1}{R_2} - \dfrac{1}{R_1}\right)$ the H.D.T. or *horizontal directive tendency* is used, and as $g = 981$ or nearly 1000, it is very convenient to measure the H.D.T. in 10^{-9} c.g.s., and this is called an *Eötvös unit*, and normally denoted by E.; readings from 5–50 E. are encountered, running up sometimes to 200 E. as a maximum.

So also in measuring the horizontal gravity gradient (which Eötvös called Gr. g and which must not be confused with the vertical gradient) a convenient unit is 10^{-9} dyne per cm. on unit mass, and this too may be denoted by E. Normal readings are 5–30 E., ranging to maxima of about 150 E.

In addition to making corrections for near and distant objects above the surface of the earth, it is necessary to remember that the rotation of the earth, and consequent spheroidal shape, modifies the value of gravity as we move northward, and that this normal northward gradient amounts to about 7 E. in latitude 30°.

MATHEMATICAL THEORY

It may be desirable first to recall a few elementary relations.

1. The curvature at any point of a plane curve is measured as for a circle passing through three points very near together.

If AB is a chord (Fig 100), and if PC is perpendicular to it at its mid-point C, then putting $AC = CB = y$, and $PC = z$, we have

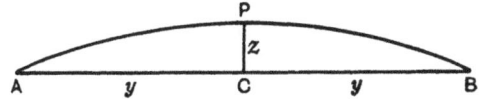

Fig. 100. Diagram for the curvature of a plane curve.

for the radius R of the circle through A, P, B, the relation

$$y^2 = z(2R - z), \qquad (5.1)$$

and, on reducing y and z indefinitely, we have R equal to the limiting value of

$$y^2/2z. \qquad (5.2)$$

This is a measure of the radius of curvature at P.

Example: If $AC = CB = 10$ ft., $PC = \frac{1}{4}$ ft., then $R = 200$ ft. nearly.

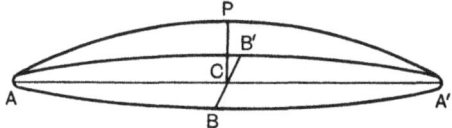

Fig. 101. Diagram for the curvature of a surface.

2. In the case of a surface the procedure is different. Consider a point P and imagine a tangent plane drawn at P.

Take a point C, close to P (Fig. 101), and on the normal at P. Through C draw a plane perpendicular to PC, and parallel to the tangent plane at P. This may cut the surface in an ellips $ABA'B'$, with its centre at C. If AA' is the major and BB' the minor axis of this ellipse, we can write down the values of the two *principal* radii of curvature at P which we will call R_1 and R_2. By the last article

$$R_1 = \frac{AC^2}{2PC}, \quad R_2 = \frac{BC^2}{2PC},$$

both in the limiting value where PC is very small.

The elliptic section close to P, as drawn in the figure, has been called the *indicatrix* because it indicates the character of the curvature at a point. This section may be *any* conic section—parabola, ellipse, or hyperbola, and it becomes of course a circle when $R_1 = R_2$, and they have the same sign. If they were equal, but had opposite signs, we should be dealing with a surface like a saddle, or a mountain pass, or the region between open thumb and forefinger, and the indicatrix would then be a rectangular hyperbola.

3. The equation of an ellipse with a, b as semi-axes and with the origin at the centre, and with axes of reference along the axes of figure, is well known to be

$$\frac{x^2}{a^2} + \frac{y^2}{b^2} = 1. \tag{5.3}$$

With other rectangular axes through the centre we have the equation for any conic section

$$ax^2 + 2hxy + by^2 = c,$$

and the centre term on the left side can be cleared by rotating the axes through an angle θ so that we have a new equation

$$AX_0^2 + BY_0^2 = c,*$$

where
$$\left.\begin{array}{l} A + B = a + b, \\ AB = ab - h^2, \end{array}\right\} \tag{5.4}$$

and the angle θ is given by

$$\tan 2\theta = \frac{2h}{a-b}. \tag{5.5}$$

* Put
$$x = X_0 \cos\theta - Y_0 \sin\theta,$$
$$y = X_0 \sin\theta + Y_0 \cos\theta,$$
in (5.3) and (5.4) follow.

Drawing a small right-angled triangle with sides $2h$ and $a-b$ as in Fig. 102, we see not only that

$$\tan 2\theta = \frac{2h}{a-b},$$

but that the square on the hypotenuse is $(a-b)^2 + 4h^2$ or

$$(a+b)^2 + 4(h^2 - ab),$$

and this equals $\qquad (A+B)^2 - 4AB$

or $\qquad (A-B)^2 \quad$ from (5.4).

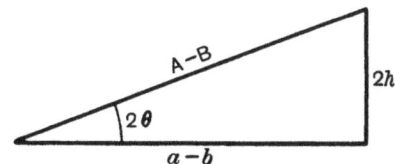

Fig. 102. Diagram showing relation between parameters
in a conic section.

Hence $\qquad A - B = (a-b)\sec 2\theta,$ $\qquad$ (5.6)

a result which will be wanted later.

4. If then we write the equation of the surface near P (Fig. 101) as

$$2z = ax^2 + 2hxy + by^2,$$ $\qquad$ (5.7)

with the axis of z along the normal PC, and with any two axes of x and y at right angles in the tangent plane, we can always simplify our equation by retaining the z-axis, and by swinging the other two axes together in the tangent plane through an angle θ given by

$$\tan 2\theta = \frac{2h}{a-b},$$ $\qquad$ (5.8)

and our equation takes the more desirable form

$$2z = Ax^2 + By^2.$$ $\qquad$ (5.9)

Putting $y = 0$, we remember that

$$R_1 = \text{limit of } \frac{AC^2}{2PC} = \text{limit of } \frac{x^2}{2z} = \frac{1}{A},$$

so that $A = \dfrac{1}{R_1}$ is the minimum curvature at P. So also putting

$x = 0$ we get $B = \dfrac{1}{R_2}$ and B is the maximum curvature at P.

In considering the whole curvature of the *surface* at P we are free to choose what we will, and

$$\frac{1}{R_1 R_2}, \quad \frac{1}{R_1} + \frac{1}{R_2}, \quad \frac{1}{R_1} - \frac{1}{R_2}$$

have all been suggested as suitable measures of curvature, but there are good reasons for preferring the first.

It is found that in torsion-balance work we want to know how the level, or equipotential, surface due to gravitation is deformed slightly in any manner from a true sphere, hence we select $\dfrac{1}{R_2} - \dfrac{1}{R_1}$ as a measure of the peculiarity of the surface at P, and we might call this the principal curvature difference. We are here assuming that R_1 is greater than R_2, so that the above is positive.

In the case of complete sphericity $R_1 = R_2$, and the measure is zero; so that $\dfrac{1}{R_2} - \dfrac{1}{R_1}$ does not really measure the curvature, but it does measure the divergence from sphericity.

It will be shown presently that the H.D.T. (horizontal directive tendency) is equal to $g\left(\dfrac{1}{R_1} - \dfrac{1}{R_2}\right)$ so that the twisting couple on a Cavendish torsion balance is intimately connected with the divergence from the true sphere of the 'level' surface at the point where the balance is set up. And the same remark holds good for the Eötvös balance, but here we have, due to the horizontal gradient of gravity, an additional torque, secured by having one of the two masses at a lower level than the other.

5. We are now ready to consider the forces acting on the balance and to take moments about a vertical line through the middle of the rod (*vide* Fig. 95). In doing so we shall define the vertical line at a point as that of a short plumb line there, necessarily perpendicular to a dish of still water or mercury at

the place or point under consideration, if unaffected by capillarity from the edge of the dish. The only force on unit mass (1 g.) at this middle point of the balance will be g dynes, the particular value of gravity there. At any point (x, y, z) a little removed from the origin, the forces on unit mass may be assumed to have linear relationships given by

$$\left.\begin{aligned} X &= Ax + Hy + Gz, \\ Y &= Hx + By + Fz, \\ Z &= Gx + Fy + Cz + g, \end{aligned}\right\} \tag{5.10}$$

where A, B, C, F, G, H will indeed have to be determined experimentally. The coordinates of m are

$$(l \cos \alpha, \, l \sin \alpha, \, h),$$

and of m' $\qquad (-l \cos \alpha, \, -l \sin \alpha, \, 0).$

The torque or couple round Oz, from x towards y, equals on unit mass,

$$l \cos \alpha (Hl \cos \alpha + Bl \sin \alpha + Fh) - l \sin \alpha (Al \cos \alpha + Hl \sin \alpha + Gh)$$
$$+ l \sin \alpha (-Al \cos \alpha - Hl \sin \alpha)$$
$$- l \cos \alpha (-Hl \cos \alpha - Bl \sin \alpha),$$

and on the actual masses

$$ml^2(B - A) \sin 2\alpha + 2ml^2 H \cos 2\alpha + mlh(F \cos \alpha - G \sin \alpha),$$

so that the torque S equals

$$K\left(\frac{B - A}{2}\right) \sin 2\alpha + KH \cos 2\alpha + mlh(F \cos \alpha - G \sin \alpha), \tag{5.11}$$

where $K = 2ml^2$ is the moment of inertia of the two masses about the suspension wire. But rotation will continue, after oscillations have subsided, until $s = \tau\theta$, where τ is the coefficient of the suspension wire and θ is the angle turned through by the balance. The reflected light from the mirror will have turned through this angle, so that if D, the distance from mirror to scale, or to photographic plate, be large, we have

$$\theta = \frac{n_\alpha - n_0}{2D}, \tag{5.12}$$

where n_0 is the 'zero' reading, to be eliminated as unknown, and n_α is the reading for the actual position with deflexion α from the x axis pointing due north.

From (5.11) and (5.12) we get

$$n_\alpha - n_0 = \frac{KD}{\tau} \overline{(B - A} \sin 2\alpha + 2H \cos 2\alpha)$$
$$- \frac{2Dmhl}{\tau} (G \sin \alpha - F \cos \alpha)$$

or

$$n_\alpha - n_0 = P\left(\frac{B - A}{2} \sin 2\alpha + H \cos 2\alpha\right) - Q(G \sin \alpha - F \cos \alpha), \tag{5.13}$$

where P and Q stand for determinable instrumental quantities

$$2\frac{KD}{\tau} \quad \text{or} \quad \frac{4ml^2D}{\tau}, \quad \text{and} \quad \frac{2Dmhl}{\tau} \text{ respectively.}$$

6. The actual readings for the six settings $\alpha = 0, 60, 120, 180, 240, 300°$ will be $n_1, n_2, ..., n_6$, and we get from (5·13) these six equations:

	$P(B-A)\sqrt{\tfrac{3}{4}}$	$PH/2$	$\sqrt{3}\,QG/2$	$QF/2$
		Multiplied by		
$n_1 - n_0$	0	+2	0	+2
$n_2 - n_0$	1	−1	−1	+1
$n_3 - n_0$	−1	−1	−1	−1
$n_4 - n_0$	0	+2	0	−2
$n_5 - n_0$	1	−1	+1	−1
$n_6 - n_0$	−1	−1	+1	+1

wherein the quantity in the left column of any row equals the sum of the quantities found by multiplying the stated figures of that row by the values of the uppermost row. Here we have six simultaneous equations to find the five unknowns n_0, $B - A$, F, G, H; so that one equation is a check.

Add the first, third and fifth rows and we get zero on the right, so that

$$n_0 = \tfrac{1}{3}(n_1 + n_3 + n_5),$$

so also

$$n_0 = \tfrac{1}{3}(n_2 + n_4 + n_6). \tag{5.14}$$

Moreover, we get the actual working formulae

$$
\left.
\begin{aligned}
B - A &= \frac{1}{P\sqrt{3}}\,(n_2 - n_3 + n_5 - n_6), \\[2mm]
2H &= \frac{n_1 + n_4 - 2n_0}{P}, \\[2mm]
F &= \frac{1}{2Q}\,(n_1 - n_4), \\[2mm]
G &= \frac{1}{2Q\sqrt{3}}(n_5 + n_6 - n_2 - n_3).
\end{aligned}
\right\}
\qquad (5.15)
$$

The instrumental quantities depend on the dimensions and material of the wire, and are calibrated by the manufacturers. If a new suspension wire is needed, the coefficient of torsion must be calculated afresh by the method of oscillations, which will give K/τ, and K was previously known. The introduction of a known heavy mass at a measured distance, first on one side and then on the other of the lower mass, will give a deflexion which, when measured, permits the measurement of τ by a deflexion method.

If two beams are used side by side, only three settings are necessary, giving $\alpha = 0°$, $120°$, $240°$ for one beam, and $\alpha = 180°$, $300°$, $420°$ for the other beam. Substitute the values obtained in (5·13), and we shall again find $B - A$, $2H$, F, G in terms of the six readings n_1, n_2, n_3 for the first and n_1', n_2', n_3', for the second beam.

In this case all six equations will be needed, because we have to find from them the *two* zero equations n_0 and n_0', one for each beam.*

Let us consider the meanings to be attached to F and G. Looking back at equations (5.10) we see

$$X = Ax + Hy + Gz,$$
$$Y = Hx + By + Fz,$$
$$Z = Gx + Fy + Cz + g,$$

* Those who are working with either apparatus will naturally consult Shaw and Lancaster Jones (n.d.).

and at unit depth below the origin, putting $x = 0$, $y = 0$, $z = 1$, we get $X = G$, $Y = F$, $Z = C + g$, as the forces on unit mass there. But because G, F, C are small the resultant force on unit mass is g, so that the direction cosines of the force R (the resultant of X, Y, Z) are at the point named $(0, 0, 1)$

$$\frac{G}{g}, \quad \frac{F}{g}, \quad \text{and} \quad 1.$$

Thus the values G, F and g give us the direction ratios of a vector 1 cm. below the origin or middle point of the rod of the balance. What is that vector? The third equation of (5.10) gives us the answer.

Since $Z = Gx + Fy + Cz + g,$

we get at once that $G = \dfrac{\partial Z}{\partial x}$ and $F = \dfrac{\partial Z}{\partial y}$,

so that G is the variation in the value of Z, and therefore of g, per unit length in the direction $y = 0$, or along Ox, while F is the variation of g per unit length along Oy, and their resultant $\sqrt{(G^2 + F^2)}$ shall be called $\Delta_H g$, which is the horizontal gradient of gravity. Hence Gr. g or $\Delta_H g = \sqrt{(G^2 + F^2)}$ and its direction is given by the angle which it makes with Ox, namely, ϕ, where $\tan \phi = \dfrac{F}{G}$.

As soon as these directions, such as ϕ, are determined at a large number of stations it is possible to draw curves at right angles to them, and along such lines or curves there will be *no* variations in the value of g. Such lines of equality are called *isogams*, a word probably carried over from magnetic measurements and derived from 'gamma', because variations of gravity have been measured in terms of units, 10^{-3} dyne per g., denoted by that Greek letter.

Immediate reference may be made to Figs. 103 and 104 (North American Exploration Company), where the isogams are drawn in full lines, and the direction and magnitude of the gravity gradients are drawn to scale in terms of 10^{-9} dyne per g. per cm., or Eötvös units.

On the same plans there are shown the curvature difference values measured as $g\left(\dfrac{1}{R_2} - \dfrac{1}{R_1}\right)$, and called the H.D.T. or horizontal directive tendency. This is now to be considered.

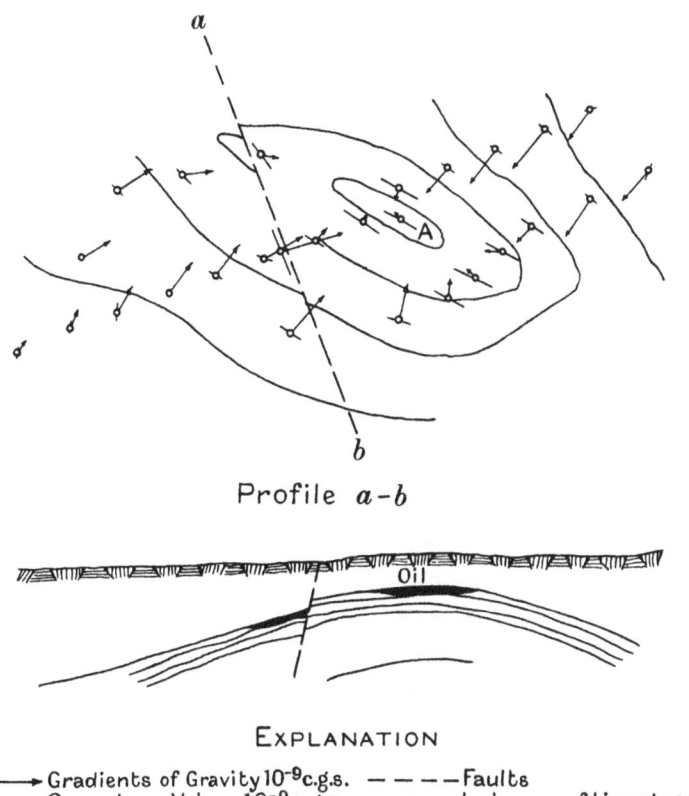

Profile $a-b$

EXPLANATION

○———▶ Gradients of Gravity 10^{-9} c.g.s.   – – – –Faults
——○—— Curvature Values 10^{-9} c.g.s.   ‿‿‿ Isohypses of Limestone

Fig. 103. Diagram showing isogams and the magnitudes to scale of horizontal gravity gradients; also the horizontal directive tendency vectors. (North American Exploration Co., *Geophysical Prospecting*, p. 20.)

At this point it becomes necessary to consider the equation of the level surface at and near our origin. Let the potential due to the whole earth at the origin, or mid-point of the rod, be U_0.

This would be defined as the work to be done in pulling unit mass, 1 g., from that point away to a great distance.

The potential at an adjacent point (x, y, z) would have a slightly different value, and we propose to assume that its value U is given by

$$U = U_0 + \tfrac{1}{2}(Ax^2 + 2Hxy + By^2 + Cz^2 + 2Gxz + 2Fyz + 2gz), \quad (5\cdot 16)$$

where the letters have the same meaning as in (5.10) because the forces on unit mass X, Y, Z are equal to the space variations of U parallel to the axes, so that

$$\left.\begin{aligned} X &= \frac{\partial U}{\partial x} = Ax + Hy + Gz, \\[4pt] Y &= \frac{\partial U}{\partial y} = Hx + By + Fz, \\[4pt] Z &= \frac{\partial U}{\partial z} = Gx + Fy + Cz + g, \end{aligned}\right\} \qquad (5.17)$$

and it is clear that the partial differentials give us the correct values of the force per unit mass at x, y, z.

Now the shape near the origin of the level or equipotential surface defined by (5.6) is dependent on terms of the second degree, namely, $Ax^2 + 2Hxy + By^2$. Let us give z some small value h, let $U - U_0$ have the corresponding small value U', then we have the equation of an *ellipse*

$$Ax^2 + 2Hxy + By^2 + Ch^2 + 2Gxh + 2Fyh + 2gh = U';$$

and we dealt with this very case in §§3 and 4 of this chapter, except that $2z$ is now replaced by $2gz$ (see (5.16)).

Turn the axes through an angle θ given by

$$\tan 2\theta = \frac{2H}{A - B}, \qquad (5.18)$$

and we shall have

$$A'x^2 + B'y^2 + Ch^2 + 2G'xh + 2F'yh + 2gh = U',$$

as a new equation to the ellipse, wherein A' will be g/R_2 and B' will be g/R_1. Hence by (5.6)

$$\left.\begin{aligned} g\left(\frac{1}{R_2} - \frac{1}{R_1}\right) &= (A - B)\sec 2\theta \\[4pt] &= \sqrt{\{(A - B)^2 + 4H^2\}}. \end{aligned}\right\} \qquad (5.19)$$

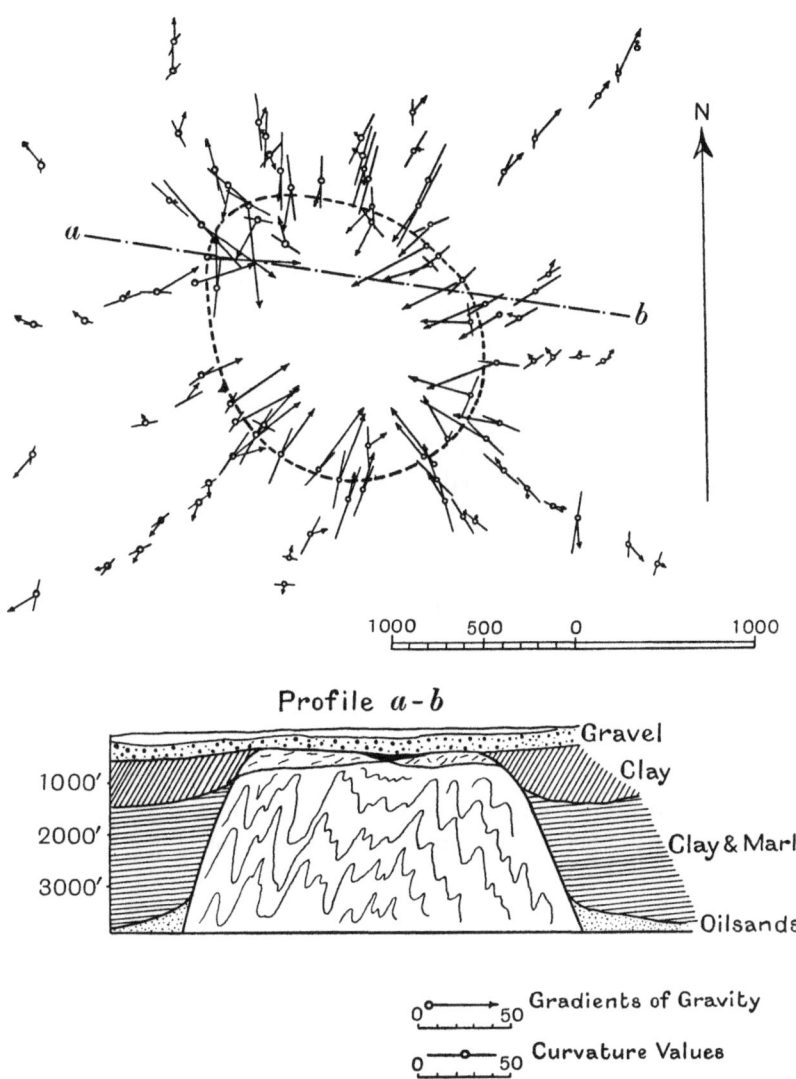

Profile *a-b*

Fig. 104. Diagram showing gravity gradients and H.D.T. over a salt dome.

But $A - B$ and H are given by (5.15). Hence (5.18) gives the direction θ and (5.19) gives the magnitude of the horizontal directive tendency (H.D.T.), also called curvature difference by some writers. On division by g we have the true 'divergence from sphericity' $\left(\dfrac{1}{R_2} - \dfrac{1}{R_1}\right)$, which lies parallel to the algebraical minimum curvature. As for the appearance of g in the statement that A and B are equal to g/R_2 and g/R_1 respectively, this is due to the fact that in (5.7) we had $2z$, whereas in (5.16) we find $2gz$.

When the H.D.T. vectors are plotted they will be found to lie along and parallel to heavy anticlines, visible or hidden, and to be at right angles to synclines or valleys, and to point to mountains or to heavy domes underground. The word 'heavy' must be inserted because the case is different if the hidden anticline or dome is composed of rocks lighter than those above them, in which case there will be effectively repulsion or negative attraction.

7.   So far, we have shown that six measurements n_1, n_2, n_3, n_4, n_5, n_6 lead to the determination of the zero reading n_0, and of the four quantities $A - B$, H, F, G. Of these, F and G lead directly to

$$\Delta_H g = \sqrt{(G^2 + F^2)}$$

and

$$\tan \phi = \frac{F}{G},$$

so that we know the magnitude and direction of the gravity gradient; whereas $A - B$ and H give us θ from

$$\tan 2\theta = \frac{2H}{A - B}$$

and

$$g\left(\frac{1}{R_2} - \frac{1}{R_1}\right) = (A - B) \sec 2\theta$$

$$= \sqrt{\{(A - B)^2 + 4H^2\}}$$

for the direction and magnitude of the H.D.T.

These four quantities, the direction and the magnitude of $\Delta_H g$, and also of the H.D.T., are all that we require, and there follow (*a*) corrections, (*b*) plotting, (*c*) interpretation. Attention is directed to Fig. 103* as a simple clear case in which to study both sets of vectors.

If the gravity gradients are added together along a straight line, selected for a series of equidistant stations; then we obtain the total change in *g* along that line and this can be termed the gravity anomaly. A good illustration is given in Fig. 105, taken from the 1926 *Summary of Progress of the Geological Survey of Great Britain*. The paper concerned, by McLintock and Phemister, describes investigations in South-west Persia, and should be read

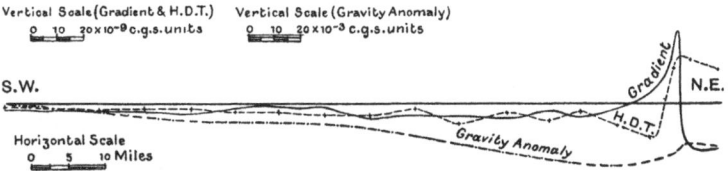

'Fig. 105. A 100-mile traverse in S.W. Persia. Gravity gradient and
H.D.T. indicate a sharp fold underground on the right.
(Geol. Survey, Gt Britain, 1926, p. 189.)

by those who are interested in the use of the torsion balance. It will be noted that, in crossing an anticline, the gradient naturally changes sign, while the H.D.T. does not (Fig. 106). In passing a vertical fault the reverse is true, and, indeed, every underground density change produces interpretable modifications of both curves. The above writers have also published (1927), an account of their survey over the Swynnerton Dyke, Yarnfield, Staffordshire.

8. If a small mass M attracts unit mass at distance r with a force F, then $F = \gamma \dfrac{M}{r^2}$, where γ is the constant of gravitation; and if F is measured in dynes, the masses in grammes, and the distance in centimetres, then $\gamma = 6 \cdot 658 \times 10^{-8}$ c.g.s. unit.

The work done in withdrawing the unit mass to a very great distance will be $\gamma M/r$, and this figure is a measure of its potential energy at distance r from M.

Moreover, if M is a large sphere, Newton showed that the above remarks still hold because a uniform sphere attracts as

if the whole mass were at its centre. Thus the equipotentials
of the sphere M are themselves spheres, and the force in dynes
on unit mass is normal everywhere to an equipotential and
equals the gradient of potential along that normal at the point
considered.

In the case of the earth we have already given a value U for
the potential (5.16) and for the three components of the force on
unit mass derived from it (5.17).

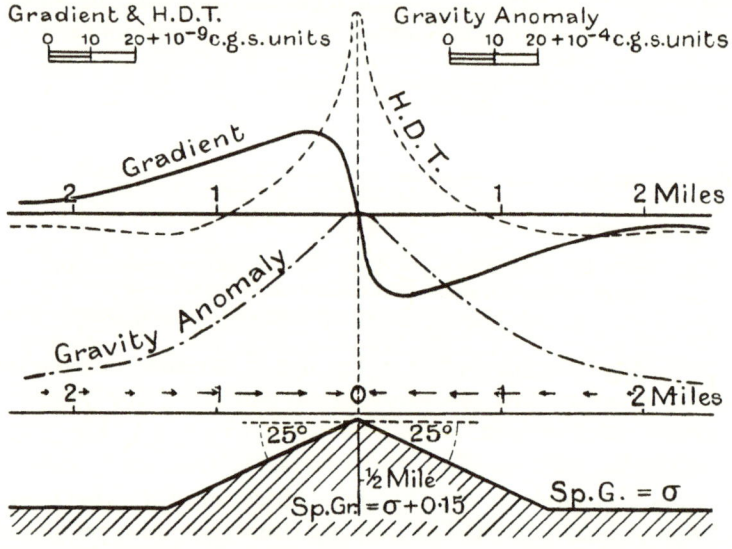

Fig. 106. Gravity gradient, gravity anomaly, and horizontal directive
tendency due to a ridge or anticline beneath the earth's surface.

If we again differentiate U partially with respect to x, we
find $A = \dfrac{\partial}{\partial x}\left(\dfrac{\partial U}{\partial x}\right)$, which is usually written $\dfrac{\partial^2 U}{\partial x^2}$, but for the
mere convenience of printing this will be written U_{xx}, so that
we have these six important relationships:

$$\left.\begin{array}{l} A = \dfrac{\partial^2 U}{\partial x^2} \text{ or } U_{xx}, \quad B = \dfrac{\partial^2 U}{\partial y^2} \text{ or } U_{yy}, \quad C = \dfrac{\partial^2 U}{\partial z^2} \text{ or } U_{zz}, \\[3mm] F = \dfrac{\partial^2 U}{\partial y\,\partial z} \text{ or } U_{yz}, \quad G = \dfrac{\partial^2 U}{\partial z\,\partial x} \text{ or } U_{zx}, \quad H = \dfrac{\partial^2 U}{\partial x\,\partial y} \text{ or } U_{xy}. \end{array}\right\} \quad (5.20)$$

The fact that in general $\dfrac{\partial^2 U}{\partial x\,\partial y}$ is equal to $\dfrac{\partial^2 U}{\partial y\,\partial x}$ justifies us in the use of six, and only six, coefficients in (5.10) and again in (5.16).

CORRECTIONS

The particular advantage of the use of U and the relations in (5.20) is that by their help we can determine the values of the corrections, for bodies either above or below the level of the torsion balance, which must be applied in order to obtain the necessary information sought for. For example, if there is a mass m at a point (x, y, z) relative to the centre of gravity of the torsion balance and at a distance r from it we have for U, the potential at the balance due to m,

$$U = \int_r^\infty \gamma \frac{m}{r^2}\,dr = \gamma\,\frac{m}{r},$$

and

$$\frac{\partial U}{\partial x} = -\gamma\,\frac{m}{r^2}\frac{x}{r},$$

and

$$A = U_{xx} = \frac{\partial^2 U}{\partial x^2} = 3\gamma\,\frac{m}{r^5}\,x^2 - \gamma\,\frac{m}{r^3}.$$

So

$$B = U_{yy} = 3\gamma\,\frac{m}{r^5}\,y^2 - \gamma\,\frac{m}{r^3}.$$

Whence

$$A - B = 3\,\frac{\gamma m}{r^5}\,(x^2 - y^2). \tag{5.21}$$

So also

$$H = U_{xy} = \frac{3\gamma m x y}{r^5},$$

$$F = U_{yz} = \frac{3\gamma m y z}{r^5},$$

$$G = U_{zx} = \frac{3\gamma m z x}{r^5},$$

and we can sum these, not only for the one mass m, but for all masses which have to be excluded from the problem in hand.

This remark brings us sharply to the question of ideals in the investigation. Values of H.D.T. and $\Delta_H g$ may be required as for a smooth terrain. But they are here required for finding minerals below the earth. The procedure may be modified with this aim in view. To be more precise, Lancaster-Jones (1928) states 'that with density 1·8 so modest a slope as 1°, or about one in sixty, leads to a correction value for U_{xz} in the instrument amounting to 13 Eötvös units'. Clearly this correction is not of such great importance if all the stations were used on such a slope, as if the stations were on different slopes. So again the presence of a mountain equidistant from a number of stations is of less importance than if it is at various distances from them.

The following corrections for causes exterior to the apparatus have to be considered:

(1) Sea level.
(2) Latitude.
(3) Terrain.
(4) Topography.
(5) Cartography.
(6) Local—ditch, bank, wall, house, trees, etc.

Helmert's formula for latitude and sea-level is

$$g = 980·617 - 2·593 \cos 2\lambda - 0·0003086H,$$

where λ is the latitude, and H the height above sea-level in metres, but in practice no correction for sea-level is made.

In the formula g is corrected for sea-level and latitude together, and the value at 45° latitude and sea-level is taken as 980·617 cm.sec.$^{-2}$. This correction includes that due to earth rotation, which adds a force per unit mass perpendicular to the axis of the earth equal to $\omega^2 r$, where $r =$ distance to the axis and ω is the angular velocity of the earth equal to 2π/sidereal day, or $7·3 \times 10^{-5}$ radian per sec. This gives rise to a term in $U = \dfrac{\omega^2 r^2}{2}$, and due to the change in this and to the elliptic figure of the earth there is the 'northing' correction referred to on p. 216. For

the shape of the earth is roughly an ellipsoid of rotation with
a semi-axis $a = 6378$ km., $b = 6357$ km., and with ellipticity

$$\frac{a-b}{a} = \frac{1}{298}.$$

The methods of correcting for terrain, topography and cartography are similar to one another in general character, but those
for terrain or due to near objects are in ordinary country of the
greatest importance. The necessary facts are obtained by taking
levels in eight principal directions such as north, north-east,
east, etc. The levels must be known on rings round the station
at 1·5, 3, 5, 10, 15, 20, 30, 40, 50, 70, 100m. distances. Now
these rings are cut up into areas by the straight lines. Every
point on each area will have its height z above or below the
centre of the instrument, and two other coordinates, an azimuth
α, and a distance ρ from the balance. Clearly we are changing
to cylindrical coordinates. Take, for example H or $U_{xy} = \dfrac{3\gamma mxy}{r^5}$
in (5.21), and consider what will develop with summation. The
mass m is replaced by an element of volume multiplied by
density σ, and we have to integrate

$$3\gamma\sigma\rho \, d\alpha \, d\rho \, dz \rho^2 \sin\alpha \cos\alpha / (\rho^2 + z^2)^{\frac{5}{2}}.$$

The ingenious devices for shortening the mathematics and
abbreviating the calculations would occupy a large volume on
the torsion balance, a book which indeed needs to be written
by an expert of threefold merit—skilled in mathematics, physics
and geology. Such calculations began with von Eötvös and
were improved by Schweydar, while Nikiforov and Numerov
simplified the proceedings for distant and more distant regions
(topography and cartography) by the use of contour bands
which cut the divided rings. The contour lines, at 5 ft. intervals,
are drawn on the map, while the circles and eight radial lines
traced on linen cloth, or transparent celluloid, may be superposed as required.

Those who do not intend to use the torsion balance would be
wearied with a further description of these corrections, and those
who intend to work with it must be or must become experts, and

they will necessarily get acquainted with the full literature. At a meeting of the A.I.M.E. at New York, three valuable summaries were presented which threw light on modern methods.*

THE GRADIOMETER

An interesting type of torsion balance has been invented and used by Shaw and Lancaster-Jones. It is called a gradiometer because it measures the horizontal gradient of gravity, while the horizontal directive tendency is not measured, and indeed is deliberately eliminated.

Three heavy masses are employed at equal intervals (120°) in azimuth (Fig. 107). Two of them, M_1 and M_2, are on a light aluminium ring, and the third, M_3, is supported by a light aluminium rod above the level of the others.

Radial connexions permit of suspension, and there is a mirror of the usual type.

Three readings are derived from three settings of the instrument, with 120° in azimuth.

It is not difficult to gather from equations such as (5.13) in the mathematical theory that the value of $A - B$, and that therefore the value of

$$g\left(\frac{1}{R_2} - \frac{1}{R_1}\right)$$

in (5.19) is balanced out in this instrument.

Whether it is better to have, or not to have, the values of the H.D.T. must depend on the nature of the problem under investigation. It would appear that, at least in some cases, the knowledge, derived from the variations in H.G.G. and in H.D.T. jointly, must add greatly to a sound interpretation of underground conditions (see Figs. 105, 106).

It seems that this type of gradiometer was invented by Marcel Brillouin about twenty-five years ago.† The overall dimensions of the modern instrument are much smaller than

---

* Barton (1928); Heiland (1928); Lancaster-Jones (1928). These were issued with *Mining and Metallurgy*, March 1928. They are fairly condensed statements from different viewpoints, and aggregate ninety pages of information.

† See also the French patent 619426 (24 June 1926) of Koybetliantz.

those indicated in the figure, as the object is to reduce the time of vibrations. Thus there is an Oertling type 4 ft. high, weighing 90 lb., with a period of 8 min. and with ribbon-torsion wire so that the balance comes to rest in 25 min. Visual reading is employed, and it is stated that seventeen stations have been read in 36 hr. The sensitivity is not, however, quoted.

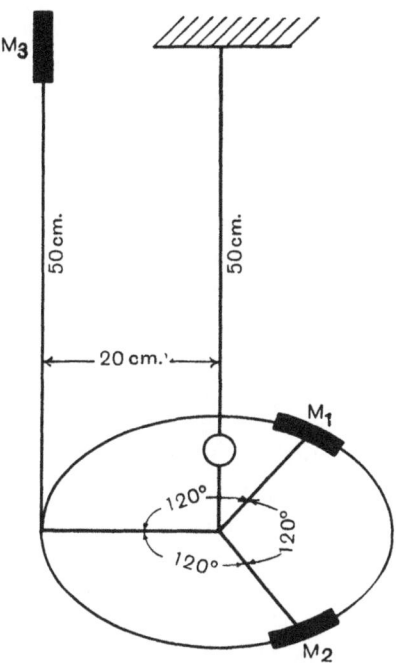

Fig. 107. Gradiometer, a torsion balance which measures gravity gradients and eliminates horizontal directive tendency.

GRAVIMETERS

During the last twenty years great advances have been achieved in the construction of gravimeters, which measure directly the change in vertical force on a given mass. The search for oil by the large oil companies has been the incentive, and it is mainly by members of their staffs that the development of such new types of instruments has been carried out, but these instruments have found application in mining and geodetic surveying also. Gravimeters are now available which are light and easy to transport, and possess very small temperature variation and high sensitivity, some instruments measuring to a hundredth of a milligal. Since gravity anomalies over such buried structures as salt domes have maximum values of the order of a milligal, it is obvious that the instruments require great precision in their construction and careful attention to their operation. When used for the detection of ore-bodies, the anomalies may only be a few hundredths of a milligal. Modern gravimeters not only have such sensitivity combined with a fairly rugged construction to withstand transport by air, motor vehicle and by hand in the field, but a reading taken at a particular point may be taken in a matter of a few minutes compared with the several hours required to set up and read the torsion balance. As a result, gravimeters have now generally replaced the torsion balance in geophysical prospecting. Gravimeters only measure differences in the gravitational attraction; they are not absolute instruments. The pendulum still remains the instrument used in making an absolute determination of the value of g at a given station. In geophysical work, however, it is irregularities in the value of g which are sought, for such anomalies are due to subsurface changes in geological formations or caused by the presence of deposits of different densities. The gravimeter serves the same general purpose for locating variations in density as the vertical magnetic variometer does in detecting changes in the magnetic properties of rocks or minerals.

The general principle involved in the design of a gravimeter is the displacement of a mass, suspended in some manner, owing to a slight change in the vertical attraction of the earth caused by a variation in the local mean density of the subsurface structure. Since any such displacements are exceedingly small, ingenious ways of amplifying such minute quantities are employed, using optical, electrical and mechanical methods. Great skill is required in the precision with which exacting tolerances are met and materials are chosen to reduce temperature and hysteresis errors. The actual displacement of the movable part of the apparatus may be only a few thousandths of a millimetre, readings of which must be repeatable under the same conditions and measurable by some magnifying device that varies with the type of apparatus designed by different companies. To be useful in geophysical prospecting, gravimeters must be capable of measuring changes of 0·02 milligal.

There are today several types of gravimeters in use which fundamentally are adaptations of the familiar Jolly spring balance or the Threfall and Pollock torsion balance, with which the gravity anomaly causes a change in the force exerted upon a mass suspended in some manner from a spring or an arm attached to a torsion wire or fibre. The change in force produces a displacement which is so small that Hooke's law holds, the actual movement being brought back to its original or null position by an auxiliary compensating force or adjustment. According to Hooke's law, the change in force ΔF resulting from a change in gravity Δg on a mass m causes a change in extension of the spring or torsion Δx, according to the equation

$$\Delta F = m\,\Delta g = k\,\Delta x \quad \text{or} \quad \Delta g = (k/m)\,\Delta x,$$

where k is a constant for a given spring or torsion suspension. Since Δg may be only 0·01 milligal, it is desirable to make k/m small so as to produce as large a value of Δx as possible. The natural period of oscillation of such a system is $T = 2\pi\sqrt{(m/k)}$; hence, to obtain high sensitivity, the natural period should be long, since $\Delta g = \dfrac{4\pi^2}{T^2}\Delta x$. Natural oscillations in gravimeters, as in seismographs, are undesirable and are eliminated by damping or by special mechanical means.

The Gulf gravimeter (Wyckoff, 1941) consists of a long vertical helical spring S of metal ribbon, at the lower end of which is a mass M in the form of a circular ring (Fig. 108). Below the ring and attached to it is a small mirror R which serves to reflect a beam of light that passes back and forth a number of times, thereby magnifying the displacement, before being directed up to the scale in an eyepiece, E. If there is a change in g, the spring will rotate as its length is altered by the resulting change in force on the mass, and the mirror will be turned through a small angle, perhaps as much as a second of arc. As a result of multiple reflexions, the image may have a final displacement of twenty or more times this amount, producing a

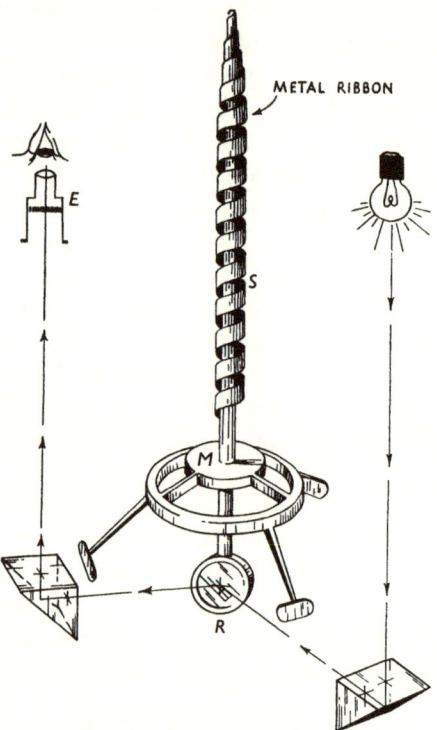

METAL RIBBON

Fig. 108. Gulf gravimeter.

shift of a millimetre as observed in the eyepiece. The temperature is thermostatically controlled by an insulating bath, the power being supplied by a storage battery. The instrument weighs approximately 25 lb. and has a sensitivity of 0·02 milligal.

Several types of gravimeter depend for their action upon the exact balancing of the force of gravity on a mass attached to a spring suspended at one end of an arm like a horizontal pendulum of long period. Fine adjustment in the restoring force is usually made by a delicate auxiliary spring. In the Thyssen gravimeter (von Thyssen, 1938) (Fig. 109), the mass M, whose position is affected by a slight change in the value of g, is attached at one end of an arm resting on a fulcrum F, like a chemical balance, the restoring force being produced by a spring

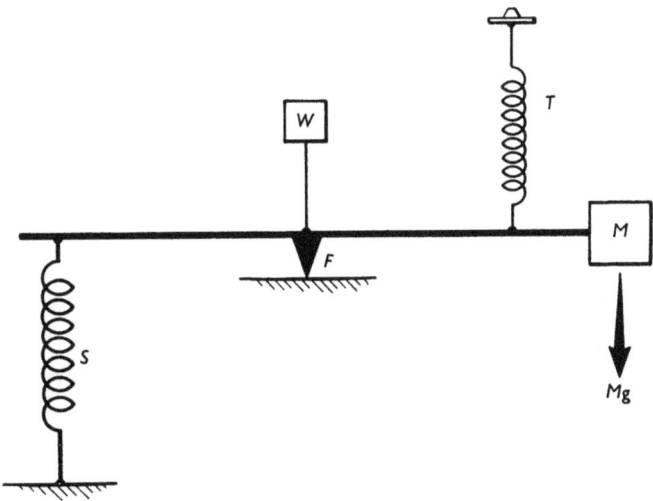

Fig. 109. Balance type of gravimeter.

S at the other end. The magnification is attained by having an auxiliary mass W attached to the beam, which is in unstable equilibrium, just above the fulcrum when the beam is horizontal. In this state it produces no turning moment. A slight displacement of the arm, however, caused by a change in g altering the force on M, results in the weight W being shifted from its position above the fulcrum, thus introducing an additional turning moment tending to increase the deflexion. The displacement is observed optically, the beam being restored to its original position by changing the tension in the auxiliary spring by a micrometer adjusting screw T. When calibrated, the gravity anomaly may be deduced from the number of divisions through which the micrometer head is turned to restore the beam to its null position.

The North American Geophysical Company have constructed a null type of instrument which is based on the long-period seismograph designed by Lacoste (1934). It is portable and is an example of what has been termed a 'zero-length spring' meter. The general principle of this gravimeter may be under-

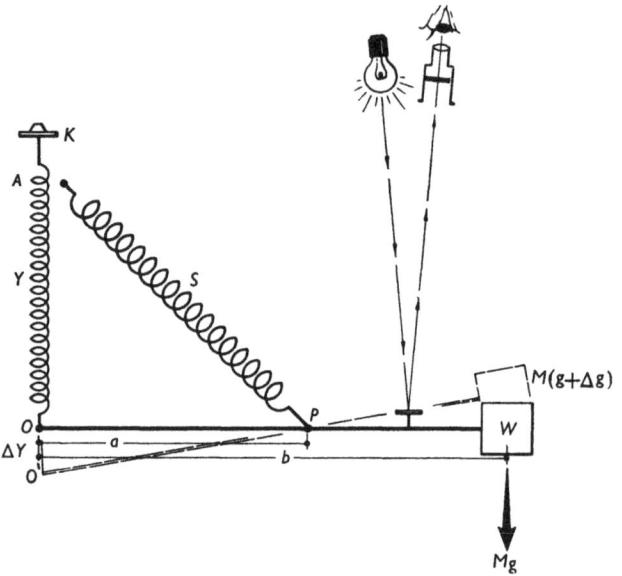

Fig. 110. 'Zero-length spring' type of gravimeter.

stood from the sketch (Fig. 110). A mass W is supported at the end of an arm at a distance a from the point O, the level of which may be lowered or raised by the micrometer screw K. A stretched spring S, which obeys Hooke's law, is attached to the arm at P, near the mass, the other end being fastened to the case. The mass is just balanced by the tension in the spring. The null position is observed in an eyepiece, the beam of light and image of the cross-hair being reflected from a mirror M. If, after adjustment at one station, the gravimeter is moved to a station at which g increases to $g + \Delta g$, causing the force on W to become $W + \Delta W$, the spring S will be stretched slightly and the weight will be lowered, owing to the increased attraction. The end of the arm at O is now lowered by means of K to a position O' when P is brought back to its original position, in which case the length of the spring S remains unaltered. The distance AO, designated by y, will thus be increased by an amount Δy, the latter being proportional to the number of turns of the knob K. If k be the constant of the spring S and m the mass of W, then it can readily be shown that

$$\Delta g = \left(\frac{kyb}{ma} - g\right) + \frac{kb}{ma}\,\Delta y$$
$$= A + B\,\Delta y.$$

The constants A and B may be determined for any particular instrument by making measurements at two stations for which the values of g are known from pendulum observations. It will be noted that the arm OP need not be exactly horizontal when the null position is established at a reference station, but the instrument must be accurately balanced, for which purpose levels are provided, before a reading is taken. The temperature must be kept constant; this is accomplished by placing the apparatus in a vessel that is thermostatically controlled with the aid of current from a storage battery. Clamping devices are incorporated to assure safety while in transport.

In making a survey with the gravimeter, one station is chosen as the standard and the apparatus adjusted for zero reading there. Readings are then taken at other points and the anomalies read in terms of the standard reading, the instrument being returned to the initial station every few hours to check it for drift correction. A survey made with this instrument over an ore-body on the property of the East Sullivan Mine near Val d'Or, P.Q. (Innes, 1947) is given in Fig. 111. The closely spaced contours would indicate a sulphide ore deposit near the surface, while the widely spaced ones are caused by a larger body which probably extends to greater depth.

A small light portable instrument which requires no external thermostatic control equipment and which possesses high sensitivity has been developed by S. Worden of the Houston Technical Laboratory. The movement of the Worden gravimeter is made entirely of quartz, thus reducing corrections for slight changes in temperature. A schematic diagram of the gravimeter is shown in Fig. 112. The mass m is supported at the end of an

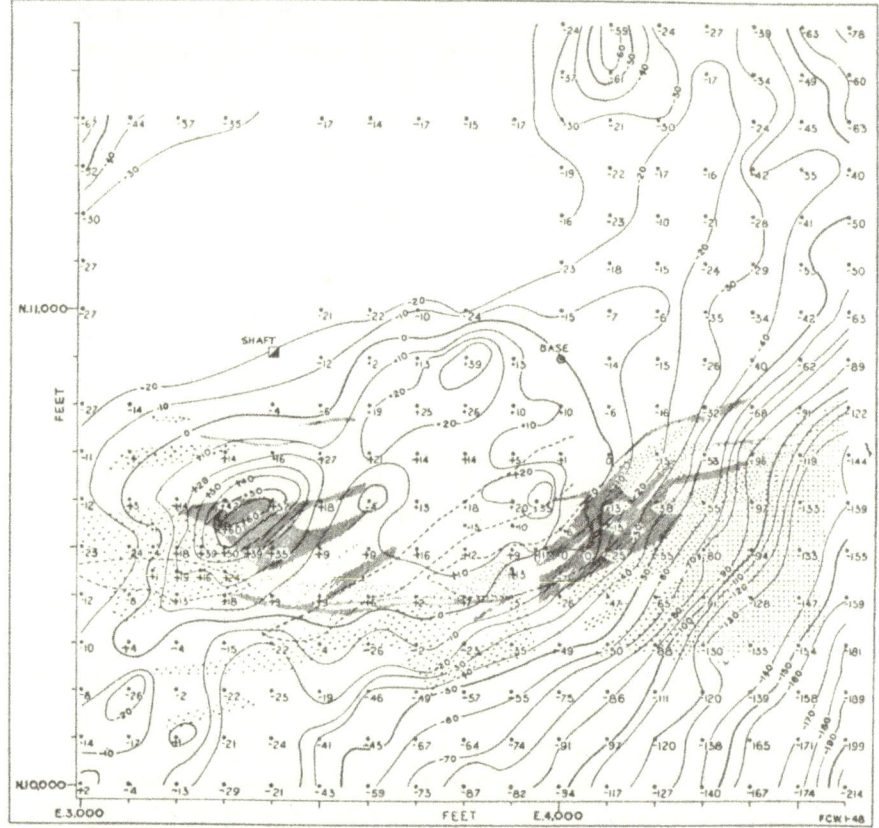

Fig. 111. Gravity survey (East Sullivan Mine, Val d'Or, P.Q.).

arm attached to a cross-beam carrying a pointer P and to which are attached the quartz springs arranged as shown. The beam may rotate about the thin hinges AA, and the mass m is balanced mainly by the spring S. The tension of the delicate spring s may be adjusted by the micrometer B so as to bring the pointer to the zero reading on the scale in the microscope M.

The movement is placed in a well-insulated thermos bottle, and operates in a partial vacuum. The outer case is provided with levelling screws and bubble levels, is moisture-proof and the arrangement is not affected by variations in barometric pressure.

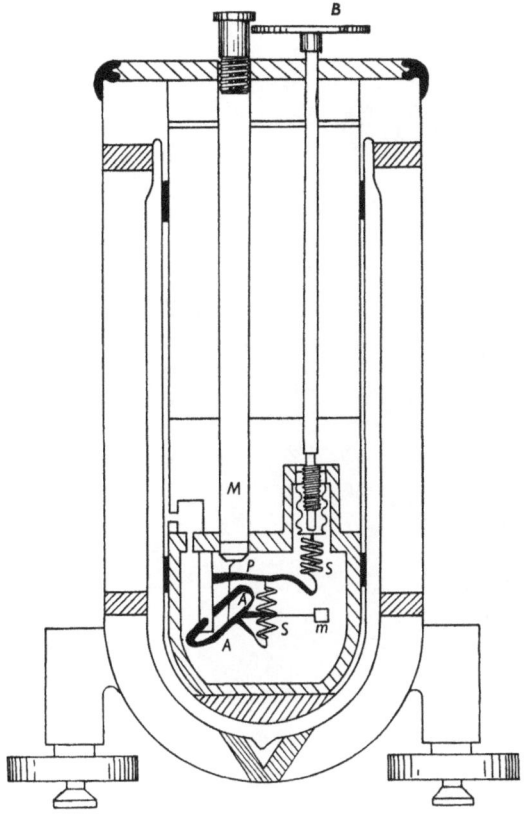

Fig. 112. Worden gravimeter.

The whole apparatus is about 5 in. in diameter, 15 in. high and weighs 5 lb. Its sensitivity is 0·01 milligal, it is non-magnetic and the correction for a temperature change of 50° F. is 0·25 milligal. Calibration is performed by either using a tilting table, by means of which the vertical component of g acting on the mass m can be changed by a known amount, or by taking readings

at two stations at which the absolute values of g have been
determined by the pendulum method. Fig. 113 shows the instrument in a car specially provided with a metal rod fitted with
feet which is passed through the floor so as to rest firmly on the

Fig. 113. Motor car fitted to take readings with the Worden gravimeter.

ground. The gravimeter can readily be carried by one surveyor,
may be flown in an aircraft (Woollard, 1950) and has been used
underground in mines (Rogers, 1952). Fig. 114 shows the results
of a survey made in the Antigonish area of Nova Scotia with
this instrument by Dr J. E. Blanchard and the Nova Scotia
Research Foundation. It was carried out to locate salt deposits
in this area.

INTERPRETATION AND RESULTS

In some simple cases a novice can obtain valuable information from a chart with isogams marked upon it, and yet further
interpretation if the gravity gradient and curvature difference

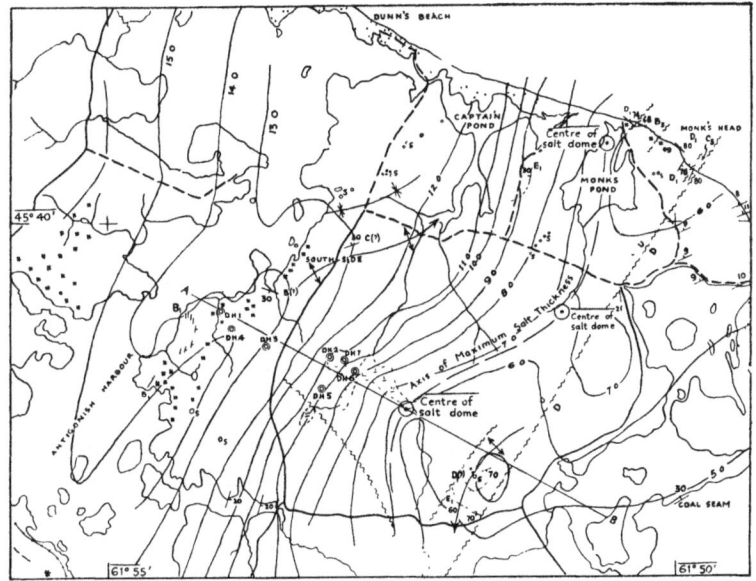

Fig. 114. Survey with Worden gravimeter (Antigonish area, N.S.).

(H.D.T.) gradient are also shown. A man accustomed to mathematics, physics and geology may very well read the information
with vastly greater skill, especially if he has had years of
experience in mining fields. Such men are rare. In many cases
it is necessary to assume an underground structure, with approximate density differences, then to calculate the resultant effect
on the surface, and to proceed step by step to a closer fit. In
other cases the indications are such that a fairly clear estimate
and close fit may be more readily made.

Thus Barton (1928) states that the Nash salt dome was discovered in 1924 by the Rycade Oil Corporation, using geophysical

methods with a 5% chance of a salt dome, but that, after a torsion balance survey, the chances were raised to 99%, and the upper rim of the salt table was correctly delineated at a depth of between 500 and 900 ft., and the position of the outer rim at a depth of 4000 ft., as subsequently verified by borings. Four other salt domes (Clemens, Allen, Long Point, Fannett) are attributed to torsion-balance surveys.

These remarkable domes may be a quarter of a mile to two or three miles in diameter. Any one of them could have paid for all the geophysical exploration so far carried out in the world. The sulphur on one underground dome, melted by superheated steam and then pumped to the surface, has already exceeded 150 million dollars in value, so that a salt dome is quite a good thing to find, and still better to hold.

In the North German Lowlands, as in the Gulf Coast, salt domes occur sometimes apparently forced up from great depths, and yet so covered with diluvial and tertiary deposits that no geologist can get a clue as to their position and extent. Here too domes and anticlines can be explored by geophysical methods, and a single boring made in view of such findings may explain the entire underground conditions and avoid the cost of many failures.

For, indeed, the whole geophysical investigation may be distinctly less in cost than a single boring. The North American Exploration Company state that, in Texas previous to and including 1924, during twenty years only six salt domes were discovered, whereas five were discovered by geophysical methods in 1925. It is claimed that in the Gulf of Mexico Coast District there were only twenty-four domes discovered previous to 1924. By 1930 there were 150 known salt domes in that district, of which 110 were found by geophysical means, mainly torsion balance and seismic methods. During the last twenty years gravimeters, owing to ease of transportation and speed with which readings may be taken, have largely replaced the torsion balance in geophysical prospecting. According to recent reviews by Hammer and Eckhardt (1952) of the Gulf Research and Development Company, there were 123 gravity crews in the field in 1951 and a somewhat greater number in 1953. Such gravity operations throughout the world amounted to 1526 crew-months, one-half of which were in the United States. If financial expenditure on geophysical surveying be a measure of the success of such work, as it undoubtedly is, it is of interest to note that oil companies alone spent over $350,000,000 during 1951 on such investigations, mostly on seismic and gravity methods.

It is well to remember that after the pressing claims of mining have received attention, there will follow a period of funda-mental investigations which will illuminate many of the half-solved problems in the pure sciences, more particularly geology. One such result is already apparent in the great increase in the application of seismic methods of geophysical prospecting to be described in the following chapter.

PROBLEMS

1. A spherical mass of rock salt, 200 m. in diameter, is embedded in limestone, the top of the rock salt being 200 m. below the horizontal surface. Find the vertical gravity anomaly and the horizontal gravity gradient on the surface at a point 400 m. from the point directly above the centre of the rock salt sphere.

2. One of the largest tunnels in the world, in which water will be brought from above Niagara Falls to the Sir Adam Beck Generating Station, is being constructed by the Ontario Hydro-Electric Power Commission, the tunnel being more than a mile long. The diameter is roughly 51 ft. and the centre is about 260 ft. below the surface over a long part of the tunnel. If the mean specific gravity of the rock removed is 2·7, find the gravity anomaly on the surface directly above the tunnel on the surface (i) before the water enters and (ii) when the water flows through the tunnel.

3. A vertical vein of pyrrhotite 20 ft. wide and extending to a depth of 1000 ft. occurs in rock of mean specific gravity 2·7. If the top of the vein is 100 ft. below the surface, what is the maximum vertical gravity anomaly due to the deposit, considering it of infinite length?

4. An iron sphere 4 m. in diameter is buried in sand with its centre 8 m. below the surface. Find the gravity anomaly due to this mass at a point 6 m. from the point vertically above the centre of the sphere, on a horizontal surface. Calculate also the horizontal gravity gradient at this point.

5. A long horizontal clay drain pipe is buried 6 ft. below the surface. If the mean specific gravity of the earth is 1·3 and the pipe is 2 ft. in diameter, what is the vertical gravitational anomaly directly above the centre of the pipe and the horizontal gravity anomaly at a point 12 ft. from the point above the pipe on the surface perpendicular to the length of the pipe?

6. A spherical cavity 100 m. in diameter occurs in limestone. The top of the cavity is 550 m. below the surface. Calculate the gravity anomaly and the horizontal directive tendency at a point on the surface above the centre of the cavity.

CHAPTER VI

SEISMIC METHODS

GENERAL SCHEME

The general scheme of prospecting by the seismic method is to fire a charge of high explosive buried a few feet in the ground. The sound, or shock, travels by the quickest route, which is under the earth and along a curved path, concave or hollow upwards, and then arrives at the receiving instrument, which is a seismograph, or earthquake recorder. The time of firing is registered on photographic paper on a rotating drum, a part of the seismograph, and the time of arrival of the shock is also recorded upon it, so that, if the distance on the surface of the earth is known, it is possible to make a measure of *distance* as compared with the *time* interval between the explosion and its arrival. Now this is not the velocity of the shock because it has gone by a longer route, which is deep in the earth.

The use of these travel times for the shock wave may be considered under two general applications of seismic exploration, identified as refraction and reflexion methods. The former arrangement was the first method employed, as it was particularly valuable in locating salt domes. But for locating anticlines and general changes in underground structures at considerable depths, the reflexion method has proved more advantageous. The great improvement in electronic equipment which has taken place in recent years and experience gained in interpretation of the resulting seismic records has contributed to the use of reflexion methods, especially in the search for oil deposits.

If in the path of sound there is a body, such as a salt dome, through which sound passes more swiftly than through the surrounding rock, then the time of the passage of the shock will be measurably shorter, and hence the detection of the salt dome is possible.

This method is generally called the *refraction* method, though sometimes it is referred to as 'the mirage sound method', because the curved path of the sound or shock ray is analogous to the curved path of a light ray in a mirage.

It will be noted that there is some resemblance here to sound-ranging, as carried out in France during the First World War, when guns were located by the sound from any one of them passing to a number of detectors or microphones, and the times of arrival recorded. But, in seismic measurement, the effect of the explosion is accelerated, and may even be reflected, by underground strata, anticlines, domes, etc. A good account of 'Sound Ranging' is given by W. L. Bragg under that title in the *Encyclopaedia Britannica* (see also Drysdale, Lamb *et al.* 1923).

SOUND

Sound will travel through àny material substances, gas, liquid or solid, owing to their elastic properties. Through a uniform substance sound will travel in a straight line, but through a non-uniform substance, in a curved path. This is true of all waves whether of sound, light, heat or radio. But for this fact there would be no long-distance wireless telegraphy, because the rays could not get round the earth. And so it comes to pass that there is sometimes a 'skip distance', or a region where the wireless waves cannot be received, because they pass far overhead, bent or reflected probably by the so-called Heaviside layer about 40–80 miles above the earth. So again the sound of big guns may be heard 20 miles away, while those same guns may be inaudible at some intermediate place, because the quickest route of the sound waves is a curved path in the upper air. Another familiar example is that of mirage, when light is bent in one of two ways, depending on whether the warmer air is below the cooler air, as over a sun-heated desert; or the reverse, as over a large region of cold water on a still summer's day, when there is not enough wind to stir and mix the layers of light and heavy air together.

Rays of sound or light passing from a light to a denser medium are bent or refracted at any point towards the normal of the face of junction of the layers. So, with the 'cold-water' mirage, light travels from a distant object having a path convex, or humped on the upper side along its passage to the observer, who looking along the tangent of the curve at its end sees the object raised up. And the contrary is true of the 'desert mirage' when the path is concave or hollow upwards. Hence, it may happen that bright light from a patch of cloud or of sky may look like a lake of fresh water on the ground at some distance away. Sound travels quicker in water than in air and yet more swiftly in the ground. Our knowledge of its speed underground is in part derived from the study of earthquakes, a branch of geophysics known as seismology.

A table of approximate velocities would not be out of place.

Table 2. *Velocity of sound and of compressional waves*

	Metres/sec.	Feet/sec.
Air at 0° C.	331	1,086
Air at 100° C.	386	1,166
Water at 4° C.	1,429	4,700
Sea water at 7° C. and 0·35 salinity	1,478	4,847
Unconsolidated sediments	1,200	3,900
Sandstone	1,400	4,600
Limestone	3,800	12,500
Rock salt	4,900	16,000
Granite	7,000	23,000
Seismic (deep in earth)	8,000 and increasing with depth	26,000
Rayleigh (*surface* waves)	3,600	11,800

WAVES

There are two main types of waves in matter: (1) compressional, where the vibrations of the molecules, or particles, are longitudinal, or to-and-fro at right angles to the wave-front; and (2) transverse, where the oscillations are more or less at right angles to the ray, and therefore the vibrations are in the plane of the wave-front, somewhat as in a wave on water, where the particles of water move up and down while the wave moves on. Unfortunately, the matter is not quite so simple as that, for there is a to-and-fro motion as well. However, in the present instance we are not interested in water waves, but in shock or sound waves in rocks and in other materials of which the earth is composed. The velocity v of such waves is given by the relation

$v = \sqrt{\dfrac{\text{elasticity}}{\text{density}}}$. There is no ambiguity about the density, which
is the mass in g. per c.c., but, in the matter of elasticity, there
are several different coefficients to be considered, and these have
unfortunately a variety of names. In the case of sound travelling
along a rod, it would be well to select Young's modulus E, but
for sound travelling in a large massive body it is necessary to
substitute $k + \frac{4}{3}n$, where k is the volume elasticity and n the
rigidity. We have to consider then

E, Young's modulus, or longitudinal elasticity.
n, Torsion modulus, or rigidity, or shear modulus.
k, Bulk modulus, or cubic elasticity, or volume elasticity.
C, The compressibility $= 1/k$.

σ, Poisson's ratio, of the ratio of lateral contraction per unit
breadth to longitudinal elongation per unit length. Thus the
velocity of sound or shock in a large mass may be taken as

$$v = \sqrt{\frac{(k + \frac{4}{3}n)}{D}},$$

or

$$v = \sqrt{\frac{(1 - \sigma)\,E}{(1 + \sigma)\,(1 - 2\sigma)\,D}},$$

which are equivalent results, owing to simple relations between
E, n, k, σ. Unfortunately, so little is known of the elastic properties of rocks, and of the speed of sound through them, that
we are forced to compare theory and observation for simpler
elements such as iron and lead.
Consider then

	D	n	k	E	Vel. $\sqrt{(E/D)}$	Vel. $\sqrt{\{(k + \frac{4}{3}n)/D\}}$
Iron	7·86	8·3	16·1	21·3	5200	5870
Lead	11·37	0·56	5·00	1·6	1180	2250
Marble	2·5–2·8	2·5	4·7	2·6	3200	5500

Now elasticity is a stress/strain, and a stress is measured as
force/area in dynes per sq.cm., while a strain is a change of
length per unit length, hence it follows that n, k, E are all
expressed in dynes per sq.cm. The above figures have, in the

case of n, k, E, to be multiplied by 10^{11}. The velocities in the two right-hand columns are given in metres per second. Sound travels slowly along a lead rod because $E = 1 \cdot 6 \times 10^{11}$ and is small, but it travels much faster in a large mass of lead, because $k + \frac{4}{3}n = 5 \cdot 7 \times 10^{11}$ is nearly four times as large as E, so that the speed is about twice as great.

When we compare these figures, which we have calculated from a theoretical basis, with speeds measured by experiment we do not seem to get very good agreement. Thus Biot, using iron rods nearly 1 km. long, found 3500 m. per sec.; Wertheim found for lead rods by theory 1250 m. per sec. and by experiment 1410 m. per sec., but for iron 4950 m. per sec. theoretically, and 4980 m. per sec. by experiment. He clamped his rods at the centre and set them vibrating longitudinally and then determined the frequency, and deduced the velocity remembering that the rod was a half-wave-length.

Some attention has been drawn to this point because it seems desirable to know more of the physical constants of the constituents of the earth.

Adams and Coker (1906) measured Young's modulus compressionally, and also measured Poisson's ratio for five marbles or limestones, seven granites, a syenite, a diabase and a sandstone, and they calculated from these determinations the shear modulus n, and the bulk modulus k for each specimen. Knott (1908) quoted these results (on p. 165), giving also the densities which he obtained from Adams. Knott adds a note (on p. 163) that the older the rock the greater the average rigidity, and from 92 specimens quotes the rigidities as

Archaean	$22 \cdot 3 \times 10^{11}$	Mesozoic	$12 \cdot 7 \times 10^{11}$
Palaeozoic	$14 \cdot 8 \times 10^{11}$	Cainozoic	$6 \cdot 3 \times 10^{11}$

The following table sets forth the results of Adams, Coker and McKergow, and there is added in the last column the mean velocity of compressional waves which we have calculated from $v = \sqrt{\dfrac{(k + \frac{4}{3}n)}{D}}$. It will be noted that the values vary from 2·9 to 6·4 km. per sec., so that at least near the surface of the earth the quickest waves travel at rates from two to four miles a second in various materials.

Happily the detection of domes or anticlines is not dependent on a knowledge of the absolute velocity of sound or shock in rocks. It is sufficient to measure a time difference in order to obtain information on the situation, which can then in general be clearly unravelled.

Table 3*

Substance	Density D	Young's modulus $E \times 10^{-11}$	Shear modulus $n \times 10^{-11}$	Bulk modulus $k \times 10^{-11}$	Poisson's ratio	Velocity compressional waves (km./sec.)
Iron, wrought	7·8–7·9	19·4	7·6	14·7	0·28	5·6
Iron, cast	7·1–7·7	10·3	4·1	6·9	0·25	4·1
Marbles and limestone (5)	2·69–2·72	5·2–7·2	2·1–3·0	3·7–5·7	0·25–0·28	5·5
Granites (7)	2·61–2·63	4·0–5·7	1·5–2·4	2·7–3·3	0·20–0·26	4·6
Nepheline syenite (I)	2·62	6·3	2·5	4·3	0·26	5·4
Diabase (Sudbury)	3·0	9·5	3·7	7·3	0·28	6·4
Sandstone (Ohio)	2·3?	1·6	0·61	1·25	0·29	2·9
Plateglass	—	7·2	3·0	4·4	0·23	—

The results of measurements of earthquakes, thanks largely to Galitzin and others, have enabled an estimate of the speed of such shocks in the earth to be made. Three types of wave are recorded, (1) compressional, (2) transverse, (3) Rayleigh, and these arrive in the order thus set forth, for their velocities are somewhat as follows:

	km./sec.	miles/sec.
Compressional or longitudinal waves	5·4–13	3½–8
Transverse waves	3·3–7	2½–4
Rayleigh waves	3·5–6	2¼–2½

As to the Rayleigh waves they travel along the surface of the earth which has an up-and-down motion like the waves of the ocean, while the Love or Q (Querwellen) waves are also surface waves, but the earth's surface has a horizontal transverse motion.

* See also Jeffreys (1952), p. 82; Heiland (1940), pp. 467–8.

Ambronn, working in Canada at Tilbury, Ontario, and in Turner Valley, Alberta, reports velocities in kilometres a second as follows:

Sand	0·6	Marl	2·5
Soft clay	1·0	Slate	3·0
Sandstone	2·2	Limestone	5·0

The large European explosions (La Courtine and Oppau) indicate for granite 5·5 km. per sec. These are by no means in ill agreement with the theoretical values of Table 3.

SEISMOGRAPHS

A large mass suspended so that it is capable of free motion will indicate an earthquake in consequence of its inertia. While the surrounding earth is in motion the large mass tends to remain at rest, and will certainly continue to do so, except in so far as motion is imparted to it by the supports, which may be suitable rods or wires. The simplest scheme is a heavy mass on the end of a wire with a light contact on smoked glass or paper beneath it. This will indicate an earthquake and give some idea of the amplitudes of waves in the ground. It is more convenient to have three recorders, one giving the north-south movement, another the east-west, and a third the vertical movement. Thus seismographs are at once divided into two classes, horizontal and vertical.

A horizontal seismograph (Fig. 113) is on the principle of a swinging gate attached by two hinges to a gate-post which is not quite upright. Everyone has swung on such a gate! If the gate is light, and you are heavy and at the end further from the post, there is a horizontal pendulum. If the post is well inclined to the vertical, the time of swing is small, and the to-and-fro swings follow rapidly; if the post is quite upright the gate will not swing to-and-fro, and it will rest in any position. For slight changes in the small inclination of the post, towards different directions, there will be large changes in the position of rest of the far end of the gate.

A vertical seismograph consists of a rod pivoted or held by a flat broad spring, at one end, and there is a heavy mass at the other. This system is supported by a spiral spring attached near and below the centre of the rod as shown in Fig. 116. The equation of motion of such a rod is found by taking moments, about the pivot, (1) of the weight of the rod and mass, (2) of the force exerted by the spring which, for oscillations of small magnitude, will be proportional to the angular displacement. These moments can together be equated to the product of the moment of inertia and of the angular acceleration of the rod about the pivot. On adding a resistance of 'damping' term, proportional to the angular velocity, the mathematical theory reduces to an equation of the same general character as that given below for the horizontal pendulum.

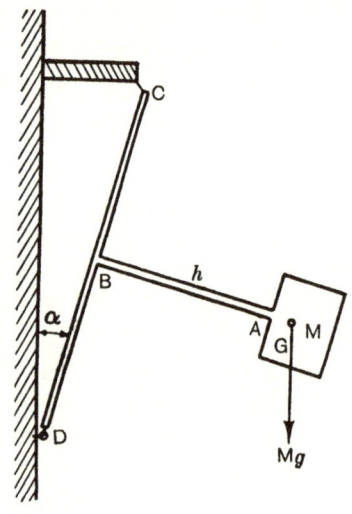

The time of oscillation of a simple pendulum is given by $t = 2\pi \sqrt{(l/g)}$, where l is its length, and g is the acceleration due to gravity. If, however (Fig. 115), the mass M is at the end of a rod AB, which in

Fig. 115. Diagram to illustrate the principle of a horizontal seismograph.

turn is fixed to CD, and if D presses on a vertical support at D, while C is attached to the same rigid support by a short wire, then we have a horizontal pendulum whose time of oscillation t is now given by $t = 2\pi \sqrt{\{l/(g \sin \alpha)\}}$ provided the mass M is small, and the rods light. Here α is the inclination of the rod BA to the horizon, or of CD to the vertical, and it can be adjusted to be very small. Suppose, for example, that $l = 1$ ft., and $\sin \alpha = \frac{1}{64}$, then since $g = 32$ ft. per sec.[2] we have $t = 2\pi \sqrt{2} = $ nearly 9 sec., which is the period of swing of a simple pendulum 64 ft. long! Moreover, the apparatus is very insensitive to a motion left to right, or right to left, but sensitive to displacement, by shock or earthquake to or from the reader, that is, from front to back, or the converse. The mass and rod will then tend to go on swinging with its own natural frequency, which of course has nothing to do with the earthquake. Such free oscillatory motion is not wanted, so that the free motion must be damped out by suitable vanes producing air resistance. A better plan is to use a copper plate between the poles of a magnet, or a flat coil of wire forming a closed circuit in a magnetic field. When the plate or coil moves in the magnetic field between the poles of two horseshoe magnets, the induced currents are, by Lenz's law, always opposed to the motion which produces them, and their reaction checks or damps the undesirable free oscillations.

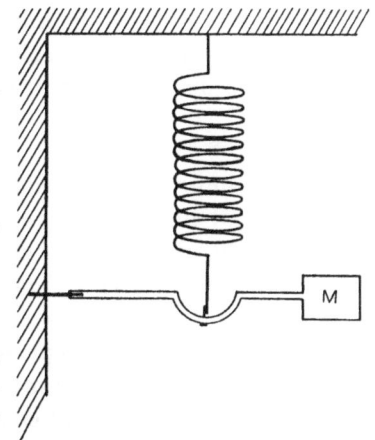

Both horizontal and vertical pendulums, in all forms of seismographs, are thus critically damped, or so adjusted that, when displaced, they tend to move slowly to the new position

Fig. 116. Diagram to illustrate the principle of a vertical seismograph.

of equilibrium without overshooting the mark, or oscillating to-and-fro about it. Their periods are usually long, about 12–14 sec. for a complete swing.

In the case of all seismographs the object in view is that the mass M shall remain at rest, while the earth and supports will move owing to the shock or to the earthquake waves. In consequence of this an attached mirror is turned through a small angle, and a reflected ray of light is turned through twice that angle, so that there is a displacement of the spot of light falling on the travelling photographic film, which may be 1 or 2 m. from the mirror.

A third method of registration is by the aid of a light coil of insulated wire attached to the light boom fixed to the pendulum. The coil is placed between the poles of a magnet, or electromagnet, firmly supported on the earth. When earthquakes cause relative motions the coil moves up and down relatively to the magnetic field, and currents are induced in the coil the terminals of which are suitably connected to a sensitive galvanometer. The galvanometer deflexions are then recorded photographically. In this case the galvanometer must be suitably damped and its period adjusted to suit the requirements of the seismograph, always remembering that we do not want the amplitude or period of the seismograph or of the galvanometer, but as far as possible that of the earthquake wave.

REFRACTION METHODS

In the region round the Gulf of Mexico the practice is to bury 150–200 lb. of high explosive about 12 or 20 ft. underground, and to fire the charge electrically. The explosion makes a large crater in the ground, which may be left unfilled! The cost is such that a single company has expended thirty thousand dollars in a month on the high explosives alone. A battery of seismographs, three or more in number, is arranged in a straight line, or in a circle with the explosion point at the centre. The time of explosion is registered on the seismograph films in one of two ways. The sound wave passing through the air will be detected by the seismograph, and a calculation involving the velocity of sound in air and the measured distance from explosion point to seismograph will enable zero time, that of the explosion, to be deduced. Now the velocity of sound in air at various temperatures is known, but there is a difficult correction to be made for the velocity of the wind, difficult of course largely owing to the highly irregular motion of the wind which may vary with time, and also from place to place at the same time. The exact time of firing is given by the explosion interrupting a continuous radio signal recorded on the observer's film, or the shot may be fired automatically by a radio signal sent from one of the recording points.

In Europe, or in places more thickly populated than the Gulf region, the use of heavy explosive charges might not be tolerated, so that the present tendency is towards more moderate charges, even say a few grammes of dynamite, which will make a detectable shock on a sensitive seismograph a few thousand feet away. It has been found, however, according to Barton, that small charges and highly sensitive galvanometers are unsatisfactory owing to microseisms and 'ground unrest' producing effects comparable to those under observation. The ordinary seismograph measures displacement or amplitude, but it is possible also to measure acceleration, by means of the change of pressure which the lever of the large inert mass will produce at a given point on a gauge. Seismographs of the Mintrop type were employed with satisfactory results in the early work, but electromagnetic induction ones are now in common use, being simpler to transport and lay out. A mass is held by a stiff spring, and a coil of wire attached to this mass is free to move in the field of a strong permanent magnet. The whole is in a pot, containing oil for damping. When the shock reaches the detector, the ground and pot move, but the inertia or the mass produces a slight motion of the coil relative to the permanent magnet, which moves with the pot. The induced e.m.f. is amplified by electronic equipment. The resulting currents are fed to the galvanometer, the amplification increasing with time to provide a record of about the same amplitude over a brief period of a few tenths of a second.

The Mintrop vertical seismograph consists of a heavy ball of lead (Fig. 117) suspended by a horizontal steel spring, passing through a hole into its centre. A long, light, conical aluminium lever, about 1 m. long, reaches upwards and ends in a light style pressing against a mirror very near its axis of suspension, which is a vertical wire controlled by a torsion head. Light passes from a lamp through a light-tight tube to the mirror and back to the film of the recording camera, about a metre distant.

An alternative plan, due to Schweydar, is the employment of a light strong thread which passes round the small circular vertical axle to which the mirror is attached. One end of the thread is attached to the end of the lever and the other to a spring sufficiently strong to keep the thread taut.

The mechanical type of seismograph, as stated above, has been largely replaced by electrical induction apparatus, wherewith a moving wire cuts a magnetic field, thus causing a current which can be amplified as desired; finally, a galvanometer and moving photographic film record the earth movements.

In Fig. 118 is a plan showing the 'shot-point' S on a leased plot, and also the six points, A to F, at each of which a seismograph was stationed. It will be noted that SA is about 7 km. and SF 9 km., or about $4\frac{1}{3}$ and $5\frac{1}{2}$ miles respectively. The records obtained showed that the shocks from S to F and S to E passed through the salt dome beneath the ground, while the other paths S to A, B, C and D did not do so; hence, not only was the salt dome detected, but its boundary was determined tangentially.

Photographs of the actual records, or six seismograms (A, B, ..., F), are shown in Fig. 119, and these have been placed one above the other in the photographic reproduction, having zero time, as determined by radio, in a common straight line on the extreme left.

In the middle of the upper strip of film A is marked a timescale for 1 sec., which on the original was 1·6 cm. or about $\frac{5}{8}$ in. long. The films, however, did not move with uniform speeds in any one camera and therefore vertical lines, most clearly seen in D, perpendicular to the sound detector line are automatically marked, five intervals to the second. The instant of explosion is shown by a break or change in the radio line. On the extreme right of the sound line is shown the effect of the noise of the explosion coming through the air. In the middle of every film is the actual seismic record of the disturbances which came through the earth by the quickest, not by the shortest, route. It is immediately seen that a clear record, with a sharp beginning, is in each case obtained of the compressional or P waves (*primae undae*). It is particularly interesting to note a similarity of type in records A to D, and a divergence from it in the character of the compressional waves in E and F. Moreover, the seismic shocks arrived in the order F, E, D, while the air shocks came in the order D, E, F. This reversal is due to the quickening of the seismic waves in the rock salt of the dome. The transverse waves

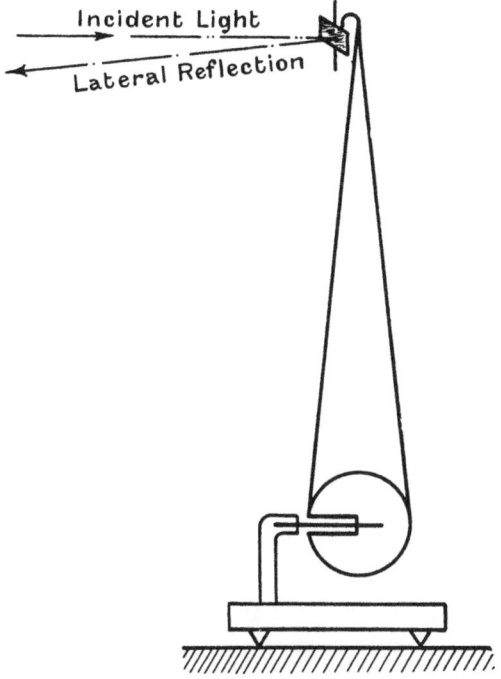

Fig. 117. Diagram to illustrate the principle of the
Mintrop field seismograph.

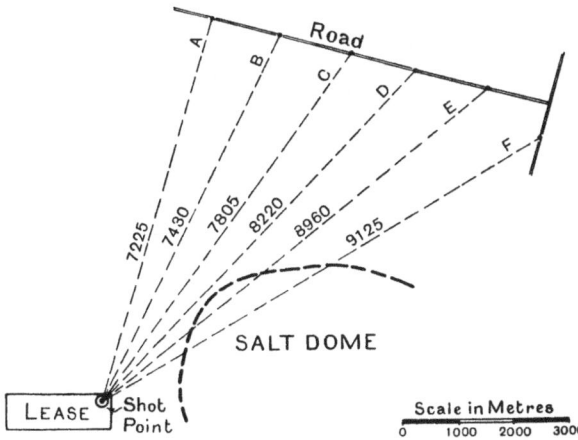

Fig. 118. Diagram showing the 'shot-point' and positions
of six recording seismographs.

travel slower and arrive later than the compressional waves. These S waves (*secundae undae*) are not used at present in geophysical prospecting, but as the science advances it may be that their interpretation will unfold, for the types of oscillation and the amplitudes are both repeatable, and in the diagram even a casual glance will show a similarity in the records of transverse waves in* C and D. However, it is premature to speak with too much confidence on the details of effects shown in the records. Any house or tree in the neighbourhood will be a centre of re-radiation. Wind will produce its microseisms, and only a prolonged study by skilled physicists of records obtained under varied conditions will secure, for the geologist and for others, the full possibilities of these interesting large-scale experiments.

Let us consider the times of travel of the shock by the two paths S to A and S to F. The velocity of sound corrected for wind was 350 m. per sec., or 1147 ft. per sec. Hence the distances SA and SF are actually 7225 and 9125 m. respectively. If the elasticity increases uniformly and in proportion to the depth, then the underground path was in each case nearly the arc of a circle. However that may be, the measured times, derived from the seismograms, were 3·501 and 3·724 sec. respectively. These are the exact times between the shot at S and the arrivals of the beginnings of the compressional waves at A and at F.

Thus the distance/time was 7225/3·503 for SA, and 9125/3·724 for SF, or 2050 and 2430 m. per sec. The actual speeds were of course much faster, because we are stating chord/time instead of arc/time. But clearly we can state that the *unquestionable effect of the salt dome was to increase the ratio 'distance/time' from 2050 to 2430 m. per sec., which is a 20 % increase, or from $1\frac{1}{4}$ to $1\frac{1}{2}$ miles per sec.*

It is stated as a result of experience from drilling that the depth reached by the path of the shock is one-third to one-quarter of the distance from shot-point to seismograph. If this

---

* In curve A (Fig. 119) the portion marked 'TRANSVERSE WAVES' is probably a surface phase. The transverse wave, if present, is perhaps under 'N' in 'COMPRESSIONAL' (Leet, 1930).

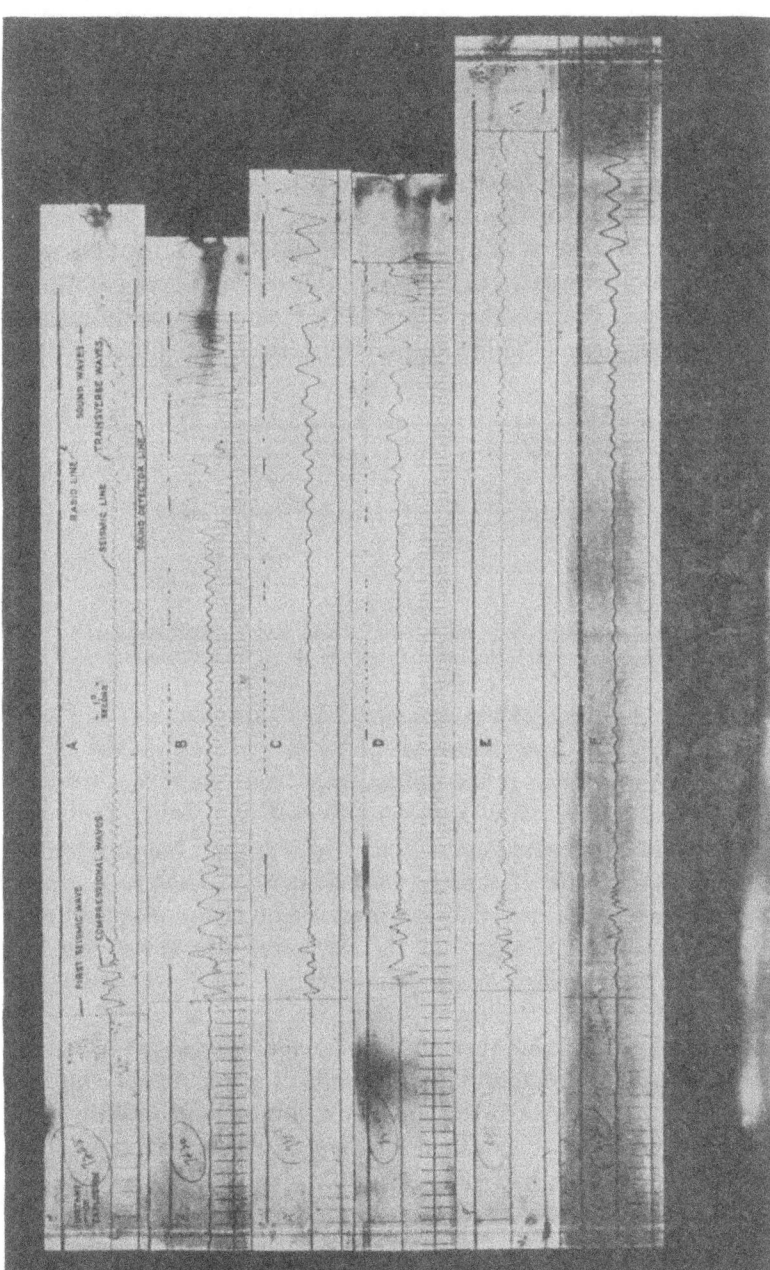

Fig. 119. Reproductions of the actual records obtained by the seismographs at the stations shown in Fig. 118.

is true we might conclude that the top of the dome was at a considerable depth, or else the passage from S to D would have found a quicker path through it. The present tendency of this art is to increase sensitivity by measuring acceleration instead of amplitudes, and to record the time of explosion by electric currents and wires rather than by sound waves in air. Enlargement of the scale of displacement or of amplitude has been achieved by Wiechert using a train of light levers to displace the mirror which reflects light from lamp to photographic film.

Thus a 4-ton steel weight falling 46 ft. (14 m.) can be detected at a distance of $1\frac{1}{2}$ miles. Again Grumach, developing the

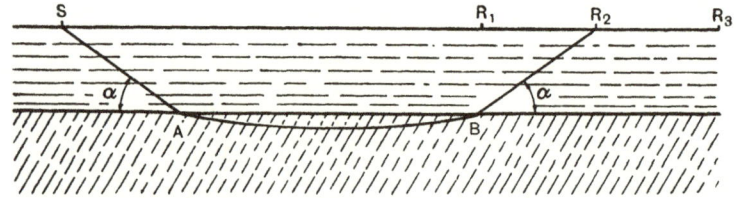

Fig. 120. A profile showing 'shot-point' S and seismograph station R_2, together with the quickest path of the shock underground.

acceleration method in 1906, measured small amplitudes (10^{-6} in.) of rapid period—three-hundredths to three-thousandths of a second. Such movement may be produced in rocks by waterfalls.

The acceleration of a mass at the end of a lever may be measured by a pressure gauge, involving a spring, but an alternative plan is the use of a piezo-electric crystal, such as a plate of quartz or tourmaline. Varying pressure on the plate produces opposite electrical charges on the surfaces, and these can be detected by a quadrant electrometer, or by a cathode-ray oscillograph.

The thickness of the overburden above denser rocks beneath has sometimes been found by the following means. A shot is fired at S, and records of the shock are obtained at a number of seismographs R_1, R_2, ... placed on the surface in a straight line which passes through S (Fig. 120). Two shocks will arrive at, say R_2, one by a more or less direct route SR_2, the other by the longer path $SABR_2$. If h is the thickness of the upper layer, α the inclination of SA to the horizon, v the velocity in the upper layer, and V the greater velocity in the lower layer, then the condition of minimum time readily gives the relation $\cos \alpha = v/V$.*

If the direct shock arrives at R_2 at the same time as the long-path shock, then, putting $SR_2 = 2d$, we have

$$\frac{2d}{v} = \frac{2h}{v \sin \alpha} + \frac{2(d - h \cot \alpha)}{V},$$

whence
$$h = d\sqrt{\frac{V - v}{V + v}}$$

gives the thickness h of the upper layer. The values of the velocities must be deduced from the records on the various seismographs, at some of which the direct shock will arrive first, and at others more remote, last. This method may also give the slope, if any, of the under stratum, or reveal the existence of faults. Slope can only be secured by shooting in reverse direction as well, unless the velocity in overburden and underlier are previously known. On paper it is simple enough, but in practice there are grave difficulties. In the first place the detection of two different compressional waves on the same seismogram is not easy, and it must be remembered too that the paths of the wave shocks will not all be straight as the diagram represents and as the theory supposes. There is ample evidence that elasticity increases with depth, so that the paths are concave

* If A' (Fig. 121) be a point near A then the times of the paths SAB, $SA'B$ must be nearly equal for minimum time from S to B. Draw AK perpendicular to SA', then the time along KA' with velocity v in the upper layer must equal the time along AA' with velocity V in the lower layer.

Hence
$$\frac{A'K}{v} = \frac{AA'}{V},$$

or
$$\cos \alpha = \frac{v}{V}.$$

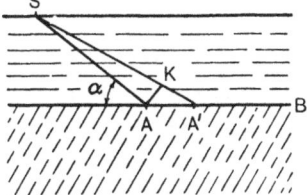

Fig. 121. A profile showing that the path of swiftest travel from one layer to another involves an angle α whose cosine is v/V.

upwards. Moreover, the path of the greater part of the shock will not follow the quickest route as given by $\cos \alpha = v/V$, which indeed corresponds to the critical angle in optics, when in Snell's Law of Refraction one angle is a right angle. This passage of sound underground has a clear further analogy with radio waves which leave a transmitter, reach to and travel along the Heaviside layer, and pass thence to a receiver. This may be seen by turning Fig. 120 upside down! The above method has actually been used for finding the thickness of the overburden in Sweden (Ambronn, *Methoden der Angewandten Geophysik*, p. 184).

Leet (1930) has given an admirable account of modern practice, and a brief summary is given in the following paragraphs (see also Ewing and Leet, 1932).

He first draws attention to the results of the large explosions at La Courtine, 10 tons of melinite, and at Oppau, 4500 tons of melinite, which gave velocities in granite and in gneiss:

	La Courtine (km. per sec.)	Oppau (km. per sec.)
P waves	5·5	5·4
S waves	—	3·15
L waves	2·8	—

In the search for salt domes on the coast of the Gulf of Mexico it has been found that the velocities of compressional waves (P) are about 1·6–3·0 km. per sec. in sedimentary rocks and about 5–5·5 km. per sec. in the rock-salt plug. Hence the possibility of detection to depths as great as perhaps 4000–6000 ft.

At the shooting station 75–1000 lb. of dynamite, 60 % quarry gelatine, is exploded 12–20 ft. underground, and round this main charge is wound a wire leading to a continuous radio signal, which is therefore stopped at the moment of discharge. A secondary charge of 16–75 lb. on the surface is simultaneously exploded whereby the observer can measure his distance from the shot-point, with due regard to wind direction and velocity. Such distance is more accurately now measured by survey.

The observers on land carry their apparatus on a half-ton

motor truck, and on water on special 35 ft. speed boats (with a speed of 35 miles an hour).

The main observing seismometer, called a tromometer, is a foot long and 3 in. in diameter, and is buried vertically about 3 ft. in the ground, or thrust with a pole in the mud.

In the metal case of the tromometer is a permanent magnet and between its poles a wire connected to a balanced weight. When the shock comes, the wire, moving across the magnetic field, generates an electromotive force, the resulting current is magnified by an amplifier a few hundred to a few million times which also filters out all frequencies but those between 20 and 60 cycles per sec., and is then taken to a galvanometer with a mirror so that light falls on a movie film, electrically driven from the signal 'ready', until the record is complete.

A synchronous motor controlled by calibrated tuning forks permits the passage to the film of 50–100 marks per sec. The observer also gets the air wave as well as the radio signal on his film. It is the final arrival of the air wave which ends his operations.

The moving mass of the tromometer has a natural period of a few hundredths of a second and is oil damped with a ratio seldom greater than four to one.

There may be three or four observers, and their films are developed and sent to the computer, who measures the records on all the films and plots the time-distance curve to the nearest 10 m. and $\frac{1}{100}$ sec.

The initial reconnaissance is made with all the observers and the shot-point in a straight line in order to find the 'ground' velocity. After that, 'fan' shooting begins with 10–15 degrees of azimuth between observers. Fan shooting has also been used to trace the course of deep metallic ore veins (Broughton Edge and Laby, 1931). The time of the first arrival of the shock wave at each detector is plotted, and from the resulting graph the centre of the vein is deduced. The results of such a survey made by Broughton Edge and Laby are shown in Fig. 124. Ambiguity may arise owing to the velocity of the shock wave varying horizontally with distance from the shot-point rather than vertically with depth. By reverse shooting, taking a record of the wave travelling in the opposite direction by interchanging the position of the recorders and shot-point, the resulting time-distance graphs will indicate a convex curvature for both direc-

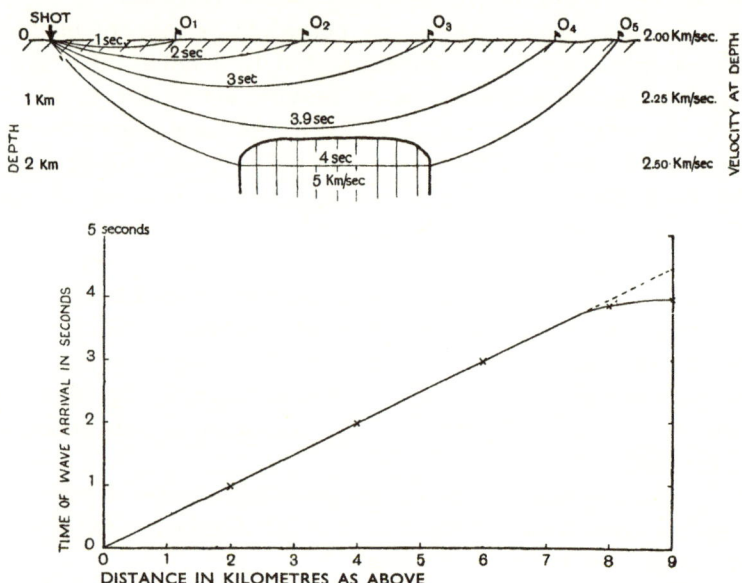

Fig. 122. The paths of shocks from the shot-point travel by circular paths to observers at O_1 to O_4. The 'travel-time' diagram (ideal) shows the speeding up through the salt plug.

tions if variation is with depth and if the variation is horizontal, one curve will be convex and that for the opposite direction concave.

It is important to remember that a compressional wave, on meeting a discontinuity underground, will give rise as a rule to both compressional and transverse waves, both reflected and refracted.

The paths of the compressional waves underground will be arcs of circles if, and only if, the velocity increases linearly with the depth.* Two ideal diagrams (Figs. 122, 123) show travel-

* See Appendix IV.

time curves for a vertical salt plug and for a hemispherical salt dome, respectively.

Refraction methods have been used sucessfully in finding depth to bedrock in highway and dam construction (Shepard

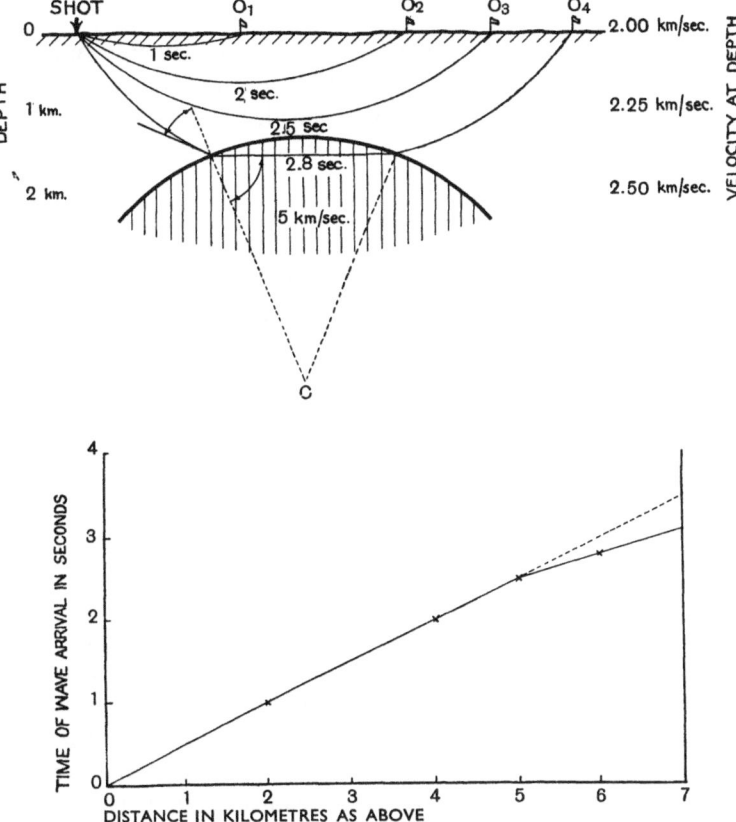

Fig. 123. Hemispherical salt dome (ideal), and paths of shocks, with 'travel-time' diagram beneath.

and Haines, 1942; Moore, 1952). When time-distance graphs are plotted, a series of intersecting straight lines are obtained, corresponding to the velocities of the shock-wave in the surface soil, clay or till and rock. Projecting these lines back to the zero distance axis, the time, Δt, for the wave to travel vertically down to the interface between two formations and back again may be found. If the velocity V in the formation is known, the depth D can be calculated as $D = \frac{1}{2} V \Delta t$.

REFLEXION METHODS

Highly successful methods were developed during and after the First World War to measure the depth of the ocean from a ship, which may be travelling at speed, by emitting a noise or

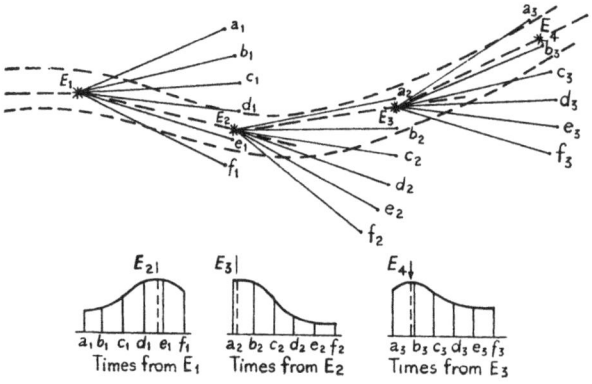

Fig. 124. Fan shooting.

sound from a plate or diaphragm fixed to the hull. The sound waves then travel in the water to the bottom of the sea, whence they are in part reflected back through the water to a microphone, or hydrophone, on the ship and in acoustic contact with the water. The effects are duly amplified or registered, and the time interval of the to-and-fro journey measured. As the velocity in sea water has been accurately determined, it is simple to calculate the depth of the ocean. Continuous records of the depth of the Atlantic Ocean have thus been obtained between America and Europe. Such methods now replace the elaborate soundings of the hydrographer from a ship at rest.

R. W. Boyle found the reflexion from icebergs of supersonic underwater waves not well marked, because the density and elasticity of icebergs are not strongly different from the corresponding properties of sea water. It is, however, possible that there may be an 'echo' from a salt dome surrounded by material of different density and elasticity.*

It has naturally occurred to many investigators to use such a reflexion method to detect the depth of any marked inequality in the strata beneath the surface of the earth.

In *The Science of Petroleum* (1938) there are articles by Burton McCollum and by J. H. Jones which give a clear picture of the methods then in use for 'exploring subsurface Geology'. Developments in instruments and in the interpretation of shock waves have since resulted in extensive use of the reflexion methods for locating oil deposits. It is customary to have a group, perhaps five to twelve in number, of electrical detectors in a row, at intervals of 100 or 150 ft., pointing at the shot-point which may be 1000 ft. away in the centre of the row, a hundred feet from the nearest recorders. The shot hole should be fairly deep—below the weathered upper layer, and bored by a rotary drill down to 50–100 ft.—and filled with water. The charge at the bottom of this hole may be of any strength from a percussion cap upwards, but a few pounds of gelignite will usually suffice. There are wires from the charge itself to the galvanometer so that the exact instance of the explosion may be registered. The galvanometer is of the Eindhoven or d'Arsonval type with five to twelve recording filaments, each connected through an electronic amplifier to a detector unit. The movements of the filaments are recorded on a moving sheet of photographic paper in a camera, the developing and fixing of the paper taking place automatically after each shot. A synchronous motor drives a rotary shutter, so that every $\frac{1}{100}$ and $\frac{1}{10}$ of a second is shown on the paper. Fig. 125 is a schematic drawing showing the field procedure in reflexion mapping and wave travel beneath the surface (Gabriel, 1936).

* The thickness of glacier ice has been found by the reflexion method (H. Mothes, *Forschungen u. Fortschritte*, vol. VI, No. 28 (1930) pp. 363–5). (See also *Bulletin of the Seismological Soc. of America*, vol. XXII, No. 3 (1932).) Actual records show the reflexions.

It is understood that the object of the undertaking is to ascertain the depth below the earth of one or more discontinuous surfaces, which may be horizontal or inclined.

On the film, one straight below the other, are the records of the explosion, arriving by wires, which gives zero time. Next come the direct compressional waves from the shot-point in order of distances. After some confusion on the records, points may be detected on each showing the first reflexions from some layer, it may be 1000 or 5000 m. below the earth, and there may be two or three such reflecting surfaces.

If the velocity of the waves has been determined in advance, then, since the records give the times, it is possible to calculate the depth.

There are, however, difficulties to be considered. There may be weathered surfaces above the main strata, and the shock waves travel at a slow rate through the looser material, for example, at 200–500 m. per sec. in the weathered beds and beneath them it may be at 1000–2000 m. per sec. Hence the desire for deep shot-holes and short shooting distances.

Since electrical seismographs and amplifiers are now used, it is possible to filter out waves of undesired frequency, and experience shows that it is advisable to permit frequencies of 40–60 cycles per sec. to pass to the recorders.

It is sometimes advisable to interchange the positions of shot-point and detector-group. By use of an old boring of great depth and an explosion charge at the bottom, it is possible to obtain a good measurement of the velocity of shock waves in the neighbourhood. If the velocity is accurately known, there is no difficulty in measuring not only the depth, but also the dip of the underground discontinuity. With sufficient shots and detectors it is finally possible to mark on a map contours showing places of equal depth (synbaths!).

Fig. 126 shows a truck fitted for seismic work in Texas, and Fig. 127 illustrates the twelve panels within the truck to be connected by wires to the seismometers. Each panel contains amplifiers and a sensitivity changer so that the recorded amplitudes of the early violent waves are diminished and those of the

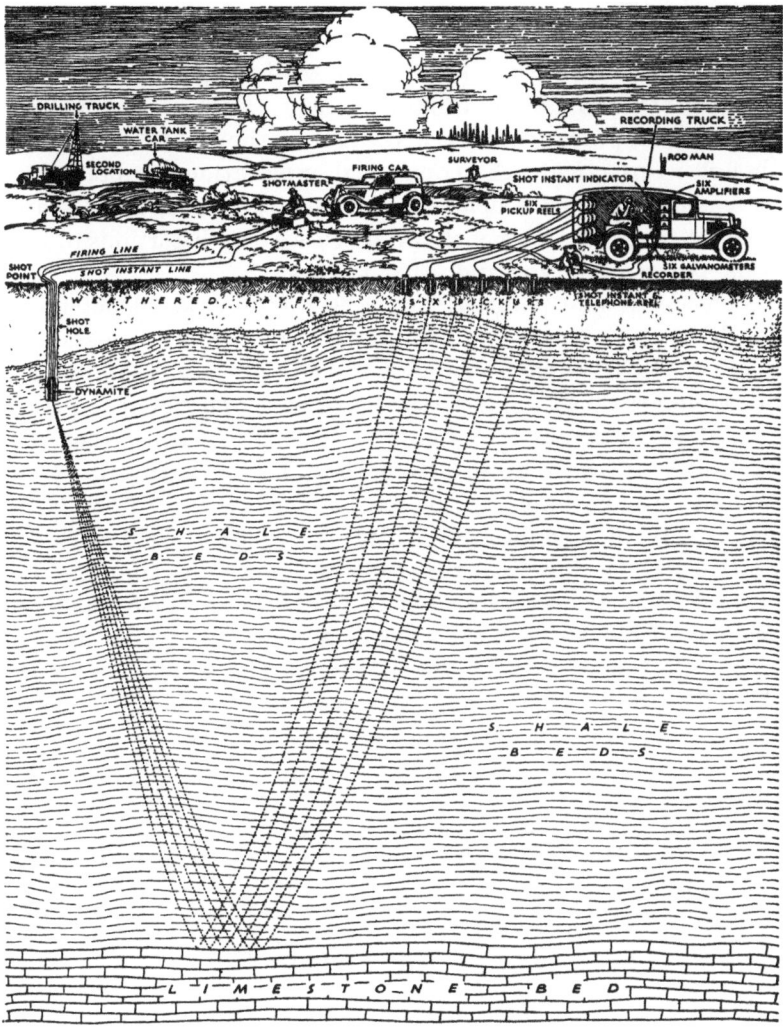

Fig. 125. Seismic reflexion surveying.

later feeble waves enhanced. Hence the excellence of the records in Fig. 128, printed by kind permission of the National Geophysical Company.

There are many methods of determining the depth below the surface of the reflecting beds and their inclinations (Heiland, 1940; Nettleton, 1940; Jakosky, 1950; Dobrin, 1952). The following simple case will serve to illustrate the general principles, though in practice each geophysical exploration company will have more detailed schemes based on experience, taking into account variable velocity of the shock waves, distortions of the beds, reverse shooting and other factors.

Consider a shot-point O (Fig. 129) and a series of detectors $S_1, S_2, ..., S_6$ set out on the surface in a row at distances $a, b, ..., f$ ft. from the shot-point to the right and a similar set of geophones $S_1', S_2', ..., S_6'$ to the left. The geophones are equally spaced at intervals of a hundred feet. The waves from the explosion will be reflected from the limestone bed, LL, considered flat for simplicity, inclined at an angle θ to the horizontal. If the medium be isotropic (which it seldom is in practice), the paths of the rays may be deduced by applying the laws of geometrical optics. Let O_1 be the geometrical image in the plane LL of O. Then the waves may be considered as coming in straight lines from O_1 to the detectors. By geometry we have the path $OR_1S_1 = O_1S_1$, and if the mean velocity in the medium be V, and the times of transits from O to $S_1, S_2, ..., S_6$ be $t_1, t_2, ..., t_6$, we have, letting p be the length of the perpendicular from O on LL,

$$(O_1S_1)^2 = (Vt_1)^2 = OS_1^2 + OO_1^2 - 2OS_1\,OO_1 \cos O_1OS_1$$

or

$$(Vt_1)^2 = a^2 + 4p^2 - 4ap \sin \theta, \tag{6.1}$$

$$(Vt_6)^2 = f^2 + 4p^2 - 4fp \sin \theta. \tag{6.2}$$

Multiplying (6.1) by f and (6.2) by a, on subtracting and transposing terms, we obtain

$$p = \frac{1}{2}\sqrt{\left\{ V^2\left[\frac{a(t_1^2 - t_6^2)}{f - a} + t_1^2 \right] + fa \right\}}. \tag{6.3}$$

Since V is known, a and f are measured, and t_1 and t_6 found from the seismic record, the length of the perpendicular p from the shot-point on the bed may be calculated. The value of the dip, θ, may be found from

$$\sin \theta = \frac{a^2 + 4p^2 - (Vt_1)^2}{4ap} = \left[(f+a) + \frac{V^2(t_1^2 - t_6^2)}{f-a} \right] \Big/ 4p. \quad (6.4)$$

Fig. 126. A seismic survey truck.

The vertical depth z_0 from the shot-point to the reflecting layer is $z_0 = p/\cos \theta$. If, however, the reflecting layer is curved, it might be advantageous to find the vertical depth z_1 to the reflecting point R_1, from A_1 on the surface, the distance $A_1 S_1$ being indicated by x_1, from the relations $z_1 = R_1 S_1 \sin \Psi$ and $x_1 = R_1 S_1 \cos \Psi$. The values of $R_1 S_1$ and Ψ may be derived from the trigonometrical expression relating one side of a triangle to the other two sides and included angle, obtaining

$$\cos \phi = \frac{4p^2 + (Vt_1)^2 - a^2}{4p Vt_1}, \quad (6.5)$$

$$R_1 S_1 = Vt_1 - \frac{p}{\cos \phi} = Vt_1 \frac{(Vt_1)^2 - a^2}{4p^2 + (Vt_1)^2 - a^2}$$

$$= \frac{(Vt_1) W}{4p^2 + W}, \quad \text{where} \quad W = (Vt_1)^2 - a^2. \quad (6.6)$$

Fig. 127. Twelve panels of amplifiers inside the seismic truck.

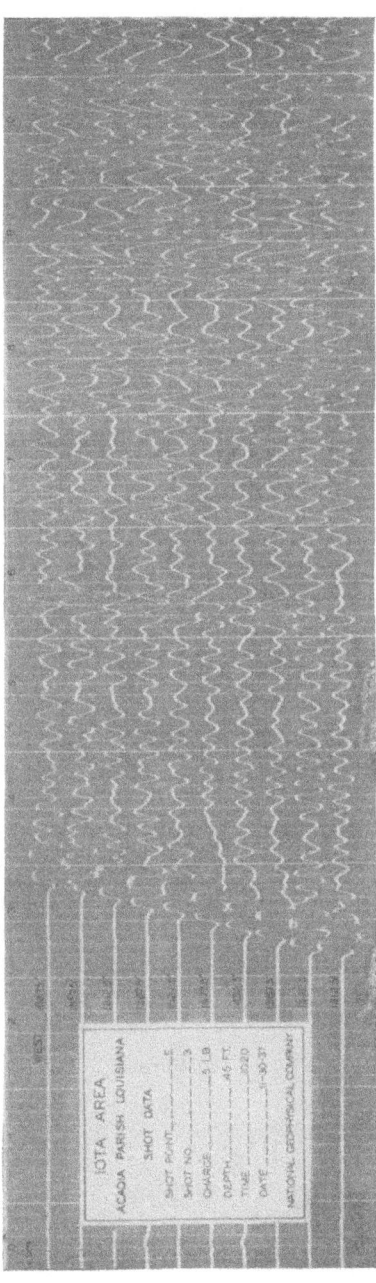

Fig. 128. Ten simultaneous records from a single shot. Less than one-third of the length of the record.

and $$\cos \Psi = \frac{(Vt_1)^2 + a^2 - 4p^2}{2aVt_1}.$$ (6.7)

By shooting both 'up' and 'down' at the two distances 'a' and 'f' say, and determining the travel times, t_u, t_d, from the seismic records, the velocity V may be found from the relation

$$V^2 = \frac{2(f^2 - a^2)}{(t_d^2 + t_u^2)_{\text{at } f} - (t_d^2 + t_u^2)_{\text{at } a}}.$$ (6.8)

It may at times be advantageous to lay out two sets of geophones along lines at right angles from the shot-point and determine the dip of the layer in the two directions. By finding the dip in the two directions, the actual strike of the reflecting bed may be determined.

As a further example of applying data secured by seismic reflexion records for determining inclination and depth of a limestone bed, the following somewhat simplified method has been suggested (though modifications and interpretations of recordings ultimately rest on field experience in any area). Five equally spaced detectors are placed in line with the shot-point S (Fig. 130), the centre one being at a distance l, which may be 1000 ft. from S. The difference in time of arrival of the first shockwave reflected from the bed at the first and last detector, Δl apart, is read from the record as Δt_1 and the travel time to the central one D is determined as t. If $D_1 A$ is drawn perpendicular to DR, then with sufficient accuracy, $AD_2 = V\Delta t_1$, where V is the velocity of the wave in the upper medium, which may be found from plotting the first arrival times after firing at the five recorders. From the diagram the angle $AD_1 D_2$ is $\theta - \alpha$, where θ is the angle of reflexion at R and α is the inclination of the bed to the horizontal. From the relation $\sin (\theta - \alpha) = AD_2/D_1 D_2$, we have

$$\Delta t_1 = \frac{\Delta l}{V} \sin (\theta - \alpha),$$ (6.9)

and by reverse shooting

$$\Delta t_2 = \frac{\Delta l}{V} \sin (\theta + \alpha).$$ (6.10)

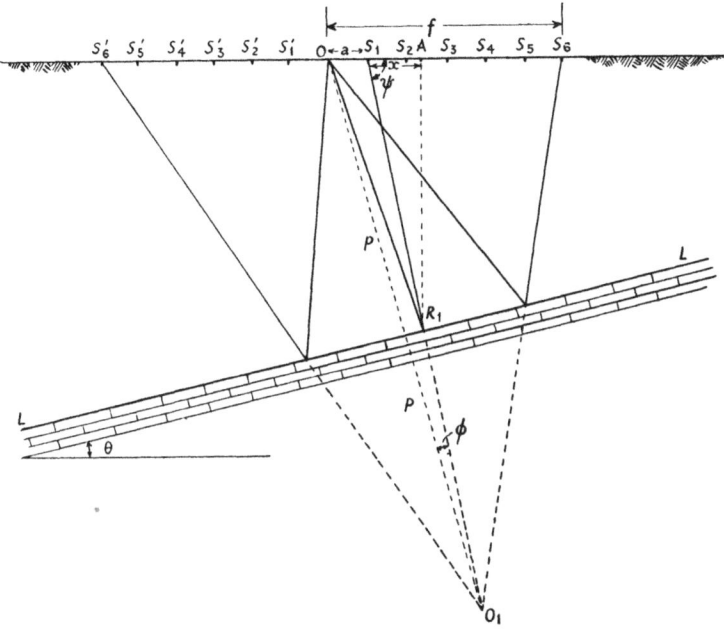

Fig. 129. Method of measuring dip by seismic reflexion.

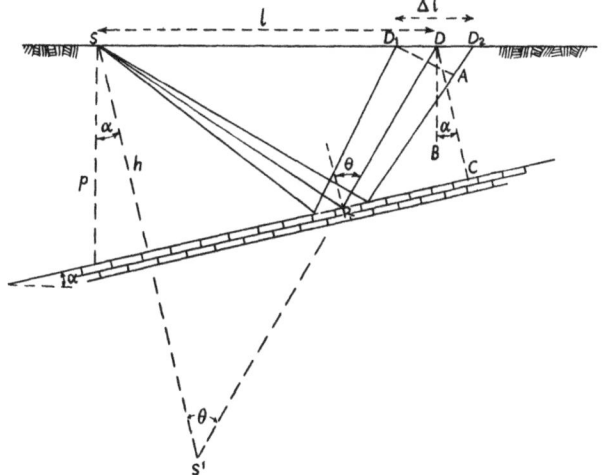

Fig. 130. Determination of the inclination and depth of
reflecting stratum.

By addition
$$\frac{\Delta t_1 + \Delta t_2}{2} = a = \frac{\Delta l}{V} \sin \theta \cos \alpha,$$

$$\frac{\Delta t_2 - \Delta t_1}{2} = b = \frac{\Delta l}{V} \cos \theta \sin \alpha.$$

(6.11)

Hence
$$\tan \alpha = \frac{\Delta t_2 - \Delta t_1}{\Delta t_2 + \Delta t_1} \tan \theta = \frac{b}{a} \tan \theta.$$

(6.12)

The waves, considered as travelling in straight lines, will appear at the detectors as coming from S', the image of S in the upper surface of the bed and $S'D = Vt$. Considering the triangle $SS'D$, we obtain by elementary trigonometry,

$$l/\sin \theta = Vt/\sin S'SD = Vt/\cos \alpha.$$

Hence
$$\sin \theta = \frac{l}{Vt} \cos \alpha = \frac{la}{t \Delta l} \sin \theta$$

on substituting for $\cos \alpha$ from (6.11).

Therefore
$$\sin \theta = \sqrt{\frac{la}{t \Delta l}},$$

$$\cos \alpha = V \sqrt{\frac{ta}{l \Delta l}}.$$

(6.13)

If V is unknown, we could determine α, since from (6.13) we have
$$\tan \theta = \sqrt{\frac{\sin^2 \theta}{1 - \sin^2 \theta}} = \sqrt{\frac{la}{t \Delta l - la}},$$

and
$$\tan \alpha = \frac{b}{a} \tan \theta = \frac{b}{a} \sqrt{\frac{la}{t \Delta l - la}}.$$

(6.14)

The perpendicular distance h from S on the bed may be deduced by considering the triangle $SS'D$, for

$$2h/\sin SDS' = l/\sin \theta \quad \text{and} \quad \sin SDS' = \cos (\theta - \alpha).$$

Hence
$$h = \frac{l}{2} \frac{\cos (\theta - \alpha)}{\sin \theta}.$$

(6.15)

The vertical depth p of the bed below the shot-point is $h/\cos \alpha$ or

$$p = \frac{l}{2} \frac{\cos (\theta - \alpha)}{\sin \theta \cos \alpha} = \frac{l}{2} (\cot \theta + \tan \alpha). \qquad (6.16)$$

Since both θ and α are known from (6.13) and (6.14), the value of p may be calculated.

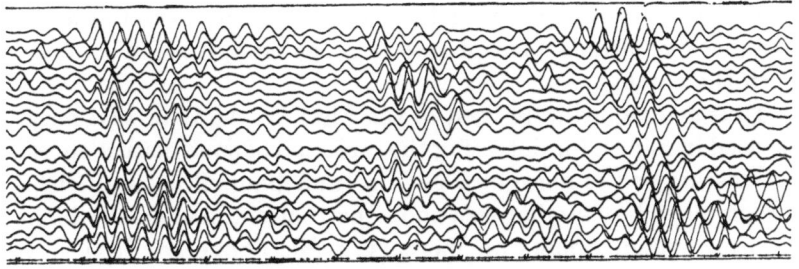

Fig. 131. Seismic reflexion record.

It should be pointed out that in practice conditions are much more complicated than the ideal premises upon which the above relations are based. There is usually a weathered stratum near the surface, in which the velocity will be different from that at depth, which will cause refraction and also affect the calculated depth. Moreover, we have considered the waves as travelling in straight lines, whereas they are slightly curved at least, since the velocity may change with depth. They serve, however, as a first approximation, it being realized that methods based on their field experience, in which the above factors have been taken into consideration, have been developed by commercial geophysical prospecting companies.

Fig. 131 is a portion of a typical seismic record from the Alberta foothills for which we are indebted to the Imperial Oil Company of Canada. It is the record obtained on twenty-four geophones with a split spread of 1320–110 ft. on either side of the shot-point.

ACOUSTIC METHODS

The fundamental difference between a 'sound' and a 'noise' is that the former can be analysed into a series of simple continuous waves, while a noise is an aggregate of shocks of short duration which do not usually submit to any such analysis. There might well be an advantage in generating a train of sound waves, which after entering the earth could be detected at a distance by suitable microphones or geophones placed in contact with the earth.

Thus geophones were used in the First World War, and may be used in mines, for detecting the direction of such disturbances as blasting or the noise of a pick in tunnelling. Two geophones are employed, of similar size and construction, in each of which an iron cylinder floats on mercury, while above the iron is a layer of enclosed air connected through a small hole with a rubber tube leading to the observer's ear. One geophone is thus connected to the right ear, the other to the left. The shock coming through the ground displaces the case of the geophone, while the inertia of the mercury and iron tends to hold them relatively at rest. Hence, the air is driven in rhythmic puffs through the holes and tubes to the observer's ears, and he hears the sound in a quite natural fashion. The remarkable physiological phenomenon of *binaural hearing* now comes into play.

If the two geophones on the rock or ground are equidistant from the source of sound, or more truly if they are both on the wave front, the hearer will have the impression that the cause of the noise is in front of his head. If, however, one geophone, say the left, is advanced a little towards the sound, the message will reach his left ear first, or rather in advanced phase, and the source of the sound will seem to come 'from his left'. Indeed, we are brought into contact with the sense of direction of sound, early developed even by quite young children and by most animals and birds, so that they can instinctively turn their heads towards any sound or noise. This is indeed one of the great safeguards of life. By moving the two geophones about until the source of sound seems to be 'in front' of the listener, it is possible to deduce that the origin of the noise or sound is in a plane perpendicular to the junction of the geophones.

Similar methods prevail at sea where a submarine bell or a Fessenden oscillator may generate a sound which can be heard, with hydrophone and amplifier, many miles away. The binaural method may in this case also be used to determine the direction of the source of sound. Were it not for the efficient use of wireless (or radio) direction-finding method there is no doubt that underwater signals would be extensively used. In war time, owing to the lack of secrecy of wireless messages, it is probable that underwater signals may be necessary in communications between ships of the same fleet.

All these ideas have been imported into geophysical prospecting, but at present with small success. Experiments tend to show that it is difficult to get signals from an oscillator into ground or rock. It is much more efficient to use the shock from a high explosive of quantity suitable to the end in view. Moreover, the ground does not transmit sound well, say, as compared with water. Thus the sounds from an oscillator in water, audible in a hydrophone many miles away at sea, are so attenuated by passage through a few hundred yards of earth that they are not audible in a microphone (or hydrophone) placed in water in a hole in the ground. For these reasons the acoustic method has not many advocates today. The question of its intelligent use in mines, whether in signalling or in exploration, must not be too speedily or rashly dismissed.

MATHEMATICAL THEORY OF SEISMOGRAPHS

If the moment of inertia of a horizontal pendulum (Fig. 115) about the line of supports C, D is MK^2, if h is the distance of the centre of gravity of the moving system of mass M from that line, and if α be the angle between AB and the vertical, then

$$MK^2 \frac{d^2\theta}{dt^2} = - Mgh \sin \alpha \, \theta \qquad (6.17)$$

is the approximate equation for small oscillations, and the period of a complete free swing is given by

$$t = 2\pi \sqrt{\frac{K^2}{gh \sin \alpha}}. \tag{6.18}$$

In all vibrating systems which swing to-and-fro without damping, or restraint, or forced movement, we have the angular acceleration proportional to the angle of displacement at any instant, or

$$\frac{d^2\theta}{dt^2} \quad \text{or} \quad \ddot{\theta} = -n^2\theta, \tag{6.19}$$

with a period

$$t = \frac{2\pi}{n},$$

where

$$n^2 = \frac{gh \sin \alpha}{K^2} \tag{6.20}$$

in the above result (6.18). When, however, the swinging system is damped by air resistance, specially when vanes are attached, or when a coil of a copper plate is used between the poles of a magnet, then an opposing force proportional to the velocity of the mass, and therefore to the angular velocity of the rod, is introduced; this force will be proportional to $d\theta/dt$ or $\dot{\theta}$ so that we now write (6·19) in the form

$$\ddot{\theta} + 2k\dot{\theta} + n^2\theta = 0, \tag{6.21}$$

and this equation will give the free swinging motion of any seismograph, or galvanometer needle, not disturbed by other agencies. Putting $\theta = ae^{ipt}$ and substituting in (6·21) we get a quadratic equation with two roots and then deduce a solution

$$\theta = Ae^{-kt}\sin\{\sqrt{(n^2 - k^2)}\,t - \eta)\}, \tag{6.22}$$

indicating a to-and-fro swing of period $2\pi\sqrt{(n^2 - k^2)}$ with amplitudes gradually decreasing to an extent depending on the value of k due to the extent of damping. Now we are not primarily interested in the swing of our seismograph, because we are really looking for the vibrations of the earth. It is therefore desirable to eliminate all to-and-fro swinging until k is as large as n, and then the solution of (6.21) becomes

$$\theta = Ae^{-kt}(t - \eta), \tag{6.23}$$

where A and η are some constants depending on initial displacement and velocity. The swinging apparatus is now said to be critically damped or *dead-beat*, and on displacement it will slowly return to a position of equilibrium like a pendulum in molasses, or a door with a spring and a dash-pot.

We must now consider *forced* oscillations, as opposed to *free*, which we have just been discussing. Suppose that the supports move to-and-fro with acceleration $\ddot{x}$ or d^2x/dt^2, where x is a small periodic displacement due to the earthquake of the type $x = x_0 \sin pt$, then equation (6.21) is replaced by

$$\ddot{\theta} + 2k\dot{\theta} + n^2\theta = \ddot{x}/l, \qquad (6\cdot24)$$

which will represent satisfactorily the motion of all seismographs. The solution (Walker, 1913, pp. 4–5) is

$$\theta = \frac{p^2 x_0 \sin(pt - \eta)}{l\sqrt{\{(n^2 - p^2)^2 + 4k^2p^2\}}}, \qquad (6.25)$$

where

$$\tan \eta = \frac{2kp}{n^2 - p^2}, \qquad (6.26)$$

and if the recorded displacement is $L\theta$ the magnification is

$$Lp^2/l\sqrt{\{(n^2 - p^2)^2 + 4k^2p^2\}}, \qquad (6.27)$$

which though complicated in appearance is instructive in character.

First of all, the oscillations of the seismograph are out of phase with the earth tremors to an extent given by the value η in (6.27), and it is interesting to note the difference between n being greater or less than p. The oscillations to-and-fro are governed mainly by p and not by n, that is to say by the period of the shocks, not of the apparatus.

If the disturbances are quick and complicated, x has not the simple recurrent sine form given by $x_0 \sin pt$, and it is necessary by heavy damping to cancel out the free vibrations so as to catch the next forced vibration. An interesting study of forced oscillations, both theoretically and experimentally, has been made by Blau (1928). The multiplication factor given in (6.27) needs to be as large as possible and consistent with the requirement ($k = n$), when the magnification factor simplifies to

$$Lp^2/l(n^2 - p^2). \tag{6.28}$$

In practice the magnification should exceed 10,000.

It is obvious that this brief discussion or analysis is quite inadequate, but further information must be gathered from a good work on seismology. The following table from Walker's book (p. 33) is instructive as to the different main types of seismographs:

Seismograph	Stationary mass m (km.)	Primary period (sec.)	Equivalent length l (cm.)	Damping modulus (k/n)
Milne	1	17	16	0·026
Wiechert	1000	12	100	0·45
Galitzin	7	24	12	1·00

A photograph of a large vertical seismograph as set up in Ottawa for recording earthquakes is shown in Fig. 132. As a contrast a photograph (Fig. 133) is also reproduced of a Wood-Anderson horizontal seismograph, without the recording gear. Between the poles of a damping magnet may be noticed the small upright cylinder attached by one generating line to a fine wire which is nearly vertical.

SUMMARY OF PROGRESS

As an indication of the progress which has been achieved in this type of prospecting, it is interesting to recall some remarks made by Barton (1928). After naming the pioneers, he states that Mintrop, in 1919, obtained a basic patent, and that in 1924 his Seismos Company discovered several salt domes. This led to an ardent invention of various types of field seismographs, some of Mintrop's type, others such as Wood-Anderson's of a different pattern, consisting of a small, massive cylinder attached to the edge of a light vertical strong stretched wire, so that a generating line of the cylinder is fastened to the generating line of the wire, constituting a horizontal pendulum. There is, of course, no real reason why seismographs should be

Fig. 132. A vertical seismograph as set up in Ottawa for
recording earthquakes.

enormous either in size or mass. Barton further states that the reflexion method was perfected by the Geophysical Research Corporation in 1926. In 1928 there were thirty or more groups or 'troops' at work in the United States and in Mexico, each consisting of three to five trained technical men, with about an

Fig. 133. A Wood-Anderson horizontal seismograph. Between the poles of a damping magnet is a cylinder attached eccentrically to a fine stretched wire, nearly upright.

equal number of helpers. Each such group or 'troop' would cost $15,000–20,000 a month, including rent of apparatus, transportation, salaries and explosives, and in a month they might explore 300 sq. miles at a cost of 5–12 cents an acre. Detailed work round a discovered dome would amount to $5–10 an acre.

A great change has taken place during the last quarter-century. As mentioned above, geophysical prospecting companies and the larger oil companies have developed their own special equipment and techniques in interpreting seismic records which they keep confidential, but their work has been very successful in providing new sources of oil and also of natural gas. For example in May 1952 there were 144 seismic parties operating in the province of Alberta, in which 3025 wells have been discovered capable of oil production and the mean daily production during a week in that month, according to the Alberta Department of Mines and Minerals, was 170,287 barrels. The number of new wells discovered and the output is increasing. The estimated expenditure in Alberta alone during 1952 on geophysical prospecting was over $36,000,000. The number of seismic parties operating throughout the world during 1952 was over 800, with a probable expenditure of over $325,000,000 in oil exploration alone.

PROBLEMS

1. A horizontal layer of limestone lies beneath a shale overburden. The direct wave and refracted wave arrive simultaneously at a detector 3300 ft. from the shot-point. If the velocity in limestone is 20,000 ft. per sec. and the travel time is 0·300 sec., find the thickness of the overburden.

2. Ten seismometers were placed in a straight line at increasing distances from the shot-point in an experimental determination of the depth to a horizontal discontinuity below. The distance from the shot-point to each seismometer and travel times to each are given in the table below. Find (i) the velocities of the waves in the two media and (ii) the vertical depth to the discontinuity.

Distance in ft.	Time in sec.	Distance in ft.	Time in sec.
500	0·108	3000	0·620
1000	0·216	3500	0·716
1500	0·300	4000	0·809
2000	0·422	4500	0·900
2500	0·530	5000	0·992

3. To find the depth to bedrock in a dam site survey the travel times from the shot-point to eight detectors which were laid out in a straight line 50 ft. apart were measured from the record. From the following readings determine the depth of overburden:

Distance in ft.	Time in sec.	Distance in ft.	Time in sec.
50	0·0190	250	0·0590
100	0·0290	300	0·0620
150	0·0385	350	0·0655
200	0·0497	400	0·0685

4. In a seismic investigation to find the depth of a horizontal limestone bed, one detector was placed 200 m. from the shot-point and a second one 1200 m. away in the same straight line. The travel times of the reflected waves to these two detectors were 0·5637 and 0·6742 sec. respectively. Find the velocity of the seismic wave in the overburden and the vertical depth of the limestone below the surface.

5. To find the depth and slope of a reflecting layer by a seismic investigation, ten detectors were placed at intervals of 50 ft. in a straight line, the first 310 ft. from the shot-point and the last 810 ft.

away. From the record, the velocity of the shock-wave in the upper medium was found to be 10,000 ft. per sec. The travel times of the reflected wave to the first and last detectors were found to be 0·0910 and 0·1130 sec. respectively. Calculate the dip of the reflecting layer, and the depths below the shot-point and the first detector.

6. A series of detectors was laid out in a straight line passing through the shot-point in a seismic investigation to find the depth to a horizontal bed of limestone. The seismic record gave the time interval for the shock-wave to travel from the shot-point to the first detector, 60 ft. away, by reflexion as 0·5637 sec. and to the last detector, 3600 ft. away, as 0·8076 sec. Calculate the velocity of the seismic wave in the upper medium and the vertical depth to the limestone.

7. In a seismic investigation in Alberta, the following readings were taken of the times of arrival of the first shock-wave at detectors placed in a line passing through the shot-point at distances from the shot-point shown in the table below. From the time-travel plot calculate the velocity of the seismic wave in the medium. Reflexion is observed later at the first detector, 55 ft. from the shot-point, the time interval being 1·809 sec. and at the last detector, 660 ft. away, after an interval of 1·814 sec. Find the vertical depth to the reflecting layer below the shot-point and the inclination of the layer.

Distance in ft.	Time in sec.	Distance in ft.	Time in sec.
55	0·016	385	0·055
110	0·016	440	0·066
165	0·021	495	0·072
220	0·031	550	0·079
275	0·038	605	0·089
330	0·046	660	0·097

8. Calculate the depth to a reflecting layer and its inclination from the following data obtained on a seismic record. The shot-point was 1320 ft. from the central of five equispaced seismometers in a line passing through the shot-point. The distance between the two extreme detectors was 360 ft. The time of travel from shot-point to reflecting layer and back to central seismometer was 0·732 sec. The difference in times of arrival at first and last seismometer was 0·022 sec., and by reverse shooting this was found to be 0·010 sec. The velocity of the shock-wave in the upper medium was 10,000 ft. per sec.

9. The following data were obtained from the record using a seismic reflexion method. The shot-point was 2000 ft. from the centre of a series of detectors, the first and last of which were 750 ft. apart. The travel time to the central detector was 0·232 sec. The interval between the arrival of the reflected wave at the first and last detector was 0·0825 sec. and by reverse shooting was 0·0207 sec. Find the depth of the reflecting layer below the shot-point, the inclination of the reflecting limestone bed, and the velocity of the seismic wave in the upper medium.

CHAPTER VII

RADIOACTIVE METHODS

RADIOACTIVITY

We have noted that mineral and oil deposits associated with certain geological formations differ in their magnetic properties, electrical conductivity, rates of oxidation, density, and the velocity with which they transmit elastic waves. Each of these properties forms the basis of several methods of geophysical prospecting described in previous chapters. There is another property associated with certain elements that has become important in recent years, not only on account of the economic value of the minerals containing these elements, but also because it has supplied means of locating other valuable deposits. This property is known as radioactivity. Elements possessing radioactivity are unstable; their atoms break down spontaneously into other atoms at rates characteristic of each type with the emission of one or more of three distinct types of radiation known as alpha (α), beta (β) and gamma (γ) rays. These types of radiation produce three different effects, in varying degrees, namely: (i) they ionize a gas, that is, they cause it to become a conductor of electricity; (ii) they produce scintillations or phosphorescence in certain minerals and chemicals; and (iii) they affect a photographic plate or emulsions exposed to their action in much the same manner as light does. All three types of rays will pass through thin sheets of material opaque to ordinary light but some are much more penetrating than others. The α-rays, which are the charged nuclei of helium atoms, are completely absorbed by a single sheet of writing paper, while the β-rays, which are electrons or in some cases positrons, are more penetrating, though even the most energetic ones are stopped by a thin sheet of lead. On the other hand, the γ-rays, which are like X-rays but of shorter wave-length, may penetrate several inches of iron. Various detecting and measuring instruments have been designed which depend for their operation on these properties of the radiations emitted, more particularly (i) and (ii) above and are now used to locate radioactive minerals and gases.

DETECTING INSTRUMENTS

Originally the loss of charge, or more strictly decrease in potential, of a gold-leaf electroscope owing to the conductivity produced in the air surrounding the leaf system, was used as a measure of the radioactivity. As a field instrument such electroscopes were delicate to handle and were subject to changes in atmospheric humidity, pressure and temperature. The invention of the Geiger-Müller counter (Geiger and Müller, 1928, 1929) and subsequent development of various electronic instruments incorporating its fundamental principle have made the gold-leaf electroscope obsolete as a geophysical prospecting unit.

A Geiger-Müller detector consists essentially of a cylindrical cathode and a fine wire anode placed axially in a sealed glass tube containing gas or a mixture of gas and some vapour at a definite pressure of a few cm. of mercury. The central anode is charged to a positive potential varying from as low as 300 V. to more than 1000 volts, depending upon the circuit and gas used, and is connected to the grid of a triode valve in such a way that any change in anode potential produces an electric impulse that may be registered after amplification by a click in a pair of headphones or by a deflexion on a meter. The particle or γ-ray which enters the tube causes ionization, and the electric field gives the ions sufficient energy to ionize other atoms by collision, resulting in a tremendous multiplication, so that the gas becomes momentarily conducting. This results in a drop in potential of the central wire anode, the change being communicated to the valve circuit. Various gases and circuits are used for quenching the discharge to reduce the detecting interval between two successive impulses and also to integrate the number of pulses in a given time, so that the instrument may be used as a ratemeter giving directly a reading of the intensity of the radiation. Fig. 134 represents in schematic diagram form a typical Geiger counter circuit, and Fig. 135 is that of a ratemeter (Franklin and Hardwick

1951). For assaying rock or uncrushed ore samples, a high-pressure ion chamber using a vibrating reed electrometer and operated with a $22\frac{1}{2}$ V. dry battery, Fig. 136, has been designed by H. Carmicheal (1946). This instrument measures the γ-ray activity from the specimen and is used by the Canadian Department of Mines for assaying the amount of uranium oxide (U_3O_8) in an uncrushed sample (Horwood and McMahon, 1950).

Scintillation counters are in general more sensitive than gas-filled Geiger counters, and may use any one of a number of artificially grown large crystals which fluoresce under the action of the rays emitted by radioactive atoms. Large crystals of

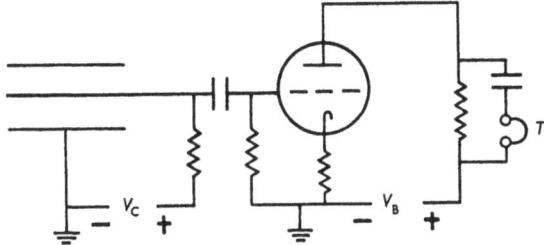

Fig. 134. Geiger-Müller counter circuit.

anthracene, stilbene and *para*-terphenyl are commonly used as phosphors, and photoelectric multipliers transform the light scintillations into corresponding electrical impulses. A typical circuit diagram of such a scintillation counter is shown in Fig. 137. Such counters are used for ground and aerial surveys as well as for drill-hole exploration. All such instruments register a background count due to cosmic rays and the low natural radioactivity which is always present, and corrections for this must be made when a quantitative measurement is required. However, the portable scintillation ratemeter (Fig. 138) designed by Roulston and Pringle (1950), (Pringle, Roulston and Taylor, 1950), called the 'scintillometer', is so efficient that the background due to cosmic rays is negligible. With such instruments readings at a station may be taken in two or three seconds.

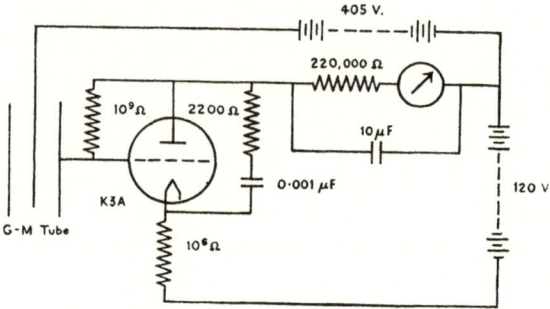

Fig. 135. Ratemeter circuit using cold cathode tube.

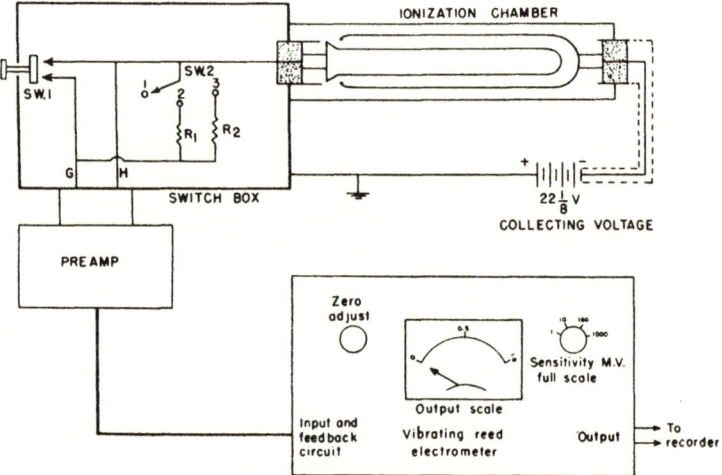

Fig. 136. Vibrating reed electrometer.

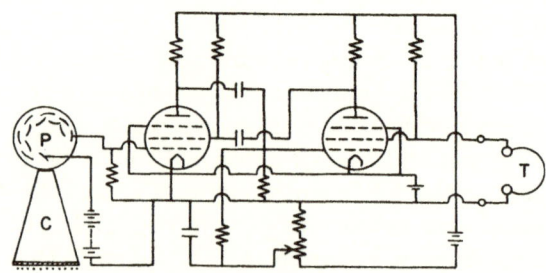

Fig. 137. Simple scintillation counter circuit.

Natural gases issue at various rates from different parts of the earth's surface. When these gases leak forth in small quantities they are for the most part hard to detect. But there is one gas which by its radioactive properties can be readily detected and

Fig. 138. Portable scintillometer.

measured. This gas is radium emanation or radon. By thrusting down into the earth 1 or 2 ft. of a hollow iron rod with holes in the bottom, it is possible with a small hand pump to extract a sample of air from the ground and pass it into a testing vessel connected with a Geiger counter with which the amount of radon present can be measured. Geiger counters, provided with a window of a thin sheet of mica through which the α-rays can readily pass, or scintillation counters, are most suitable for making such measurements.

RADIOACTIVE ROCKS

All rocks may be divided into two classes, those which are radioactive and those which are not. Those which are radioactive contain a fair percentage of uranium or thorium minerals or both; examples of these minerals are pitchblende, carnotite, thorianite. Samples of such rocks brought near a Geiger counter or placed inside the testing chamber of an electroscope will readily and rapidly be identified as radioactive.

It has been proved, however, by the late Lord Rayleigh that all rocks, primary and secondary, contain a small amount of the element radium, about one part in a million million. Hence true radioactive rocks are those which contain about one part in a million of radium (or radiothorium), while ordinary rocks contain about one part in a million million. Radium is a transformation product of uranium, from ores of which it is extracted, and most of the activity associated with pitchblende and other radioactive ores is due to the radium content. Ores containing 0.05% of U_3O_8 are readily assayed by radiometric methods. Many radioactive substances also give rise to the gas helium, which might be used as an indicator but for its inertness, so that the only procedure is collection from a large quantity of natural gas at low temperatures; indeed, this procedure is followed in obtaining helium for commercial purposes, but has no direct connexion at present with prospecting work.

In the search for radioactive rocks or minerals, such as pitchblende or certain pegmatites, it is sufficient to have Geiger or scintillation counters which can be battery-operated and carried along over the ground or flown in an aircraft for regional surveys. Since the α- and β-rays are absorbed by only a few inches of overburden, except in the case of exposed outcrops, it is the γ-rays which will reveal the presence of radioactive ores. Cosmic radiation, which is ever-present, is also detected by counters, and so the background reading must be considered in searching for radioactive deposits, unless the ore is near the surface. When a ratemeter is used the normal background reading is readily observed, and any pronounced increase due to the presence of ore is at once detected. If a pair of earphones is used, the number of clicks per second will greatly increase. The nearer the instrument is carried to the ground, the better the chance of discovering a deposit.

The early discoveries of pitchblende, carnotite and other highly radioactive mineral deposits were made by prospectors recognizing outcrops from their colour, density and general mineralogical properties and sending samples to Government laboratories where their activity was tested with gold-leaf electroscopes. Indeed, that procedure is still followed,* except that the detecting instruments are ion chambers and electron counters, and in Canada at the present time the service is provided free. The improvement in portable instruments has simplified identification of such ores in the field, but the rare occurrence of high-grade commercial deposits, to be found only by ground exploration, did not attract many prospectors into this field, since until ten years ago the radium content was the main value of the ore. With the successful construction of the chain-reacting nuclear or atomic piles and the possibility of so many useful applications of atomic energy, the situation has completely changed. Now that atomic piles require natural uranium metal as a basic fuel, the demand for minerals containing uranium has become so great that a more rapid method of locating possible sources is extremely important. During the last few years successful experiments have been carried out in making rapid surveys over large areas from aircraft in which special γ-ray detectors have been installed. Such regional surveys indicate favourable areas for prospecting by ground crews with portable equipment.

Radon, derived from the parent radium in the rocks in the ground, is produced at a rate depending on the amount of radium present. Solutions may seep up towards the surface with the upward currents of water. The radon has first to escape from the place of generation, and such part of it as is not occluded will then diffuse through the more permeable rocks, but in the case of wet clays, for instance, it may collect beneath them and only escape round the edges, through more or less vertical

* *Prospecting for Uranium* (1951).

faults, or from the summit of anticlines. The rate of migration must be faster than the rate of decay, and the radon disintegrates to half value in less than four days, and practically disappears in a month. In its migrations radon imitates the behaviour of natural gases, and like them it may be an indicator of oil deposits. Hence there will be certain regions, limited in area, where a radioactive method based on the detection of radon may have great value and usefulness, and there will be other vast areas where it would be perfectly futile to use it.

AERIAL PROSPECTING FOR RADIOACTIVE DEPOSITS

Aerial prospecting for uranium deposits appears to have been first seriously considered in 1944 (Nat. Res. Council Report CRR-495, 1952). Since that time an ever-increasing number of surveys have been made, using various means of recording the activity in aircraft during passage over large areas in Canada, United States, Australia and other countries (Brownell, 1950a, b; Lundberg, 1950, 1952; Lundberg et al. 1952). Since safe flying over hilly wooded country requires the aircraft to be at least 500 ft. above ground, it is only the γ-rays which are detectable. Special equipment designed by members of the Canadian National Research Council at Chalk River, Ontario, in which arrays of large Geiger and ion-chamber counters were used, was first flown in an aircraft in 1949. The circuits were arranged to eliminate the effect of cosmic rays or at least reduce it to a low level, so that the radiation from the ground became more obvious. The instrument was calibrated by flying over Round Lake, Ontario, in the winter and noting the response received from radioactive sources of known strength placed at intervals on the ice in the same straight line. An actual record is shown in Fig. 139. It is noteworthy that the background decreases very definitely when the aircraft passes over a lake, river or marsh, because of the absorption of the γ-rays in the water.

Owing to the necessity of obtaining high sensitivity when the apparatus is flown so high above the ground, scintillation counters were found to be more efficient. Specially grown large crystals of sodium iodide activated with thallium iodide have given excellent results. Good optical contact with the photo-electric multiplier used to transform the light intensity of the scintillations into corresponding electrical impulses is accomplished by using a thin film of Nujol oil. The impulses are amplified and passed through a special filter network designed to pass the shape of the response as the effect from a point source on the ground when the aircraft is flying at a height of 500 ft. The response is printed on a moving paper strip by Brush Development Company recorders. In the Chalk River

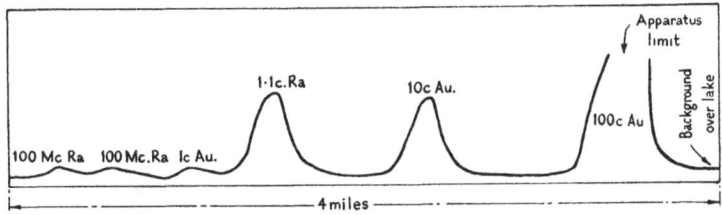

Fig. 139. Record of airborne detection of radium.

detector, three partially shielded crystal units (Fig. 140) are used, connected through amplifiers to special counting rate-meters operating three Brush pens. One records the combined signal from all three scintillation units, the other two the signals from the port and starboard side of the aircraft only. Records must be corrected for variations in the height of the plane above ground during its flight.

An aircraft fitted with this apparatus and a recording radio altimeter as well as a camera taking a continuous photograph of the terrain on a strip film by which the survey is correlated with the navigator's log, has been flown over a large area in the Canadian North-West. In one survey covering a grid of 64 sq. miles in the Beaverlodge area of northern Saskatchewan, twenty anomalies were recorded, all but two of which corresponded to known deposits. One of the two others was due to a deposit of sandstone, which often simulates a uranium deposit, owing to the sandstone usually containing a higher concentration of uranium than the ordinary rock formations. A typical record with the correction for altitude is shown in Fig. 141, an indication being evident at A. To check the value of such an indication, a ground survey is made and if the result is favourable, trenching and drilling would follow. The record of a recent aerial survey

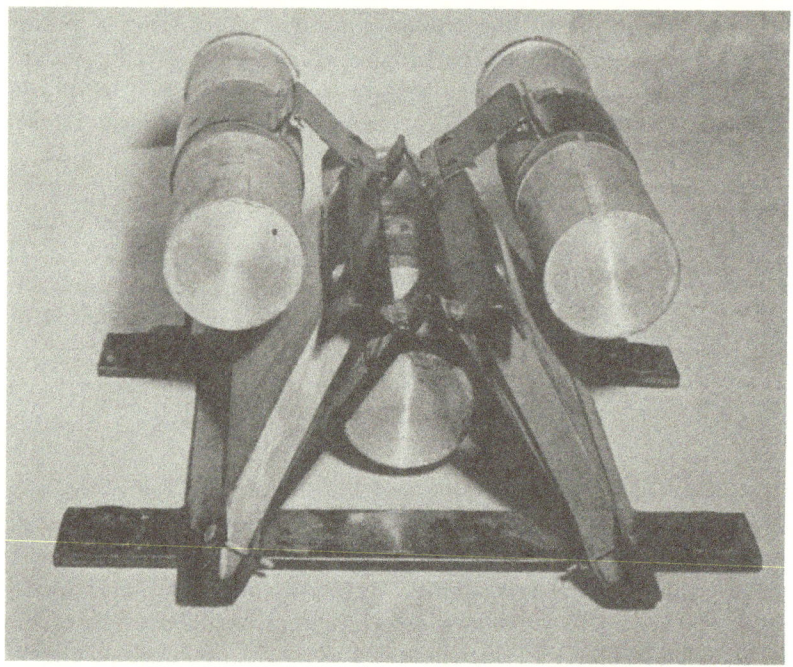

Fig. 140. Crystal units for aerial prospecting.

along a line over the Goldfields area of Saskatchewan with a scintillometer detector is shown in Fig. 142. Three uranium ore deposits are indicated on this part of the record, which was supplied through the kindness of Professor G. M. Brownell.

RADIOACTIVE ASSAYING

It is also possible to examine specimens of rock from drill holes, in fact the presence of any appreciable amount of radioactive material in the core will be indicated by merely passing a counter along it. A special type of doughnut Geiger counter

(Fig. 143) has recently been invented by members of the Canadian National Research Council, Ottawa, for measuring the activity of the core from a drill-hole. The core is simply passed through the doughnut, and the activity is recorded as it passes along.

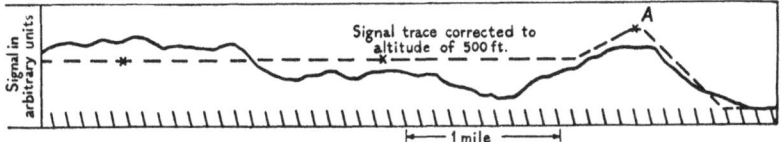

Fig. 141. Record of γ-ray activity.

It has almost 100 % efficiency and has been applied for measuring activity along tubes containing radioactive solutions. To assay a specimen of ore, a special procedure exists, using two Geiger counters, one for measuring β-radiation and the other γ-radiation simultaneously from the same specimen (Lapointe, 1950; Wilmot and McMahon 1951; Lapointe and Williamson, 1952). The rock

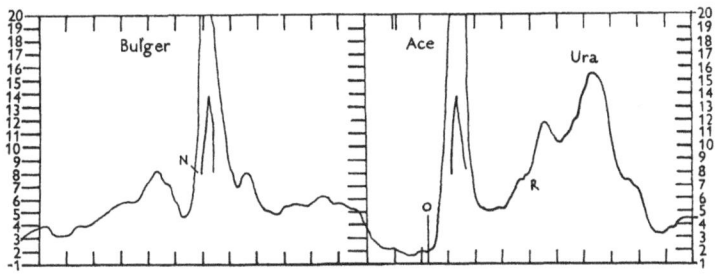

Fig. 142. Aerial survey with scintillometer
(Goldfields area, Saskatchewan).

is ground up, and in this process some of the radon gas and thoron (if there is any thorium present, as is often the case) will escape. There may also be some leaching of the soluble radioactive salts. Corrections must be made for such losses in assaying for uranium oxide, since the method is based upon the uranium being present in equilibrium with its disintegration products. To determine whether there has been any loss in the grinding, the specimen, say 300–500 g. of powdered material, is placed in a tray midway between the two counters, one above, measuring the β-ray activity, the other below, with a sheet of lead interposed to absorb any β-rays, which registers the γ-ray activity. The apparatus is calibrated by first taking the natural background count without any specimen present, and then samples containing known amounts of uranium oxide in equilibrium with the decay products placed in the tray and the readings taken. The ratio of the γ-ray to β-ray activity for such a standard specimen is then known. The more oxide present, the

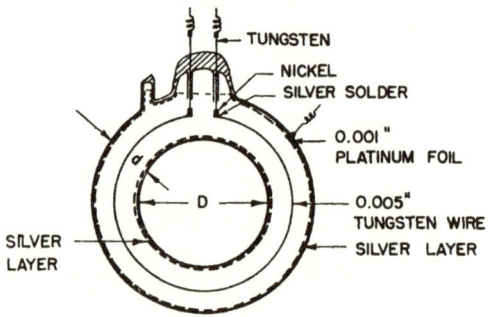

Fig. 143. Doughnut Geiger counter.

higher the count-rate, but the ratio will be constant. If there has been any leaching or escape of radon in grinding a specimen, the ratio will be altered, the γ-ray assay will read higher than the β-ray and a correction must be made. If it should prove that the β-ray assay is higher than the γ-ray, then either potassium is present (the β-radiation from which would increase the β-ray count above the normal) or some γ-ray emitters of the uranium series have been leached out. The ground ore is laid in the pan so that the thickness is that given by 1·5–2·0 g. per cm.^{2}. This radioactive assaying of minerals for uranium oxide (U_3O_8) is used with success at the Department of Mines and Technical Surveys, Ottawa, and at the uranium mines of Eldorado Mining and Refining Limited. Hilborn (1951) has constructed a β- and γ-ray measuring equipment in which the normal gas-filled tube for measuring the γ-ray activity is replaced with a scintillation counter. Samples containing from 0·005 to 2 % U_3O_8 can be accurately assayed with this equipment, a photograph of which is shown in Fig. 144. Radioassaying with such counters agrees within 5 % with the much more tedius chemical method and is sufficiently accurate for the general evaluation of the ore.

It is sometimes desirable to make a quantitative analysis of the amount of uranium in a specimen of ore when the equipment described above is not available or when uranium oxide standards

Fig. 144. Apparatus for radioactive assaying.

for calibration are lacking. The simplest method is then to determine the amount of radium present and to calculate the uranium which is in equilibrium with the radium. In order to measure the amount of radium present it is necessary to carry out the rather elaborate operation of reducing the whole specimen, say 100 g. of rock, to a complete state of solution. This solution is boiled to release the radon and then kept in a well-sealed vessel for a month, allowing the radon which is produced from the decay of radium to come into equilibrium. The solution is again boiled, and the gas transferred to an ionization chamber and the activity measured with a Geiger counter or an electroscope, not neglecting to find first the natural background count or leak and subtracting this from the observed value. This value is compared with that of a standard solution, that is, one in which a given amount of radium, say a few milligrammes, has been dissolved.

The amount of radium in the specimen being thus found, the amount of uranium may be determined from the equilibrium equation $N_u \lambda_u = N_{ra} \lambda_{ra} = N_{rd} \lambda_{rd}$, where N and λ are the number of atoms and the transformation constants respectively of the uranium, radium and radon in equilibrium. This relation holds for all the decay products of a radioactive series when in equilibrium. Hence

$$N_u = N_{rd} \frac{\lambda_{rd}}{\lambda_u} = N_{rd} \frac{2 \cdot 097 \times 10^{-6} \text{ sec.}^{-1}}{4 \cdot 9 \times 10^{-18} \text{ sec.}^{-1}} = 4 \cdot 3 \times 10^{11} N_{rd}.$$

It is only necessary then to multiply the amount of radon by the factor $4 \cdot 3 \times 10^{11}$ to determine the number of atoms of uranium in the specimen. Other methods are employed (particularly for very small quantities) in which a millilitre of the active solution is evaporated to dryness on a stainless steel disk and the activity measured with a scintillation counter or fluorometer.

Besides the direct detection of radioactive substances by the radiations they emit, there is the further point that radioactive change is associated with a measurable evolution of heat, so that there is a simple relation between the amount of radium or thorium present and the heat evolved. The consequent local rise of temperature must depend on the specific heat of the containing rock and on the conductivity of the rock and of the surrounding medium, and on the convection of heat by water or air, if they pass near or through the rock.

To sum up: indications may be obtained by radioactive methods:

(1) By measurement of the radon and thoron contents of rock, water or oil specimens.

(2) By collection of radon or thoron with other gases from the earth, possibly deducing the position of faults, or of anticlines.

(3) By examination of the temporary suspension of radon in oil or water.

(4) By measurement of the γ-radiation from radioactive matter, made either above the earth or in mines or tunnels underground or in drill holes.

(5) By comparing the γ- and β-radiation from ground specimens of rock using coincident counters.

(6) By measurement of the temperature and temperature gradients of rocks as modified by their radioactive constituents.

Although some of these methods are now widely used and have proven value in prospecting for radioactive ore, others are not yet so far developed as to be in a reliable state for ordinary prospecting, and hence they lend themselves readily to fraud and quackery. Before embarking on any such exploration methods, it is well to take advice from those who are familiar with electronic equipment and the possibilities of radioactivity and its applications. The use of Geiger-Müller counters and the scintillation counter is well established and reliable. They are simple to use, but the quantitative evaluation of the amount of uranium oxide in a mineral requires much skill and experience.

The above brief summary indicates that radioactive methods have considerable flexibility, and their possibilities are still by no means exhausted.

THEORY OF RADIOACTIVITY

Radioactive atoms of one type tend to break up and transform into radioactive atoms of a successive type or into stable atoms in some cases, at a time rate proportional to the number of atoms present. So also the number of persons killed per annum by motor-cars in a given city is proportional to the population of that city; the proportionality factors may vary from city to city according to the intelligence, sobriety and care of the citizens, and to the efficient enforcement of wise traffic regulations. In radioactivity the proportionality factor is independent of the exterior physical conditions such as temperature and pressure. It is regulated from within the atom by conditions not yet understood, which may be attributed to providence or chance, and of which we have, like a large insurance office, only statistical measurements, dealing with great numbers.

All of these statements are summarized mathematically by the relation $\dot{N} = -\lambda N$, where N is the number of atoms existent at time t, $\dot{N}$ or dN/dt is the rate of their transformation, and λ is the proportionality factor, whose inverse $1/\lambda$ is the *average* life of an atom. On integrating the above equation we have $N = N_0 e^{-\lambda t}$, in which N_0 is the number of atoms present at a time $t = 0$ and N the number at the end of a time t. Half the atoms will have decayed in a time T, when $N = \frac{1}{2}N_0$; the value of T, known as the *half-life* of the radioactive atoms, is related to λ by the relation $T = \log_e 2/\lambda = 0.693/\lambda$. The values of the half-life T and decay constant λ are characteristic of every type of radioactive atom, some kinds of atoms such as polonium 212 (thorium C'') having a half-life of only 3×10^{-7} sec. and others, uranium 238 for example, having a half-life as long as 4.51×10^9 years.

The process of transformation appears to be a violent one, and to be accompanied by the ejection, from the nucleus or central citadel of the atom, of a high-velocity projectile, which may be an electron (sometimes positively charged, called a positron to distinguish it from the negatively charged electron) or β-particle; or a helium atom stripped of its two outer electrons, forming an α-particle. There are also electromagnetic radiations, the γ-rays, strictly resembling X- or Röntgen rays, but generally more penetrating because their wave-lengths are shorter.

There are more than forty naturally occurring radioactive elements which disintegrate in the manner described, and these may be grouped into three main families: (1) uranium-radium, (2) thorium, (3) actinium. In addition, more than 800 radioactive isotopes of stable atoms of the non-active elements have been produced in atomic piles and by cyclotrons during recent years, and some of these have found useful application in geophysical work. Such isotopes have the same chemical properties as the corresponding inactive atoms, but differ in mass number, owing to the increase or decrease of the number of neutrons in the atomic nuclei of the elements concerned. Ordinary cobalt, for example, consists of atoms of mass number 59, but when a piece of this metal is placed in an atomic pile and subjected to intense neutron bombardment, many of the atoms pick up an extra neutron in their nuclei, forming the cobalt isotope of mass number 60, which is radioactive. It emits a γ-ray and has a half-life of 5·3 years.

The nucleus of the hydrogen atom of mass number 1 is called a *proton*, and has a positive charge equal to the negative charge on an electron. The neutron is a particle having no charge and a mass only slightly greater than the proton. Protons and neutrons form the nuclei, or central citadel, of all atoms, isotopes of any element all having the same number of protons but different number of neutrons, which explains the different mass numbers which the isotopes of any one element possess.

The heaviest element, uranium, is essentially a mixture of two isotopes, the amount of a third one being negligible. One isotope has a mass number 238, the other, which forms one part in 140 of natural uranium, has a mass number 235. The isotopes are indistinguishable chemically, but the lighter one is of special interest as being the only element found in nature which is fissionable and consequently of great importance as a nuclear fuel for atomic piles. Uranium after several changes gives rise to an important series of radioactive elements:

	Half-life	Radiation	Transformation constant λ (sec.$^{-1}$)
Uranium (238)	$4\cdot55 \times 10^9$ years	α	$4\cdot9 \ \times 10^{-18}$
Uranium (234)	$2\cdot69 \times 10^5$ years	α	$2 \ \ \times 10^{-14}$
Ionium (230)	$8\cdot22 \times 10^4$ years	α	$2\cdot9 \ \times 10^{-13}$
Radium	1590 years	α	$1\cdot37 \times 10^{-11}$
Radon	3·825 days	α	$2\cdot09 \times 10^{-6}$
Radium A	3·05 minutes	α	$3\cdot79 \times 10^{-3}$
Radium B	26·8 minutes	β and γ	$4\cdot31 \times 10^{-4}$
Radium C	19·7 minutes	α, β and γ	$5\cdot86 \times 10^{-4}$
Radium D	22 years	β and γ	$1\cdot00 \times 10^{-9}$
Radium E	5·0 days	β	$1\cdot60 \times 10^{-6}$
Radium F	136·3 days	α	$5\cdot89 \times 10^{-8}$

Radium F is polonium and its decay product is an isotope of lead.

Radon gas is particularly serviceable in prospecting because it breaks up slowly, so that half the atoms in a given quantity disintegrate in 3·8 days, whereas the corresponding inert gases from the other two radioactive families, namely, thoron and actinon, have half-periods of 54·5 and 3·92 sec. respectively. The amount of radon in a given sample of either water or oil may be in suspension like carbon dioxide in soda-water, in which case it will all disintegrate in less than a month; or it may be due to radium salts in solution, in which case a supply of radon will be continually replenished, at least for thousands of years, by the disintegration of its parent radium.

UNITS

There was often some confusion in the record of radioactive measurements, owing to the diversity of units formerly employed. The Eman, Mache and electrostatic units are no longer used. On account of their high activity, radium and its product radon were used as a basis for measurement of activity, but with the production of hundreds of new radio-isotopes of the elements in atomic piles and by accelerators it became desirable to have an established unit for measuring the activity of any radioactive substance. The unit now accepted is the *curie*, named in honour of Madame Curie, and is the activity corresponding to $3·7 \times 10^{10}$ disintegrations per second. The millicurie (mc.) and microcurie (μc.) are used to designate a thousandth and millionth of a curie respectively. Another unit sometimes used, but not officially adopted, is the *rutherford*, corresponding to an activity equivalent to 10^6 disintegrations per second. The *specific activity* of any substance is the number of curies (or millicuries) per gramme of material.

Radium contents in solids, such as rocks, should be expressed in grammes of the element radium per gramme of rock measured.

Radium contents in liquids are conveniently expressed in grammes of the element radium per litre of the liquid examined, but some might prefer grammes per cubic centimetre of the liquid.

In the case of radon, its activity is measured in curies. The amount of radon in equilibrium with 1 g. of the element radium will be approximately 1 curie. Such equilibrium may be regarded as established in about a month, since the radon in a sealed flask will, if initially zero, reach 50 % value in rather less than 4 days, and 75 % in less than 8 days. The exact value can be calculated for any stated time, or it may be taken from a table given as an appendix in Rutherford (1915).

Radium appears to have been distributed with singular uniformity throughout primary rocks and to have been subsequently redistributed among secondary rocks, the amount being of the order of one part of radium to a million million parts of rock by weight. Thus we have the following:

Name of rock or mineral	Mass of radium in g. per g. of rock ($\times 10^{-12}$)
Granite	4·78–0·68
Hornblende	1·2
Basalt	0·52
Olivine	0·33
Marble	1·93
Quartzite	5·0
Sandstones	2–4
Slates	3–8
Dolomites	8
Chalk	0·39
Canadian rocks	4·3–0·16

The amount of radium actually present in solution in various waters may be gauged from the following:

Source of water	Mass of radium in g. per c.c. ($\times 10^{-14}$)
Saratoga, N.Y.	1–10
Bath, England	14
Carlsbad, Czechoslovakia	4–10
Atlantic Ocean	3·4–1·4
Indian Ocean	0·7
St Lawrence River	0·025

The amount of radon in suspension as a gas is indicated by the following list which gives larger values obtained from sources of exceptional strength:

Source of water	10^{-12} curies/litre
Valdemorillo, Spain	218,000
Aix-les-Bains, France	20,500
Carlsbad, Czechoslovakia	11,500
Manitou, Colorado	30,500
Hot Springs, Arkansas	9,000
Bow Park Well, Canada	800

Nearly all well and river waters contain a measurable small quantity of radon, but in the case of most waters, even highly charged, the amount of radon present is such that a pint will not contain a medical dose, and it is necessary perhaps to drink a barrel or two at a time to achieve any result. Hence we are subject on the one hand to the vagaries of quacks, and on the other to our ignorance of the benefit of small doses of so powerful a radioactive substance as radon, together with its three swiftly changing descendants, radium A, B and C.

For measurement of the quantity of γ-radiation the *röntgen* (r.) is the standard unit. It is defined as the quantity of radiation which will produce in a cubic centimetre of air at 0° C and 76 cm. of mercury pressure, one electrostatic unit of either positive or negative charge. Alternately, it is the quantity of radiation which will produce $2 \cdot 083 \times 10^9$ pairs of ions per cubic centimetre of air under standard conditions of temperature and pressure. The milliröntgen (mr.) and microröntgen (μr.) are the thousandth and millionth part of this unit respectively and are the units most commonly used in geophysical work. According to present medical practice, the maximum permissible dosage which a human can receive without any apparent harmful effects is 300 mr. per week.

When Geiger or scintillation counters are used in geophysical exploration, it is the γ-rays which are detected and the intensities are recorded either in so many counts per minute or in mr. per hour. Figs. 145 and 146 show the results of surveys made by Brownell (1950*b*) over pitchblend deposits in Saskatchewan using the scintillometer showing lines of equal radioactivity, called 'isorads'. The authors are indebted to Professor Brownell for supplying these examples.

The α- and β-rays from radioactive atoms are ejected with various maximum velocities, depending upon the particular type of nucleus from which they are emitted, and the energy associated with such particles is expressed in millions of electron-volts, designated MeV. One MeV. is the energy an electron would acquire if it were accelerated by a difference of potential of one million volts, equivalent to $1 \cdot 6$ millionths of an erg or $3 \cdot 8 \times 10^{-14}$

calorie. The energy of the γ-rays is similarly expressed in MeV. units. Charts of all stable and radioactive nuclei of known atoms and their isotopes, such as that published by the Research Laboratory of the General Electric Company, give the energies of the various radiations emitted in terms of MeV. units. Thus

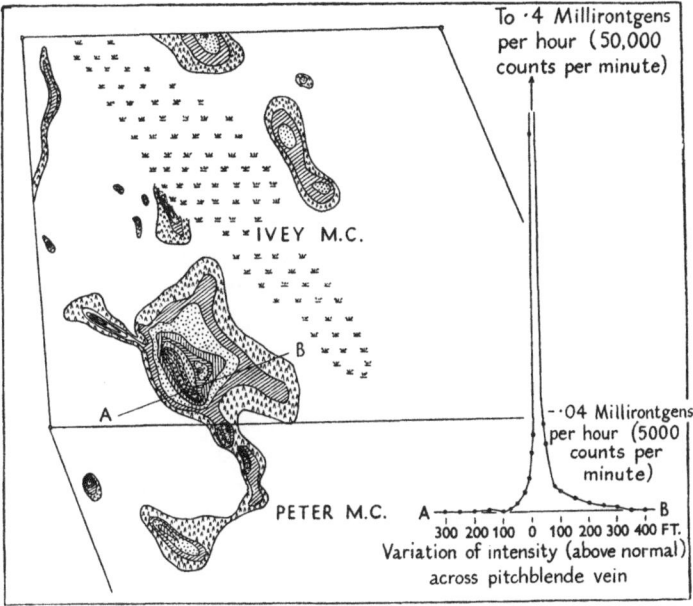

Fig. 145. Radiation survey (Goldfields area, Saskatchewan).

the α-rays from radon have a maximum energy of 5·49 MeV.; the β-rays from radium B have 0·7 MeV., and its γ-rays range from 0·053 to 0·35 MeV. On the other hand, the pile-produced radio-isotope of cobalt of mass number 60 emits γ-rays having an energy of 1·33 MeV. The shorter the half-life of a radioactive atom, the more energetic in general are the particles it emits.

DRILL-HOLE EXPLORATION

Today the use of radiations from radioactive atoms has found direct application in drill-hole explorations as well as in the location of oil and gas deposits (Russell, 1941, 1951; Pringle et al. 1952;

Lundberg and Isford, 1953). The earliest radioactive exploration was probably that developed by Ambronn and others by the second method quoted above, which still seems capable of varied applications.

A sharp pointed probe is driven into the ground, and about the middle there is a thicker portion which tightly plugs the hole. Below the plug are holes leading up the hollow shaft to airtight rubber tubes connected to a hand pump by which the air is drawn out of the ground and driven into the ionization chamber of an electroscope or Geiger counter. The air thus pumped from the ground will be found in general to contain a measurable quantity of radon, derived from the radium in the earth. The amount of radon measured by the Geiger or scintillation counter will be found to vary from place to place, depending on the local content of radium, or the radon conveyed to the surface by water, or by underground air circulation, and the amount of radon may be enhanced by the presence of faults and other geological unconformities.

Proceeding along a selected line, measurements may be made every 100 ft. or less, and any marked divergence from the average reading of the counter or electroscope, will, on repetition, require an explanation relating to underground structure. Similar procedure using a sensitive scintillation counter will indicate irregularities due to radon seeping to the surface.

Such work may also be conducted down mines and in tunnels, as well as on the surface, in connexion with exploration for water, brine, ore, oil, natural gas, or radioactive minerals. In drill-hole surveys, which will be discussed in the following chapter, radioactive sources are used as depth markers and neutrons are used to gather information regarding the nature of adjacent formations useful in locating water, oil and gas deposits. Neutrons are particles possessing no charge with a mass slightly greater than the proton, which form a part of all nuclei with one exception of the lightest nucleus, namely that of hydrogen of mass number 1, which contains only a proton. Neutrons used in drill-hole surveying are produced by bombarding beryllium with energetic α-particles, the beryllium nucleus absorbing the

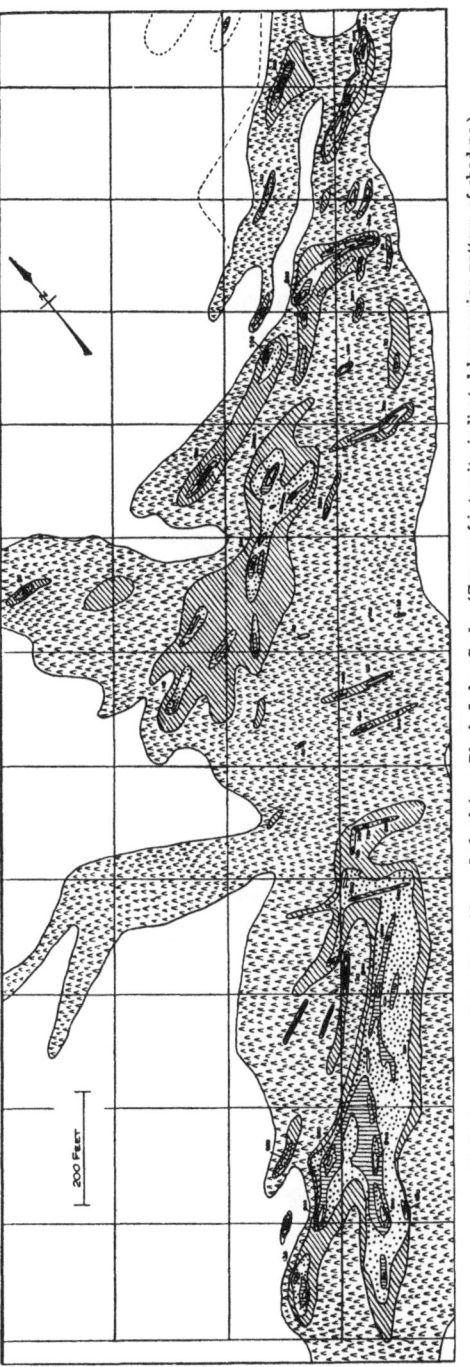

Radiation survey of the Nisto Mines Ltd. claims, Black Lake, Sask. (Zones of intensity indicated by varying pattern of shading.)

Fig. 146. Radiation survey (Black Lake, Saskatchewan).

α-particle and then disintegrating to form a carbon nucleus by ejecting a neutron with high velocity. This reaction may be represented by an equation similar to that used by chemists but actually being a nuclear transformation:

$$^9_4\text{Be} + ^4_2\text{He} \rightarrow ^{12}_6\text{C} + ^1_0n + \text{energy}.$$

The indices represent the mass numbers of the nuclei (sum of the number of protons and neutrons of the atom) and the subscripts the atomic numbers of the atoms (number of protons in the nucleus). Such neutron sources are made by mixing powdered beryllium with radium, polonium or actinium in an airtight capsule. The neutrons, being uncharged particles, can penetrate through considerable thicknesses of matter.

Neutrons released by the above reaction have velocities of several miles per second. By collision with other nuclei their velocities are gradually reduced, particularly if the impact is upon nuclei of about their own mass, such as hydrogen or hydrogenous material. After many collisions they finally reach a velocity which is in equilibrium with the energy which atoms and molecules have at ordinary temperatures and are then spoken of as 'thermal neutrons'. Under such conditions they have velocities of about 1·7 miles per sec. and are more readily absorbed or captured by other nuclei. This capture process results in the formation of an isotope of the same or some other atom, with the release of radiation. Neutrons do not produce any appreciable ionization in a gas so special detectors are needed to record them. This is usually accomplished by filling Geiger-Müller tubes with boron fluoride gas or lining the inner walls of the tube with boron carbide. The neutrons react with the boron, transforming it into lithium and releasing an energetic α-particle which produces ionization. In atomic piles (nuclear reactors), neutrons in large numbers are released and any stable atom exposed to such high flux density of neutrons becomes by capture transformed into an unstable atom of either the same kind or some other type, which are known as radioactive isotopes. Some of these new isotopes have found application in geophysical surveys which will be described in following chapters.

PROBLEMS

1. The half-life of radon is 3·825 days. What is its *average* life?

2. The *average* life of radium C is 28·1 min. Calculate its transformation constant λ.

3. The half-life of the radio-isotope of sodium of mass 24 is 14·9 hr. Calculate its transformation constant λ.

4. One milligramme of radon is placed in a radon-beryllium capsule. What mass of radon is left at the end of 10 days?

5. If 10^{13} α-particles are emitted per second from 1 mg. of polonium, which contains $2·87 \times 10^{18}$ atoms, calculate its half-life.

6. The following readings were made with a scintillometer rate-meter over a claim in the Goldfields area of Saskatchewan by Professor G. M. Brownell. The stations were 25 ft. apart and the lines 50 ft. apart. From the results of this survey, where are the best places to drill for locating pitchblende?

Lines

Station	1	1½	2	2½	3	3½	4	4½	5	5½	6	6½	7
1	29		31		31		32		30		30		30
2	30		31		31		32		32		33		32
3	33		30		32		33		35		37		31
4	30		30		31		33		38	41	40		34
5	33		30		33		38		54	51	43		34
5½	—		—		—		42		201	52	41		—
6	30		30		36	42	48	800	170	44	38		32
6½	—		—		40		88	706	51		—		—
7	29		36		43		470	800	45	41	36		33
7½	—		—		52		121		—		—		—
8	30		39	49	151	615	57	48	48		41	38	35
8½	—		45		202		—		61		49		—
9	31	40	48	61	87		62	101	191		63		38
9½	—		—		68		54	607	224		49		—
10	42		44		47		45		45		42		37
10½	43		—		42		—		39		—		--
11	37		36		38		36		36		38		31
12	33		31		36		32		31		31		32
13	29		32		38		31		31		32		30
13½	—		39		39		—		—		—		--
14	31	36	44	46	47	39	32		30		30		30
14½	—		37		—		—		—		—		—
15	30		30		32		32		31		31		30

7. A sample of gas gives an activity of 2·5 μc. If this activity is due to radon, how many atoms of this element must have been present in the sample?

8. If the half-life of radium is 1600 years, what fraction of the radium atoms decay per second?

9. Scale readings were taken with a scintillometer at stations 25 ft. apart on a set of parallel lines 25 ft. apart, over a claim in Saskatchewan with a view of locating uranium ore. Part of the results of this survey, supplied by Professor G. M. Brownell, are given in the table below. Draw the 'isorads' and deduce the strike of the ore.

Station	Lines							
	1	2	3	4	5	6	7	8
1	11	11	11	12	11	12	14	15
2	12	12	11	12	12	15	15	17
3	14	14	12	14	14	17	17	17
4	17	15	15	15	15	24	17	17
5	19	18	18	18	17	34	22	16
5½	—	30	21	24	22	—	—	—
6	16	25	24	23	28	47	27	18
7	17	23	25	28	34	47	30	24
8	16	17	30	36	46	66	47	38
8½	—	—	—	70	200	—	66	60
9	15	22	36	1200	200	310	260	40
10	14	18	22	50	35	35	46	25
11	13	15	18	22	26	26	33	17
12	12	12	15	18	17	17	14	13

CHAPTER VIII

WELL-LOGGING METHODS

INTRODUCTION

Following the discovery of a favourable prospect, it has been the general practice for many years to determine the extent, dip and amount of mineral by diamond drilling and examination of the core material. Drilling for minerals and oil gives information about the lithology of the formations as well as an estimate of the metal, and in some cases the oil, content of the rock through which the drill passes. But the examination of the cores and mud or sludge only reveals what was in the path of the drill and in many cases a single hole has failed to cut the ore-body or strike oil. Dry and blank holes are common occurrences. It is consequently desirable to develop means of surveying the region adjacent to any bore-hole at points throughout its depth with a view to detecting ore or oil in proximity to the hole, and also to enable the geologist or engineer to arrive at a more reliable decision where the next hole is to be drilled or at what point in the bore-hole the porosity and other conditions are most favourable for obtaining oil. The Schlumberger brothers were the first to devise successfully the means of making such investigations, and in recent years many methods have been developed for surveying or 'logging' such bore-holes by geophysical methods. It has already been mentioned on p. 142 how the region round a drill-hole or between two such holes may be surveyed electrically. It is now generally accepted practice to log every hole as it is drilled by one or more of the geophysical methods to be briefly outlined in this chapter.

It is convenient to group the different geophysical well-logging methods under the following headings:

Resistivity. Neutron.
Self or spontaneous potential. Thermal.
Radioactivity. Acoustic.

RESISTIVITY METHODS

The exploration of the region round a bore-hole is carried out by using lead or copper electrodes exposed to the sides of the uncased drill-hole or to the mud and water in the hole.

These electrodes are connected to three (or in some cases four) insulated copper wires near the centre of a cable which has rubber covering and may for the greater part of its length have armour protection. Two of these wires lead to the potentiometer and the other one to one terminal of the generator or battery, the other terminal of which is usually connected to the ground at the surface. Any of the methods listed on p. 108 and described in Chapter III may be employed, and the region investigated will have a radius estimated to be roughly equal to the interval between the electrodes, though the penetration may be somewhat less. There is sometimes difficulty in lowering the cable, and if the armour is spirally wound it is apt to kink. For this reason, readings are usually made as the electrode system is withdrawn from the hole, rather than when lowered.

In one arrangement (Fig. 147) one current electrode C_1 is above ground, while P_1, P_2 and C_2 are at the bottom of the cable. P_1, P_2 are the pick-up electrodes connected to the potentiometer and C_2 the current electrode. The spacing of the electrodes will depend upon whether it is desirable to detect narrow strata on the walls of the hole or survey farther back from the surface of the bore-hole. If the three electrodes are equally spaced, their distance apart may vary from a few feet to as much as 40 ft. Often the potential electrodes are more closely spaced than the distance from the current electrode C_2 to the potential electrode P_2. Since the mud on the walls may tend to shield the more remote regions, it is sometimes useful to have a fourth electrode or to change the spacing so as to obtain the mean resistivities of the regions near the wall and more remote. Exploration may be carried out in stages as the cable is unwound and lowered into the hole or withdrawn. If conducting ore-bodies are in the vicinity of any part of the bore-hole the mean resistivity will decrease; that is, if the current between C_1, C_2 is kept constant, the potentiometer reading will then indicate less potential difference between P_1 and P_2.

If the three electrodes pass in the neighbourhood of oil sands, which are bad conductors, then the resistivity will increase, and in such a case a graph may be drawn indicating mean resistivity with depth (Fig. 147). With modern equipment a graph is

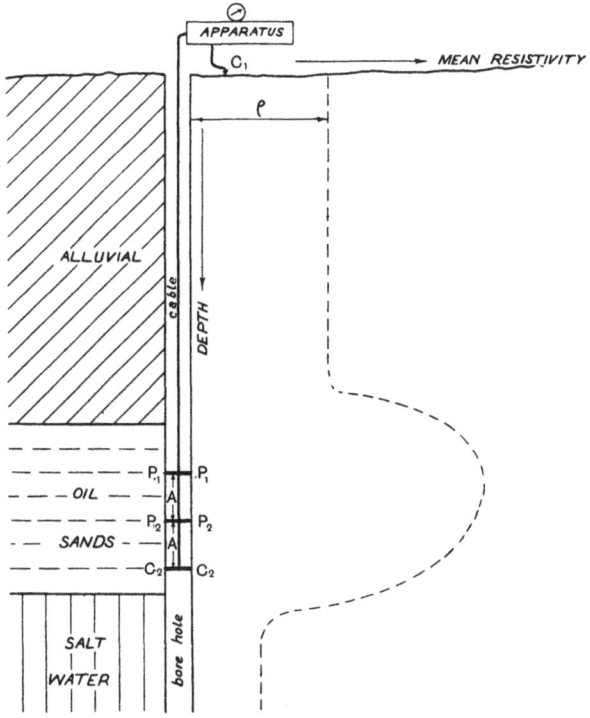

Fig. 147. Resistivity exploration down a bore-hole, indicating the nature of the region around it.

recorded on a moving paper strip, indicating directly the variation in mean resistivity with depth. In this simplified drawing, which is a rough approximation to an actual case, the alluvial beds had a fairly uniform resistivity, while the oil sands showed a much higher value, but the salt water beneath was a better conductor than either.

The mean or apparent resistivity measured by the simple system described above may not represent the true resistivity of the adjacent strata owing to the infiltration of water which becomes mixed with the mud in the hole. In the case of oil bore-holes, salt water and oil seeping into the hole may obscure the true resistivity of the region measured. To overcome such difficulties various modifications have been introduced. Among these is 'laterolog' (Doll, 1951) in which seven electrodes are used, a central current electrode, two sets of potential electrodes placed symmetrically above and below this central one and two additional current electrodes at equal distances from the central one. The seven electrodes are in a vertical line and the current introduced through the two outer electrodes is adjusted to keep the potential difference between the two pairs of potential electrodes zero. For a steady current through the central electrode, the variation of potential of any one of the potential electrodes is proportional to the resistivity of the bed adjacent to the central bed.

A method of applying the change in mutual induction between a coil in which an alternating current from an oscillator is passed and a second detecting coil held on a sonde a short distance above it, has been used to measure changes in conductivity of strata as the apparatus is lowered down a bore-hole. The exciting coil and detecting one are held at a fixed distance apart on the sonde and are connected by wires in the cable to oscillator and amplifier on the surface. The mutual inductance between the coils is dependent upon the conductivity of the medium adjacent and between the coils. The method has been used in oil-well logging and according to Doll (1949) gives a better record of the actual resistivity of the bed than the conventional electrical resistivity arrangement.

In such drill-hole logging, the true resistivity of the adjacent bed, which is the result desired, is related to the apparent resistivity measured, but this relation will depend upon a number of factors, such as the electrode separation compared to the thickness of the bed. The conductivity of the mud in the hole as compared to that of the bed affects the results deduced from the field readings. The Schlumberger Well Surveying Corporation have issued a number of handbooks, discussing such variations and containing graphs of departure curves for typical conditions.

THE SPONTANEOUS POTENTIAL METHOD

Since the first application of resistivity methods to drill-hole surveys were made by C. and M. Schlumberger in France in 1929, they have not only improved the technique of this process, but they made highly important additions to it (*The Science of Petroleum*, 1938, pp. 351–63). If a non-polarizable electrode is lowered down an uncased bore-hole, and connected by an insulated wire to a potentiometer, the other side of which is suitably earthed, then it will be. found that a varying potential may be indicated and recorded. This is known as the spontaneous or self-potential method and is usually taken simultaneously with the resistivity survey, using one of the electrodes for the self-potential electrode.

The conductivity of a given specimen of rock is largely dependent upon its porosity, upon the quantity of water and the nature and amounts of salts therein contained. This fact may readily be tested by measuring the resistance of a piece of limestone, first dry, then moistened with distilled water, and finally soaked in salt water. It may be admitted that every rock has not only a certain *porosity* but also a degree of *permeability* to water, and that water may have various conductivities according to its purity or the nature of its solutes. While it is easy thus to understand the resistivity of a rock specimen it is more difficult to account for the *spontaneous polarization* or self-potential observed in bore-holes, which, at the junction of, say, shale and water sands, may amount to about 50 mV.

The causes of this self-potential are said to be two in number:

(1) *Electro-infiltration* due to a difference of pressure causing the mud and sludge of the bore-hole to enter into the rock, in which case the electric current passes also in that direction.

(2) *Electro-osmosis* when the potential difference is proportional to the logarithm of the ratio of the resistivities, so that

there may be a self-potential between drilling mud and the salt water of the surrounding rocks. This is said to be less effective than (1).

The procedure in the field is to have a motor-car with a powerful winch in order to lower and raise a cable in the uncased bore-hole. The cable has a heavy plummet (50–600 lb.) at the lower end, and its breaking strain should be 2–8 tons. The cable should have a core of three or four wires exceedingly well insulated, for it may be lowered in salty water to depths ranging up to 10,000 ft.

The same cable is used for measurements of the resistivity and of the self-potential concurrently and the values of these are continuously recorded when the cable is lowered, and again when it is raised, at speeds of about 6000 ft. an hour. This method has been tried in France, Venezuela (1929), U.S.S.R., Rumania (1931), and the United States. The method is extensively used, and by using potential electrodes only a few inches apart to measure 'micrologs' of resistivity, the variations in potential when passing even thin layers may be related to the resistivities. The self-potential is probably due to chemical action—the more shaly the sand, the less the potential. Watery sand shows a higher negative potential than oily sand, while the corresponding resistivity for the former is low compared with that for the latter. The results are useful for correlating the lithology in neighbouring holes in a given district. Industrial currents and earth currents may in some cases be troublesome. The results of an actual survey, shown in Fig. 148, have been included here by the kindness of Mr Leonardon.

RADIOACTIVE METHODS

As mentioned in Chapter VII, different rocks contain varying amounts of radioactive materials which emit radiations detectable by suitable instruments. Since both α- and β-rays are readily absorbed by matter, it is the γ-rays which are effective in producing indications of the presence of radioactive material. In drill-hole surveys, γ-ray detectors are lowered down the

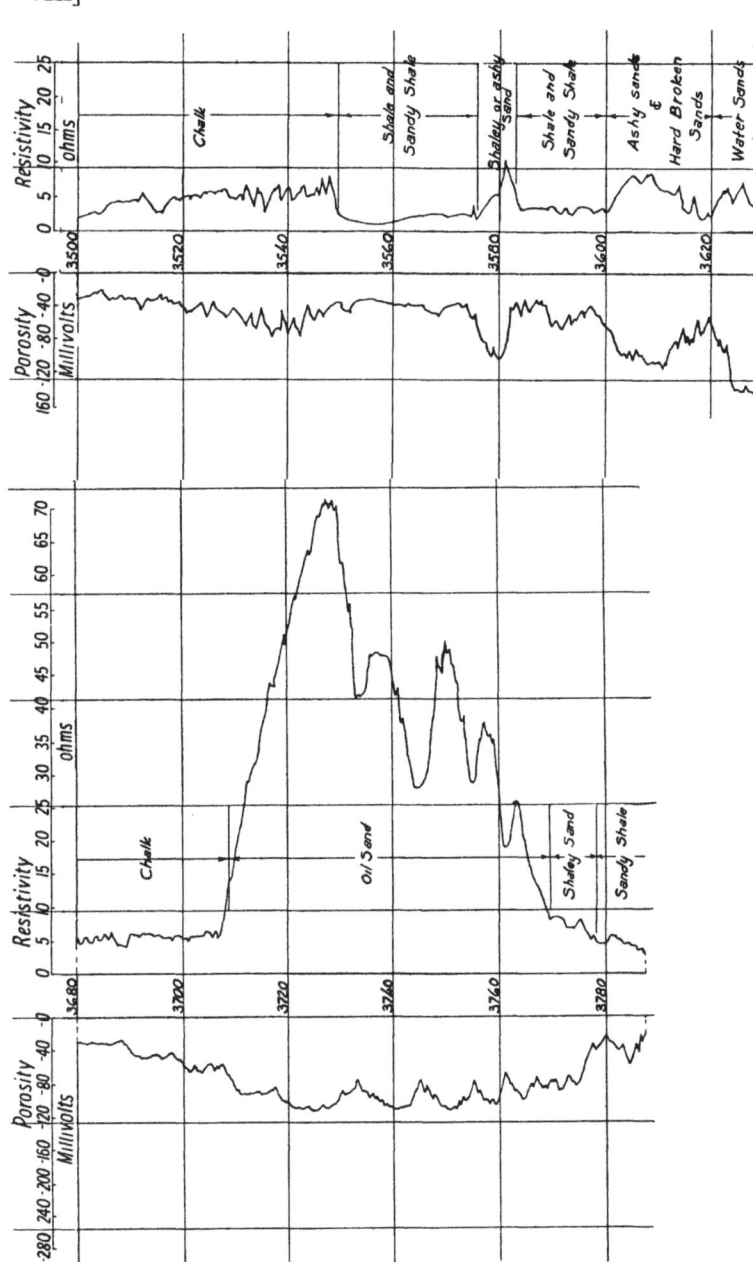

Fig. 148. Electric coring, East Texas. Resistivity and porosity.

bore-hole by a cable through which insulated copper wires pass
to recording instruments on the surface, a complete survey of
the hole being thus obtained on a moving sheet of paper showing
the variations of the radioactivity with depth. When the variation in intensity of the γ-rays has been correlated with the
formations down one drill-hole, the log of neighbouring holes,
using the same apparatus, will indicate similar structures. One
advantage of this method of surveying is that it may be used
in drill-holes in which the casing has been inserted. Furthermore,
there are certain similarities between the graphs obtained by
electrical resistivity methods and those found in the same hole
with a γ-ray detector. The first types of such detectors used an
ionization gas-filled Geiger counter, the batteries to operate it
being carried in the sonde apparatus, and the variations in
current produced by the γ-rays were being amplified by the
equipment on the surface. To obtain sufficient detail, the gas-
filled tubes were 18 to 36 inches in length and the motion of the
sonde was about 5 ft. per minute, owing to the quenching time
required by the gas. The position of the detector is recorded by
measuring the amount of cable let down the drill-hole.

A definite improvement in γ-ray surveying has been achieved
by using crystal detectors, such as the scintillometer, since there
is no need of quenching and small crystals may be used, resulting
in better definition of the layers. Moreover, the intensity of the
response is proportional to the energy of the γ-ray, affording
a more accurate and sensitive record of the activity. The
McCullough Tool Company of Houston have developed this
instrument, which has a survey speed of 200 ft. per minute.
A comparison between the gas-filled Geiger counter response
and that obtained in the same drill-hole with the scintillometer
is shown in Fig. 149. It will be noted that shales have high
radioactivity, while limestone, sandstone, dolomites and salt
are less radioactive. The scintillometer gives an accurate definition of the bottom A and top B of the stratum, while the Geiger
detector, owing to its greater length indicates the position as
C and D. The crystal, photomultiplier tube and sub-surface
amplifier are all contained in the sonde. Thallium-activated

sodium iodide crystals are used and the equipment is calibrated by using radioactive sources of known strength. A recent article (Dr Giovanni *et al.* 1953) describes a similar type of scintillation unit for drill-hole surveying which has been used with success at rates of 20–30 ft. per minute.

An additional advantage of the γ-ray method is that it may be applied in a cased drill-hole, since the γ-rays will penetrate through the casing. This type of detector has been used to

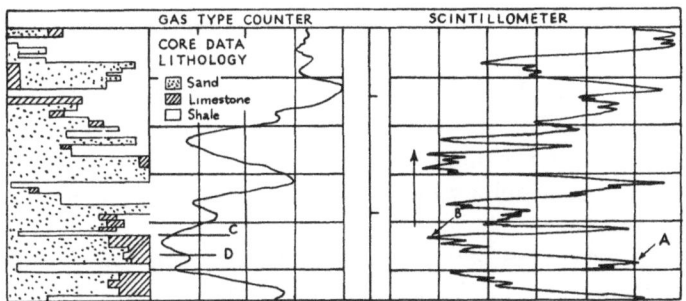

Fig. 149. Comparison of records with gas type of detector and scintillometer.

determine the height to which the cement used behind the casing has risen (Jackson and Campbell, 1945). The hole is first surveyed before the cement is introduced and some radioactive isotope, such as iodine 131, is mixed with the cement before it is inserted. A second survey is made after the cement has been introduced and the position of it can be deduced from the extra activity observed with the detector.

NEUTRON METHODS

The use of neutrons for drill-hole surveying appears to have been first suggested and tested by Pontecorvo (1941). A neutron source produced by a mixture of radium and beryllium (see p. 314), is placed in a solid lead block to absorb the γ-radiation which comes from the radium. The neutron source is at the lower end of a plummet, above which is the γ-ray detector and usually an amplifier and batteries, the impulses being led to the surface through insulated wires in the cable. These impulses are further amplified and recorded on moving strips of paper similar to the system used in radioactive surveying methods. Any direct effect of the γ-rays on the Geiger counter will be constant, if arising from the radium source. Neutrons are slowed down by collision with hydrogen nuclei to thermal velocities, and many will be captured by the hydrogen, giving rise to γ-radiation which will be detected by the Geiger counter. The gas-filled tube detector responds to γ-rays and the presence of hydrogen in the neighbourhood of the cased or uncased wall of the drill-hole can thus be detected. Thus the neutron methods indicates the presence of oil or water, both of which contain hydrogen as a constituent. The similarity of surveys made down adjacent drill-holes with the neutron and resistivity methods and the relation to the lithology may be inferred from the results shown in Fig. 150.

Certain advances have been made in instrumentation during the past few years in the use of neutron well-logging. Polonium can be substituted for radium in the beryllium mixture, thus avoiding any γ-radiation directly from the neutrons source, and reducing the shielding. Furthermore, the detector may be filled with boron trifluoride gas and its inner walls lined with boron, which will respond to neutrons and give little response to γ-radiation. Since hydrogen scatters fast neutrons and slows them down to velocities at which they are more readily captured, the number which arrive at the neutron detector from the source, which is well shielded from the detector, will be less when hydrogen is present than when it is absent. A decrease in response is an indication of the presence of hydrogen in the formation or stratum, and it may be assumed that this is due to the presence of either water, oil or gas. Consequently, surveys made in the same hole using a γ-ray detector and a neutron one will normally result in records which resemble mirror images of each other. Limestones, sandstones or dolomite, if non-porous, give a large response on neutron detectors while these formations, containing little radioactive matter, give low γ-radiation. If the crystal type of detector, such as the scintillometer, be

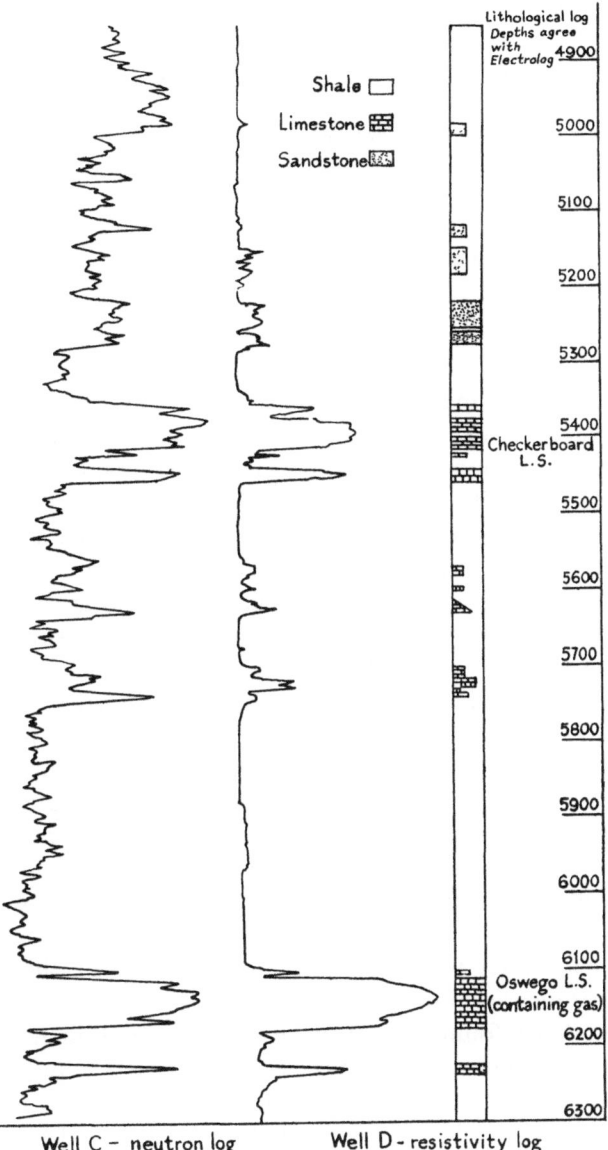

Fig. 150. Results of neutron and resistivity survey
(Oklahoma City).

used, in place of the boron-filled Geiger tube, the response to fast neutrons is measurable and the rate of survey may be greatly increased. Neutron surveys are particularly useful for gaining knowledge of the porosity of limestone and dolomite layers, since if porous, they will usually contain water, gas or oil, all of which have an abundance of hydrogen in their chemical composition.

GEOTHERMAL METHODS

The investigation of the temperature gradients in various regions may give some indication of the geological history as well as structure of the ground beneath. Rapid local variations in temperature may be brought about by several causes: (a) chemical action, (b) presence of radioactive materials such as radium, uranium, thorium or their products, (c) differences in the conductivity of the rocks, (d) geological transformations brought about by pressure, friction or the presence of water.

The procedure is usually to place delicate thermometers down bore-holes and provide suitable means of adjusting the thermometers so that they will make good contact with the surrounding rock and so attain the actual temperature of the rock. It has been found that in general the temperature rises 1° C. for an increase in depth of from 30 to 45 ft. When disturbing influences are present, or when the conductivity of the rocks changes, these values will be altered considerably. By taking a number of readings at various depths and also at different points over an area, the isothermal lines may be drawn. A map constructed from such temperature measurements, either in plane or profile, will give indications of any abnormal variations. The interpretation may then be made from a knowledge of the general geological character of the region.

It has been found that the basic rocks contain less radioactive material than the acid rocks. Holmes has found that the heat produced per c.c. per second in the acid rocks is nearly four times as much as that produced in basic rocks from the uranium and thorium content alone. The rate of leakage of this heat to the surface will depend upon the thermal conductivity of the rock and also the temperature gradient. The following table gives typical values for the thermal conductivity of different rocks, measured in calories per sq.cm. per °C. gradient.

Material	Thermal conductivity × 10³
Sand, carbon, quartz sand	0·3–0·8
Granite, gneiss, porphyry	4·0–5·5
Sandstone, quartz, rock salt	6·0–6·6
Magnetite	30

According to Jeffreys (1952) the thermal conductivity of the ground at the surface is 0·006 and the temperature gradient 0·00032° C. per cm., which gives the leakage of heat from the earth as $1·9 \times 10^{-6}$ calories per sq.cm. per sec. This loss would just be balanced by a depth of 19 km. of acid rock by its radio-active content.

Many interesting results were obtained during the construction of the two great tunnels, the St Gothard and the Simplon. In the case of the latter, the maximum temperature found was 55·4° C. at a point 1700m. below the surface of the mountain. From a study of rock temperature in the St Gothard, a prediction of 47° C. for that in the Simplon was made by Stapff, while others made much lower estimates. In the St Gothard, when passing through gneisses and schists, the temperature gradient averaged 1° per 46·6 metres, while when traversing granite at the north end, the gradient became 1° per 20·9m. From an analysis of the rock specimens (Joly, 1909), it was found that the granite contained unusually high proportions of radioactive material which would account in part for the higher temperatures recorded. The variation in temperature due to radioactive causes was definitely demonstrated in these two cases.

Temperature gradients have been measured in a number of mines in Ontario and Quebec by Misener and Thompson (1950). At the Malartic Gold fields, Delnite and Kerr-Addison Gold mines the gradient varies from 170 to 197 ft. per ° F., while in the New Calumet, Lake Shore and Asbestos mines, the gradient is 116 to 136 ft. per ° F. A demonstration oil well drilled in Toronto gave a gradient of 1°F. per 114 ft. According to Wilson

(1949) the group with high gradients lie in the Grenville province and those with the lowest within the Superior province.

Geothermal methods are not as yet used extensively in searching for ore but, nevertheless, they have a useful and definite place in applied geophysics.

ACOUSTIC METHODS

It has already been stated (p. 251) that a shock-wave travels with different velocities in different formations. A method of acoustic logging in drill-holes has been described by Summers and Broding (1952), in which this fact is applied in a unique manner. They use a pulse of high-frequency acoustic waves produced by an electronic generator and arranged by a system of gating to cut off the pulse and detect the reception by a suitable device attached below the source on the same plummet. The receiver is 5·3 ft. below the generator and is separated by rubber insulation to prevent direct reception through the plummet. The wave or impulse travels through the formation and is received a fraction of a millisecond later, the time interval being registered on an oscillograph or Esterline-Angus recorder. Results described by the authors appear to give the correct velocities in the various formations as the instrument is lowered or raised in the drill-hole at a rate of approximately 100 ft. per min. The velocity logs indicate the lithology in much the same manner as resistivity logs, according to these preliminary experiments. The method is not applicable to cased wells.

CHAPTER IX

OTHER METHODS AND APPLICATIONS

In previous chapters simple descriptions of the principles involved in the various more direct methods of applied geophysics have been given, but there are other helpful means of gaining information about underground structures which are of interest. Furthermore, some of the methods described have application in locating pipes and other metal structures underground. Simple lecture and laboratory demonstrations illustrating the principles of geophysical methods which have been used by the authors may be suggestive. Such general information is discussed in this chapter.

PALAEONTOLOGY

The application of micro-palaeontology to the petroleum industry has become of considerable importance in recent years. In America the micro-palaeontologist commenced his work on petroleum geology about forty years ago, and so rapid has the development of this branch of the science become, that there are at least 300 micro-palaeontologists investigating small fossils with the object of identifying stratigraphy and structure. In correlating structures the presence of Foraminifera furnishes often the only criterion. Such fossils have been studied until as many as 3000 different species are classified.

Micro-palaeontology may be included as a branch of sedimentary petrography and as such is intimately associated with other microscopic investigations. The work is particularly useful in oil fields where one drill-hole has been made and the rock from the different levels investigated under the microscope. From such a study, the different geological layers and the stratigraphy are identified with the presence of certain forms of fossils. Thus, the results from one drill-hole will generally serve as a key to the surrounding neighbourhood. When other holes are bored, the positions of the various layers are correlated by the presence of the same type of fossils, and thus an idea of the expectation of petroleum may be ascertained as the hole or well is in process of drilling. The results of such correlation work have been so successful that in 1927 a Journal of Palaeontology was founded, and in this journal many new specimens of fossils have been described, classified and named. Naturally much of this work still remains in the hands of the different oil companies, but, as with other methods and aids for locating oil and ore, the process is gradually being disclosed to the public.

RADIOACTIVE DEPTH MARKERS

The problem of accurate depth measurement in bore-holes is an important factor in perforating the walls for production or in cementing. As the hole becomes deeper, temperature changes and the tension on the cable make accurate measurements difficult. Geophysical logging may indicate the proper zones which it is necessary later to identify. Such points can be indicated by a permanent marker using a radioactive element contained in a steel bullet which is shot into the wall from a special type of gun attached to the plummet. A fraction of a milligramme of radium in a brass capsule attached to the steel-pointed rivet, provides a source of γ-rays which can readily be identified by a γ-ray sonde detector. It is possible also to use the radioisotope of cobalt of mass number 60, which gives a very penetrating γ-ray. This particular isotope is produced in nuclear reactors and has a half-life of 5 years. Doll and Schwede (1947) have given an account of tests using such markers and point out the assistance they provide to engineers in solving the problem of depth measurement.

THE DIPMETER

A method of determining the angle and direction of the dip of strata by a modification of the usual arrangement of electrodes in a bore-hole survey has been designed by the Schlumberger Well Surveying Corporation. Three electrodes spaced 120° apart and in the same horizontal plane make contacts with the sides of the bore-hole and each measures the spontaneous potential produced at its respective point of contact. From the magnitude of the separate readings as the cable is lowered past a stratum, the dip of the layer may be inferred (Doll, 1943).

SALT DOMES

So much has been written in preceding chapters on salt domes that it seems desirable to give a detailed account of two famous domes.

The following is a summary of an article by Bevier (1925):

Damon Mound [see Fig. 151] is a conspicuous rounded elevation in north-western Brazoria County, Texas, 38 miles south-west of Houston. It is oval in shape, covers an area of 1670 acres and rises 83 ft. above the surrounding prairie. This elevation is the result of an intrusive salt plug which has raised the surface above its normal position.

The salt plug is composed of almost pure rock salt and is capped by deposits of gypsum, anhydrite and limestone. Formations surrounding the salt plug are inclined at angles of about 45° away from the salt plug. Mineralized water and gas seepage in shallow water wells, sour dirt, sulphur, and lime rock on the surface are indications of the salt dome.

Two hundred and ninety-one wells were drilled prior to the year 1924, 85 producing oil, 154 being dry, and the remainder being classed as sulphur tests. Total production for this period amounts to 5,008,870 bbls., all of which was obtained from an area of 280 acres. There are two producing areas, one on the north-eastern side, the other on the south-western side of the dome. The major portion of oil was obtained from the south-western side. Production per acre averages 19,265 bbls. Average production of producing wells is 58,927 bbls., and the average depth 2416 ft. The oil is of asphaltic base with gravities averaging about 22·5° B. Production is obtained around the sides of the dome in sands and limes of Oligocene age.

The following is a description by Kelly (1925) of one of the most remarkable mines in the world—a sulphur mine—which, over a small area of 75 acres, has produced a gross value of $150,000,000, then about thirty million sterling:

The sulphur salt dome in Louisiana [Fig. 152] is a typical Gulf Coast salt dome, but is exceptional in its small area, about 75 acres, in the very great thickness of the cap, about 1000 ft., and in the

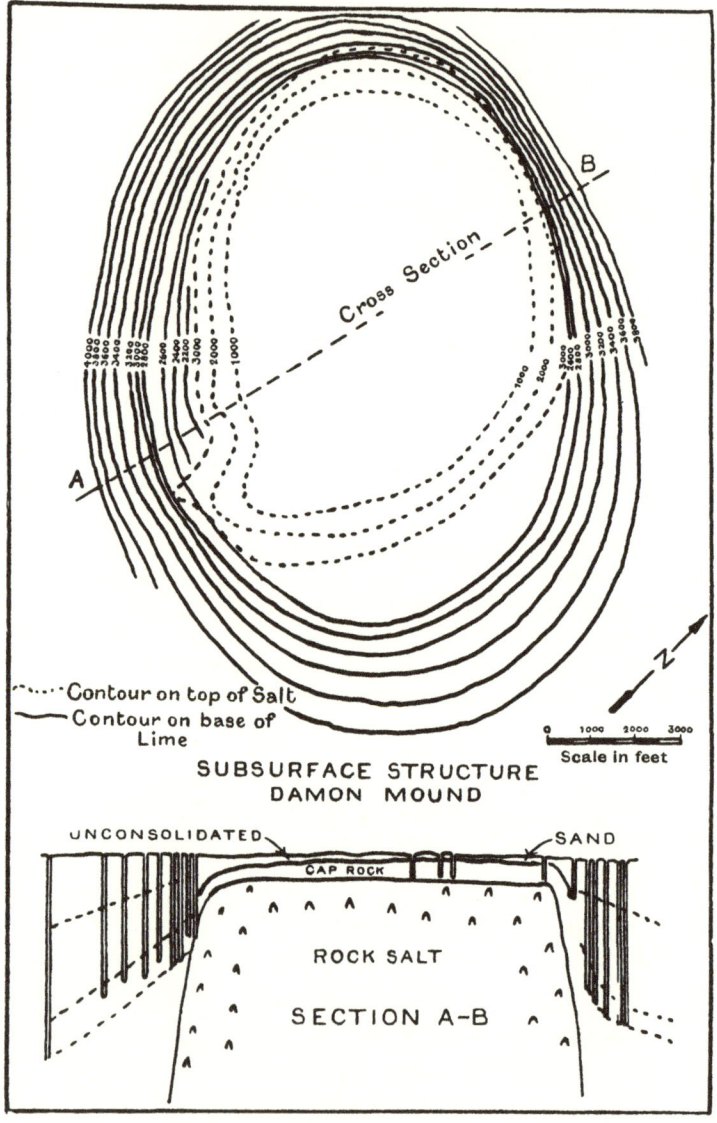

Fig. 151. Damon Mound. A salt dome shown by underground gradients, and in profile.

richness of its deposit, of native sulphur. The cap is composed of anhydrite and a mantle-like mass of 'lime' rock covering the top and flanks of the anhydrite.* The sulphur is found as a secondary mineral in the transition zone between the 'lime' rock and the anhydrite and has probably been precipitated from hydrogen sulphide or metallic sulphides in solution. After a long history of vain expensive attempts to mine the sulphur, the French process was perfected by which the sulphur is melted in place and pumped in liquid

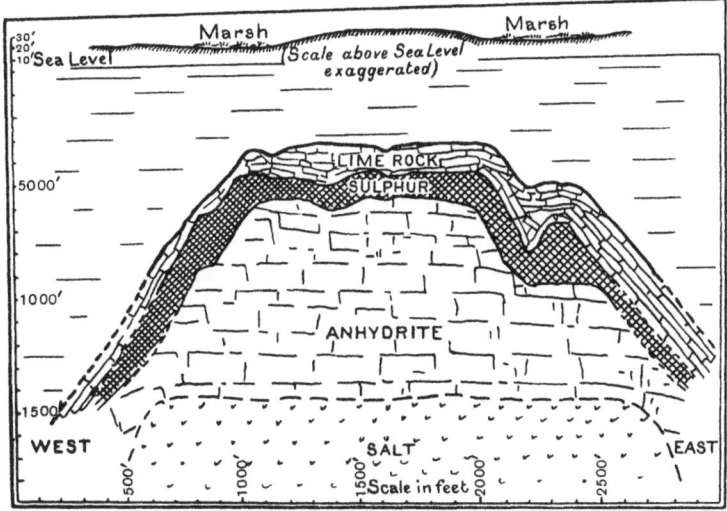

Fig. 152. Sulphur salt dome, Louisiana, shown in profile.

form to the surface. The total production to date has been about 9,000,000 tons of sulphur with a gross value of $150,000,000. Small amounts of heavy oil were encountered in the early sulphur wells, but not in commercial quantity. The flanks have not been well tested for oil.

The search for domes is not confined to land areas; for example, several parties of the Geophysical Research Corporation, operating along the delta regions of Louisiana, have developed a technique for prospecting areas covered by water. Large sections of that country are at elevations of less than 10 ft.

* Anhydrite $CaSO_4$; Gypsum $CaSO_4$, $2H_2O$.

above mean sea-level, and the entire region is intersected by a network of bayous representing former distributaries of the depositing river. Encroachment of the Gulf on parts where deposition is no longer taking place has resulted in the formation of many shallow bays with depths of the order of 6–10 ft., which have become the scene of this novel form of seismic prospecting.

To cite one party as a specific example, the men, numbering from 22 to 30 at different times, lived on two house-boats anchored miles from the nearest habitation. Observers operated in specially designed motor boats, about 35 ft. in length, capable of making a speed of 35 miles per hour. Forward compartments housed the instruments, and permanent masts carried the wireless aerials. The seismograph used is so designed and water-proofed that it can be 'planted' a few feet beneath the bottom by means of a pole manipulated from the boat deck. Otherwise, procedure is much the same as in land shooting, except that the ease of moving from station to station and rapidity with which instruments can be located have greatly increased 'production' in areas covered. Needless to say, the development of these methods opened a large field of untouched acreage which yielded some surprisingly successful results.

GEOPHYSICAL EXPLORATION IN AUSTRALIA

The Commonwealth of Australia and the British Empire Marketing Board have cooperated in a scheme for exploring by various methods the less known and more promising regions of Australia. The working parties were under the combined guidance of geologists and physicists, and their reports contain material of great interest. It is known from other workers that electrical methods are carried on with great difficulty on that continent owing to arid surface conditions, and strongly conducting zones beneath. The results of this important investigation are embodied in *The Principles and Practice of Geophysical Prospecting* by Broughton Edge and Laby (1931). These investigations have been extended in recent years, particularly with magnetic and radioactive aerial survey equipment, with a view to locating ore deposits.

BRIEF NOTES ON DIAMOND DRILLING*

The essential parts of a diamond drill are: a cutting bit set with six or more black diamonds, or other very hard material, so arranged that when revolved they cut an annular hole around a central core of rock; hollow rods to connect the bit to the driving mechanism; a driving engine to revolve the rods and bit at a high rate of speed, and mechanism for feeding the rod and bit forward either automatically or by hand. Water is forced down through the rods, out of the bit and up through the space between the rods and the rock to the surface. This water serves to keep the diamonds cool and to remove the rock cuttings in the form of a sludge which may be collected for examination and analysis. At intervals the core is broken off, by an ingenious arrangement situated just back of the cutting bit, and is removed. Fig. 153 shows the cutting bit,

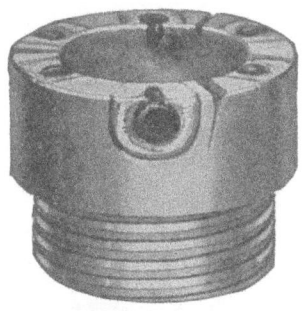

Fig. 153. Cutting bit of diamond drill showing black diamonds packed in position.

and Fig. 154 the general arrangement of a steam-driven drill located on the surface and with the hole protected by casing through the surface soil and the partly decomposed rock immediately underlying it.

Diamond drills may be driven by hand, steam, electricity, compressed air or gasoline engine. They may be transported over difficult mountain trails, taken down into mines or operated from rafts. Holes may be drilled vertically, horizontally or at any desired angle. Very great depth may be obtained, 5000 ft. is comparatively common, and one hole has been put down to about 6700 ft. Almost any material from sand to the hardest rock, may be penetrated, though speed and cost are largely dependent upon the nature of the material encountered. Other core drills are practically limited to vertical holes in fairly soft rock.

* We are indebted to our colleague, Professor W. G. McBride, for these notes on diamond drilling.

The recovery of the core is an important function of the diamond drill, since it gives a solid section of the formations, which may be examined in the same way as other rock specimens, and preserved in suitable boxes for future reference. The pro-

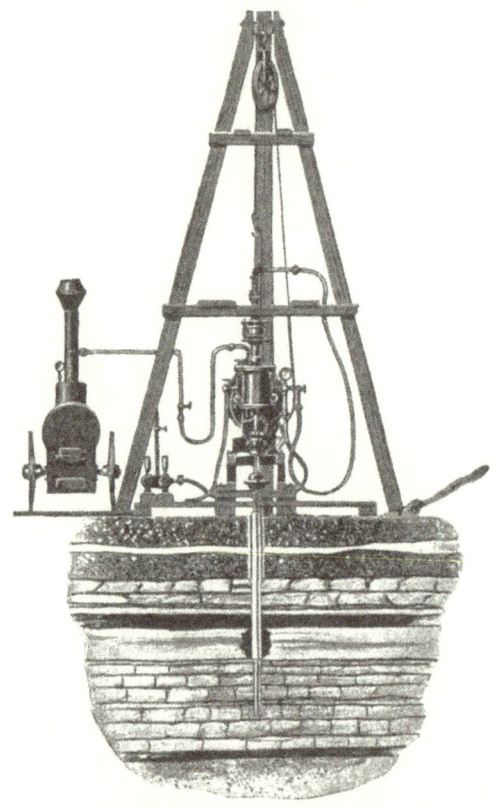

Fig. 154. Arrangement of typical diamond drill outfit, showing the principle of operation.

portion of core recovered will vary from almost 100 % in solid homogeneous rock to nothing in unconsolidated material. In most cases sufficient is obtained to enable the engineer, assisted by an examination of the sludge, to make a complete geological section along the bore. The diameter of the core ranges from

$\frac{13}{16}$ to $6\frac{3}{8}$ in., the largest sizes being used for oil or gas exploration only.

Speed and cost of diamond drilling vary within wide limits, depending principally upon the nature of the material to be drilled and the depth of hole. Short holes of 100–200 ft. are expensive because of the amount of time spent in setting up and starting the hole, while in deep holes, 2000 ft. and over, the time consumed pulling up and lowering the rods cuts down the speed and increases the expense. Hard rock reduces speed and increases diamond wear. Unconsolidated material, decomposed rock or brecciated material makes it necessary to case off or cement the hole, increases the danger of losing diamonds or even the entire bit, and sometimes limits the depth of the hole. Speed and costs can be estimated with fair accuracy by practical engineers when the conditions are known, but otherwise they are a matter of guess-work, in which experience plays a very important part. Counting from beginning of set-up to completion of hole, a diamond drill will bore, under reasonably good conditions, from 0·5 to 2 ft. per hour of working time, and costs will run from $2.00 to $6.00 per foot. Under unusual conditions wide variations from this range will occur. Diamond drilling is usually done by contract, as it is a highly specialized operation requiring an initial investment which is high in proportion to the operating costs on the ordinary operation.

The diamond drill is most useful for testing geological formations. In this, it is particularly valuable in that it gives actual specimens of rock formations large enough for microscopic or other optical examination, physical tests and chemical analysis. Bedrock soundings for dams, bridge piers and large buildings may be made by it where the rock is concealed by soil, gravel or water, and not only the depth of the overburden, but the characteristics of the underlying rock may be accurately determined. The formations through which long railway tunnels must pass may also be correctly found, and this not only assists in deciding upon the location, but enables the engineer to prepare for difficulties that will arise in driving the tunnel. Tests for ore-bodies are carried out by the diamond drill in areas which give indications of mineral deposits, and in mines prospecting is done to delimit ore-bodies already discovered and to discover other associated deposits. The compactness of the diamond drill and its ability to drill holes at any angle, make it particularly useful in underground work. A drift will cost as much as two to four diamond drill holes, and a shaft as much as eight to twelve. With proper care, good samples may be obtained, and ore values determined with fair accuracy. Diamond drills are also used for oil and gas prospecting in unproven or 'wild cat' fields. In these, they give a better geological record than the churn drill at about the same cost. For holes large enough for commercial extraction of oil, the standard well-drilling outfit using churn or rotary drills is generally preferred, though the diamond drill has distinct advantages under certain conditions.

Diamond drill-holes do not follow a straight line, but are diverted from it by the rotational friction of the bit and drill rods, by faults and by alternating hardness in the rock. Where the bit passes through faults or bedding planes at a sharp angle with the line of the hole, deflexion is quite rapid, and in long holes becomes very pronounced. In a 4000 ft. hole a deviation of 48° from the vertical has been found. In short holes the deviation is usually not sufficient to affect the results, and for ordinary purposes can be neglected up to 500 ft. Surveys of drill holes are made, with special equipment, of sufficient accuracy to serve the purpose of the engineer or geologist. When a hole has departed from its proper course, it may be brought back again by the use of a deflecting wedge inserted in the bottom of the hole. Where it is desirable to cut a vein, or other formation, in two places at considerable depth, much drilling may be saved by using the deflecting wedge to make a compound hole shaped somewhat like the letter Y inverted.

The diamond drill has certain disadvantages, particularly for mine development. The hole, on account of its small section, is seldom of value for subsequent operations, while shafts and drifts are usually available as a base for further work. Extreme care must be taken if samples from diamond drill holes are to be used as the basis for estimating ore-bodies, since the drill bit will naturally follow the line of least resistance, which may be a streak of soft ore. Cases are on record where this actually occurred, and the results were very misleading. Moreover, an engineer examining a mine where the ore has been prospected by diamond drills has no way of checking the sampling except by re-drilling the ground at great expense. When ordinary mine shafts, drifts and raises have been run, the ore may be re‑sampled any number of times at a small expense. The diamond drill, however, is a valuable aid in the preliminary prospecting of mineralized ground, if its limitations are realized, proper precautions taken and the results interpreted by a competent engineer or geologist. (See also Peele, 1927; Poston, 1928; Wright, 1928; Banks, 1929.)

BURIED TREASURE

The question is frequently raised as to the possibility of finding small conductors buried a few feet underground. In 1930 a method was described which proved successful in finding a bomb, weighing seventeen pounds, whose centre of gravity was 2 ft. below the earth's surface (Theodorsen, 1930). Three coils at equal intervals, wound on a light wooden drum or cylinder, with its axis vertical, are carried by two men holding suitable handles so that the lowest coil is a few inches from the ground. The centre coil has a curent of a few amperes flowing through it due to a stationary 500-cycle generator connected to the coil with long double flex. The upper and lower coils are connected in opposition to headphones worn by an observer who walks alongside the cylinder as it is carried to-and-fro over the area which is being searched. These two coils must be balanced for resistance, inductance and capacitance, so that there is almost silence in the phones. When, however, the cylinder passes over any underground metallic conductor of sufficient size, the current induced in that conductor by the central excited coil so distorts the electro-magnetic field through the lowest coil that a note is heard in the headphones, and precise location is then possible.

No doubt buried treasure—gold, silver, copper—in mass or in specie could also be found in this manner, or by analogous devices such as the well-known Hughes Induction Balance, but the authors regret that they have no practical experience in the finding of buried treasure!

LABORATORY EXPERIMENTS

It is often necessary to illustrate and explain large-scale field work by small scale experiments in lecture room or laboratory.

A few suggestions on this matter may be helpful:

(a) *Magnetic.* A small dip needle may be balanced with a small piece of soft wax on one end, so that it is in equilibrium when horizontal. It may then be moved slowly above and across a bar magnet or a block of magnetite, and the changes of dip noted and explained.

A student may then be asked to find a piece of magnetite which is hidden in a box of sand a few inches deep. He can quickly do so by noting the deviation of a compass as well as by his observation of the dip needle as just explained (see also p. 21).

(b) The self-potential method may be illustrated by burying a piece of sulphide ore under a few inches of *damp* sand in a large rectangular wooden tray. It is wise first to dip the ore into weak acid to increase the oxidation. Small-sized porous pots, as described in Chapter III, may be used as electrodes and connected to a microammeter. The survey may be carried out, and small pegs used to mark out the equipotentials on the top of the sand. The whole process is precisely the same as that used in the field, but on a smaller scale. The values of the currents may be measured, the galvanic effects of iron stakes determined, and practice in the use of a potentiometer also obtained.

(c) A shallow box of wood, without any metals, may be made water-tight with marine glue and filled with water about 1 cm. deep. The box may be about 2 ft. × 1 ft. 6 in. in area. Copper wires or rods are placed in the water on either side from end to end. These are connected through a 40 W. lamp to the direct-current lighting circuit (110 V.). Two search electrodes are placed in the water and connected to a suitable galvanometer. It is readily shown that the equipotentials are parallel to the two copper rods at the edge. Introduce a piece of metal, such as copper, and it will be found that the equipotentials are bent away from the conductor. If salt or a little acid is added to the water it becomes a better conductor, and inferior conductors such as ebonite or solid paraffin may be placed in the water, and then the equipotentials will tend to crowd into them. The lines or current flow may in every case be drawn perpendicular to the equipotentials, and it is instructive to note the changes due to the size, position and conductivity of the bodies placed in the water.

(d) All the experiments in the above section may be carried out using alternating current from a buzzer, induction coil or an oscillator. The current is again introduced into the water by the two copper rods along two edges. The reception electrodes may be connected to headphones, and the galvanometer is no longer required. Equipotentials are now plotted by obtaining silence or null points. The lines of flow at right angles to them may be plotted as before, when sundry good or bad conductors are introduced into the water.

It is very convenient to have the bottom of the box made of glass, beneath which is pasted section paper on which the chief lines are marked A, B, C—one way, and 0, 10, 20—in a perpendicular direction. The position of each electrode can be rapidly read, and the result can be plotted on another piece of section paper.

It must certainly be remembered that experiments in shallow water are a surface, or two-dimensional, problem, while experiments on the surface of the earth involve depth and present a three-dimensional problem.

(e) Take a tank of water with vertical glass sides about 4 ft. long and 2 ft. high, with the plates of glass about 3 in. apart.

The rest of the tank should not be made of metal, or if it is, paraffin wax or marine glue must be used to prevent contact of the metal with the water. Four pieces of wood may bridge the glass plates resting on their upper edges and four copper rods are passed through tight-fitting holes, one in each piece of wood, so that the rods are vertical and their ends just dip into the water. The rods are equally spaced with intervals A cm. between them. Direct or alternating currents are introduced to the water by the two outer electrodes, and the potential differences between the inner electrodes are detected or measured by galvanometer or headphones.

Do not introduce iron into the water but use only a single metal such as copper, otherwise electrolytic action will spoil the experiment, at least for direct current work.

Introduce a long bar of copper lowered by two strings to the bottom of the water. It will lower the potential difference of the two inner electrodes. Raise the bar to a higher level and note the potential differences for different levels. Also, keep it at one level, and vary the interval A between the four electrodes. A large variety of convincing experiments can be carried out in this manner. Next use a Megger and read E/I direct in ohms and repeat the previous experiments. All the work described in Chapter III can thus be illustrated and more fully understood. Again it is necessary to remember that a narrow tank furnishes a two-dimensional problem, while the earth gives us a three-dimensional problem.

(f) Instead of parallel wires and point electrodes it is possible to have a loop of one or many turns in air, or insulated and placed under water. A search coil of several turns may be connected to headphones, with or without amplifier, or to a loudspeaker. A great series of useful experiments is possible involving magnitude, direction and phase for various frequencies of alternating currents, with or without secondary conductors, especially if the search coil is connected to a cathode ray oscillograph, the other terminals of which are connected into the loop current so as to show the phase charges and elliptical polarization.

One example will serve as an illustration. A loop consisting of ten turns of ordinary lamp flex may be placed on a table, in the form of a rectangle 12×7 ft. This is excited by a current of about 2 amp. from a 500-cycle generator. A small search coil placed with its plane vertical in the central portion of the loop will produce no sound when connected to an amplifier and loudspeaker. This indicates that the magnetic component of the field is practically vertical. A small tilt of the search coil brings forth a note. When a piece of copper about 2 ft. × 1 ft. × $\frac{1}{2}$ in. is placed in the centre of the loop and the search coil, kept with its plane vertical, passed over it, there will be a distinct increase in intensity of the sound as the coil passes over the edges of the conductor and a minimum in the centre. If the coil is held horizontally it will be noted that the maximum change now takes place when the coil is over the centre of the conductor. Such a demonstration will illustrate clearly the variations in the horizontal and vertical components of the resultant vector when crossing a conductor excited by a horizontal loop.

It is impossible to summarize here investigations of so wide a scope. A demonstration of seismic methods may be mentioned as an interesting example. Using a cathode-ray oscillograph connected to a piezo-electric crystal pickup and giving a tap to the table, a record similar to that from a seismic field investigation is produced.

SEARCHING FOR UNDERGROUND PIPES

It is sometimes necessary to locate iron pipes underground, pipes which may convey water, gas, steam or oil. At other times it is desirable to trace underground cables, the positions of which are not known. The common practice is the laborious one of digging trenches until the pipe or cable has been found. It is much more simple and economical in many cases to find and to trace the pipe or cable by quite ordinary electrical methods.

The first case is that when two points on the metallic conductor are already known and can be reached. Take a 'buzzer' or induction coil, or better still a 500-cycle generator driven by a quarter-horse-power motor. Attach two insulated wires, laid on the ground, with good contacts to the two known points on the lost conductor, while the other ends of the wires are led to the generator. The insulated wires on the ground should be at least 20 or 30 yards from the supposed line of the underground conductor.

The observer takes a light wooden frame, about 4 × 4 ft., round which are wound about two hundred turns of no. 28 cotton or silk covered wire. The ends of this coil are joined to head-phones. The alternating magnetic field due to the current in the underground conductor, a few amperes it may be at most, enables the observer, carrying the coil vertically above the ground, with its plane parallel to the conductor, to trace the pipe or conductor almost as fast as he can walk. Clearly this is the Leader Gear Method of the First World War (Drysdale *et al.* 1923, p. 318).

When the lost conductor has been staked out roughly it is well to retrace it with the coil quite horizontal, and now it can be marked out within a few inches. For when the centre of the horizontal coil is exactly over the underground conductor there will be complete silence, but the note due to induction in the coil will become audible on moving even a foot to right or left.

By tilting the coil on either side it is, moreover, possible to estimate the depth of the lost, now found, conductor; but we must not expect too great precision in this case, because there is refraction or bending of the circular fields of force around the conductor as they enter the air from the moist earth (see Fig. 87).

When only one point is known on the conductor it is still possible to proceed, with somewhat more difficulty, by con-necting one wire from the generator to the available point, and the other wire from the generator to one or more iron stakes driven together well into the ground at the place most likely to be near the lost conductor. In this respect the first trial may not be very successful, but different places can be tried and the improvement, or reverse, readily noted. Much of the current will find its way back along the conductor which may be traced from the known point. Nor need one despair of finding a lost conductor when no points are known upon it. In this case two earth electrodes are used and one can be swung right or left, keeping a good loop with the insulated wires on the earth, until the listener with his coil finds some response from the sought conductor. Then the second electrode can also be swung until a maximum response is obtained.

All this depends on the current between the two electrodes on the earth finding a more ready path through the underground conductor than through the earth surrounding them. In difficult cases it may be advisable to use an amplifier with the observing coil and headphones. In simpler cases the ease of exploration is rather astonishing even to those familiar with the method.

Recently at the Sacred Heart Convent, near the Back River, a few miles from Montreal, engineers required quickly the precise location of the iron water pipe from the St Lawrence to the Convent. The work was carried out in four hours from the first notification, and the engineer in charge stated that the work was 'very convincing and of great practical value'.

With this simple equipment, all the underground pipes on both the McGill and Macdonald College campus were mapped by the maintenance staff. Many such devices are now available for locating metal objects buried a few feet underground, among which is a small portable instrument manufactured by the Detectron Company. This consists of a high-frequency electronic oscillator operated with flashlight dry batteries that acts as the transmitter; these oscillations set up currents in the buried conductor and a second tuned receiver with amplifier and meter as well as headphones is used to detect the secondary field from the conductor. The transmitter is held close to the ground, 20 or 30 ft. from the receiver, which is held vertical for a maximum response when over the pipe or conductor, and horizontal for a minimum indication. If the position of one end of a pipe is known and the transmitter placed vertically above and parallel to it, the pipe can be traced with the receiver for a distance of 100–500 ft. By holding the receiver coil at 45° to the vertical and moving at right angles to the length of the pipe, the depth of pipe will be the distance one has moved to obtain a minimum, which is obvious from the geometry.

COMPARISONS OF SURVEYS OVER THE SAME AREA BY
DIFFERENT COMPANIES

Independent surveys over the same area by different companies, employing quite dissimilar methods, must always prove of great interest to geologists and to mining engineers. In

Fig. 155 is an example from Northern Ontario where the Self-Potential (Porous Pot), and a Resistivity Method were employed by one company, and an electro-magnetic survey with

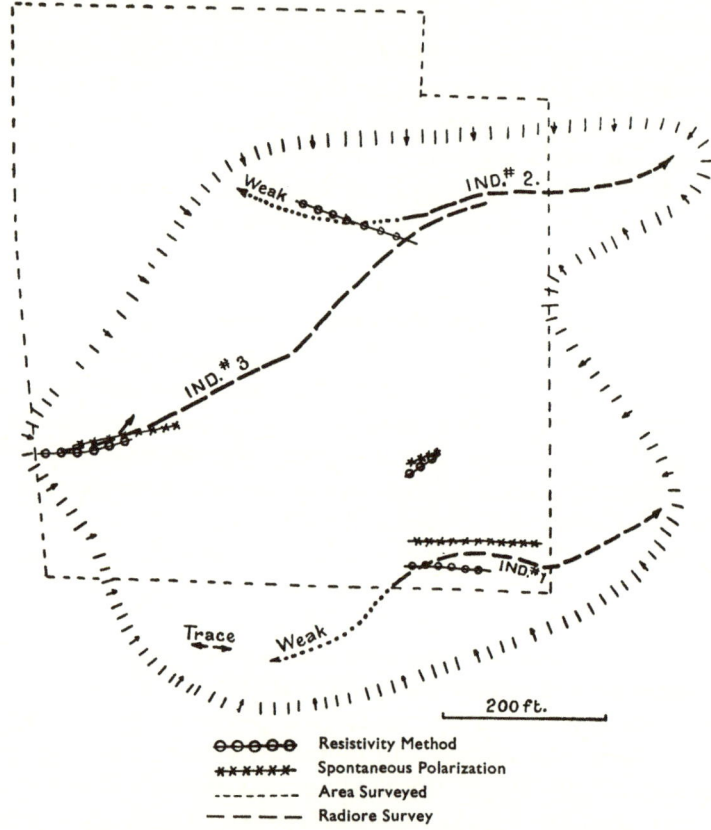

Fig. 155. Comparisons of surveys in Northern Ontario.

vertical loops and radio frequency was later used by a different company.

It will be noted that the agreements are satisfactory, but that the Radiore method determines the axis of the vein in a more prolonged manner. A special point to be observed is that the self-potential is a surface indication while the other methods tend to search deeper, and evidence of the dip of the ore-body may be deduced from their relative positions on the map, when plotted.

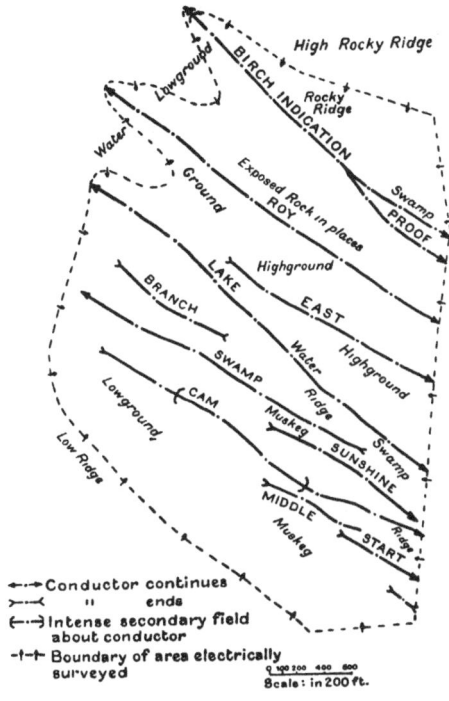

Fig. 156. Radiore survey, Arno Mine, Canada.

The Radiore method, with vertical coils, appears to detect conductors large and small, and the axes of these conductors throw much light on the geological structure of the area. An example may be seen in Fig. 156 where nine conducting axes are shown on the Arno mine.

The Birch Indication was discovered by the method employed and it has been amply verified by drilling, which cut through 40 ft. of pyrite and pyrrhotite with disseminations of chalco-pyrite. This is a good example of the efficiency of prospecting for sulphide conductors over areas which are deeply overburdened, and where trenching operations would prove very expensive.

It appears desirable to check the values of these abundant indications by alternative methods. Thus a magnetic survey would trace the pyrrhotite, a self-potential test would reveal relative intensities, and a Megger measurement (Gish and Rooney), along and at right angles to each strike, would rapidly throw light on the relative values of the indications.

APPENDIX I

WENNER'S FORMULA

We owe the following rigorous proof of Wenner's formula to our colleague Dr L. V. King.

Let V be the potential at any point due to current flow between electrodes C_1 and C_2.

V must satisfy $\nabla^2 V = 0$ in an indefinitely extended homogeneous medium.

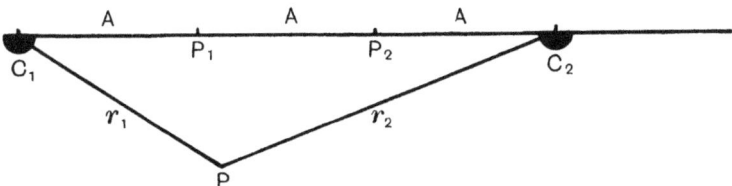

Fig. 157. Diagram for proof of Wenner's formula.

At a point P (Fig. 157) distant r_1 and r_2 from electrodes of small dimensions (compared to r_1 and r_2) a solution of $\nabla/V = 0$ is

$$V = \frac{A}{r_1} + \frac{B}{r_2},$$

where A and B are constants.

The surface of the semi-infinite plane is easily seen to be everywhere at right angles to the equipotential surfaces. Consider electrode C_1 to be a small hemisphere; then if ρ is the specific resistance, the normal current flow is $-\dfrac{1}{\rho}\dfrac{\partial V}{\partial n}$, so that out-flow of current from C_1 is $I = \dfrac{1}{\rho}\displaystyle\int \dfrac{\partial V}{\partial n}\,dS$ over the electrode.

Neglect the term $\dfrac{B}{r_2}$ and write $dS = r^2\,d\omega$, where ω is a solid angle, then

$$I = -\frac{1}{\rho}\int \frac{\partial}{\partial r}\left(\frac{A}{r}\right) r^2\,d\omega = \frac{A}{\rho}\,2\pi.$$

Hence $A = \dfrac{\rho I}{2\pi}$ and by symmetry $B = -\dfrac{\rho I}{2\pi}$.

Thus at any point $\quad V = \dfrac{\rho I}{2\pi}\left(\dfrac{1}{r_1} - \dfrac{1}{r_2}\right).$

If P_1 and P_2 be electrodes so that $C_1 P_1 = P_1 P_2 = P_2 C_2 = A$,

$$V_P = \frac{\rho I}{2\pi}\left(\frac{1}{A} - \frac{1}{2A}\right),$$

$$V_Q = \frac{\rho I}{2\pi}\left(\frac{1}{2A} - \frac{1}{A}\right),$$

$$\therefore V_P - V_Q = \frac{\rho I}{2\pi}\frac{1}{A},$$

which is Wenner's formula.

By images, solutions may be obtained similarly for semi-infinite media with two specific resistances ρ_1 and ρ_2 bounded by a plane parallel to the surface.

APPENDIX II

RESISTANCE AND CAPACITANCE

An interesting and helpful analogy exists between the capacitance (or capacity) of conductors in a dielectric medium, and the resistance between similar, and similarly situated, good conductors in a resisting medium.

Thus the capacitance between two conducting spherical shells with an insulator (dielectric constant k) between them is $kab/(a-b)$, when a, b are the radii of the outer and of the inner spheres. And the resistance between two good conducting shells separated by a medium whose specific resistance, or resistivity, is ρ, is $\dfrac{\rho}{4\pi}\left(\dfrac{1}{b}-\dfrac{1}{a}\right)$. The reason for this similarity is that in both cases the integral $\displaystyle\int_{b}^{a}\dfrac{dx}{4\pi x^2}$ is involved in the calculations. Thus the potential difference

$$v=\int_{b}^{a}\frac{q\,dx}{kx^2}=\frac{q}{k}\left(\frac{1}{b}-\frac{1}{a}\right),$$

so that the capacitance

$$C=\frac{\text{quantity}}{\text{potential difference}}=\frac{q}{v}=\frac{kab}{a-b}.$$

And the resistance between the shells is equal to

$$\int_{b}^{a}\frac{\rho\,dx}{4\pi x^2}=\frac{\rho(a-b)}{4\pi ab}.$$

It will, however, be more useful to consider the reciprocal of the resistance $(1/r)$ and to call it the conductance, and to call the reciprocal of the resistivity $(1/\rho)$ the conductivity σ.

Then for the concentric spheres we have capacitance $=\dfrac{kab}{a-b}$ in electrostatic units, and conductance $=\dfrac{4\pi\sigma ab}{a-b}$ in electromagnetic units. We can quickly derive one from the other by an interchange of k and $4\pi\sigma$.

Moreover, there is great generality in this proposition, a few examples of which may be cited.

Conductors	Capacitance in electrostatic units	Conductance in electromagnetic units
Concentric spheres	$\dfrac{kab}{a-b}$	$\dfrac{4\pi\sigma ab}{a-b}$
Parallel plates (large)	$\dfrac{k \text{ area}}{4\pi \text{ thickness}}$	$\dfrac{\sigma \text{ area}}{\text{thickness}}$
Concentric cylinders	$\dfrac{k}{2\log_e a/b}$ per cm.	$\dfrac{2\pi\sigma}{\log_e a/b}$ per cm.
Thin parallel wires, radius each a, distant d apart	$\dfrac{k}{2\log_e d/a}$ per cm.	$\dfrac{2\pi\sigma}{\log_e d/a}$ per cm.
Two small spheres, radius a each, a large distance c apart	$\dfrac{k}{2}\dfrac{ac}{c-a}$	$2\pi\sigma\,\dfrac{ac}{c-a}$

This table can be extended, and the method of images can be employed for conductance as well as for capacitance problems.

There is the further advantage to the geophysicist that, by help of this analogy, he can often the more readily visualize his results.

Thus an electrode, assumed hemispherical, placed just in homogeneous ground will have an effective conductance approximating to $2\pi\sigma a$, dependent on its radius a, and on the conductivity σ of the surrounding medium. If, however, a good conducting ore-body exists in its neighbourhood, then the conductance will be enhanced in a similar manner to the corresponding capacitance, and to its increase due to the presence of the conductor.

Suppose now that two equal hemispherical electrodes are placed in contact with the surface of the ground at a given distance r apart, what would be the effect of moving them to a greater interval? Clearly their joint capacitance would be decreased, and therefore the conductance between them would be diminished—a little.

Moreover, if the electrodes are fixed the conductance between them would be increased by a good conducting body near them in the same manner as the capacitance between similar bodies in electrostatics.

If we take the resistance R (see Jeans (1925), §§ 116, 386–8), between two hemispherical electrodes, each radius a, at an interval c between them, as given approximately by

$$R = \frac{\rho}{\pi} \frac{c-a}{ca}$$

$$= \frac{\rho}{\pi a} - \frac{\rho}{\pi c},$$

then we get, for values

$\rho = 30,000$ ohm-cm.,

$a = 10$ cm. $= 4$ in. about,

$c = 1500$ cm. $= 60$ ft. about,

that the first term is of the order 1000 ohms and the second of 7 ohms; and we may consider the first term as due to the two stake or electrode resistances combined, while the second is mainly due to the intervening earth. Such a division is somewhat artificial. It is advantageous to use the Megger from time to time to measure the stake or electrode resistance, which is the object for which that apparatus was originally designed.

APPENDIX III

ATTRACTION OF VERTICAL MAGNETIC DIKES BENEATH THE EARTH

Let the dike have thickness h, breadth b, and let the top be at a depth D below ground. Let $-m$, $+m$ be the magnetic surface density per unit area on the top and bottom of the dike.

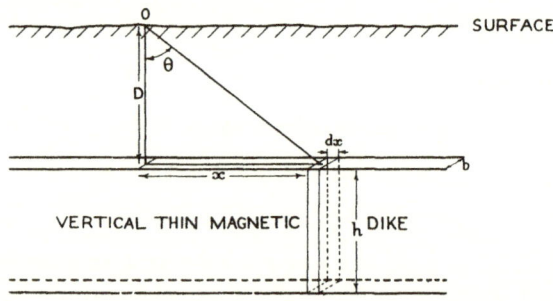

Fig. 158. Diagram to determine the vertical intensity at the surface due to a long, thin, vertical, magnetic dike.

Then considering the element $b\,dx$ (Fig. 158) we find the vertical force on unit pole at O to be

$$V_t = \int \frac{mb\,dx}{D^2 + x^2} \cos\theta$$

$$= mb \int \frac{D\,dx}{(D^2 + x^2)^{\frac{3}{2}}},$$

and putting $x = D \tan \phi$, it follows that

$$V_t = \frac{mb}{D} \int \frac{\sec^2 \phi\,d\phi}{\sec^3 \phi} = \frac{mb}{D} \int_{-\frac{1}{2}\pi}^{+\frac{1}{2}\pi} \cos \phi\,d\phi = \frac{2mb}{D}.$$

So also the vertical force on unit pole at O due to the bottom layer is

$$V_b = -\frac{2mb}{D + h}.$$

Hence the total vertical force due to the whole dike, on unit pole at O, is

$$V = V_t + V_b = 2mb\left(\frac{1}{D} - \frac{1}{D+h}\right)$$

$$= \frac{2mbh}{D(D+h)}.$$

Measurements can next be taken on a raised firm wooden platform at a height H above the ground, and then the vertical intensity due to the dike will be

$$V' = \frac{2mbh}{(D+H)\,(D+H+h)}.$$

The ratio of these is $(D+H)\,(D+H+h)/D(D+h)$ and if D and H are known, and V, V' measured, the vertical thickness of the dike can be found.

Dr L. V. King has extended this theory for the case of a thick, very long, vertical dike, and a summary of his results may be briefly stated.

(1) The magnetic potential due to a long line, with pole density σ per unit length, at shortest distance R from the observer, is

$$2\sigma \log R.$$

(2) The vertical force in dynes on unit pole due to the horizontal top of a broad dike (Fig. 159) with surface density m, is

$$2m\left(\tan^{-1}\frac{X+b}{Y} - \tan^{-1}\frac{X-b}{Y}\right),$$

where $2b$ is the breadth of the dike, Y the height of the observer above it, and X the horizontal distance from the top centre line.

This vertical intensity may also be written $2m(\phi_A - \phi_B)$, or $2m\phi$, where ϕ_A, ϕ_B, ϕ are the angles APN, BPN, APB. The horizontal force due to the magnetism on the top of the dike is

$$2m \log \frac{\cos \phi_A}{\cos \phi_B},$$

or

$$2m \log \frac{PA}{PB}.$$

(3) Similar results may be written for the bottom of the dike. In Fig. 159, *ABCD* is a vertical section of an inclined dike which extends to infinity both ways, running at right angles to the section *ABCD*.

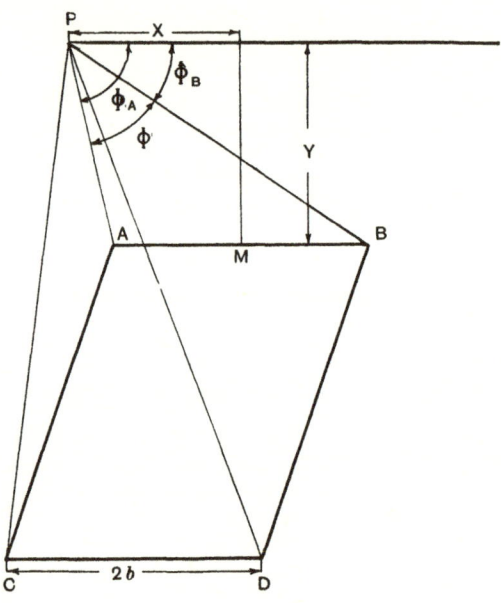

Fig. 159. Diagram for the determination of horizontal and vertical intensity on the earth's surface, due to a very long, broad, inclined, magnetic dike.

(4) Hence the vertical and horizontal intensities at any point *P*, due to the whole dike, are

$$2m(\text{angle } APB - \text{angle } CPD)$$

and
$$2m \log \frac{PA}{PB}\frac{PD}{PC} \text{ respectively.}$$

By measuring both vertical and horizontal intensities over the ground, and on a raised platform a few feet (say 10 or 20 ft.) high, and subtracting these from the normal magnetic components at a distance from the dike, it is possible to deduce the breadth *b*, the vertical thickness *h*, the depth *D* and the surface

density m. It must be admitted that theory has at present out-
stripped practice, but at the Falconbridge Nickel Mine some
preliminary experiments by Dr F. W. Lee and the authors
indicate the possibility of a successful estimate of the vertical
extent of the main ore-body, which has a fairly straight strike
about 2 miles long, a dip of 70°, and an overburden of drift about
120 ft. thick.

APPENDIX IV

VELOCITY OF A SHOCK UNDERGROUND

If the path of a shock underground is a circular arc then the velocity must increase linearly with depth; and conversely.

If the path of the shock is a minimum in time we must have as in optics,

$$\sin \phi = \frac{v}{v_0},$$

where v is the velocity underground at a point P on the path, v_0 the velocity at the lowest depth reached by the wave, and ϕ the angle between the tangent at P and the vertical.

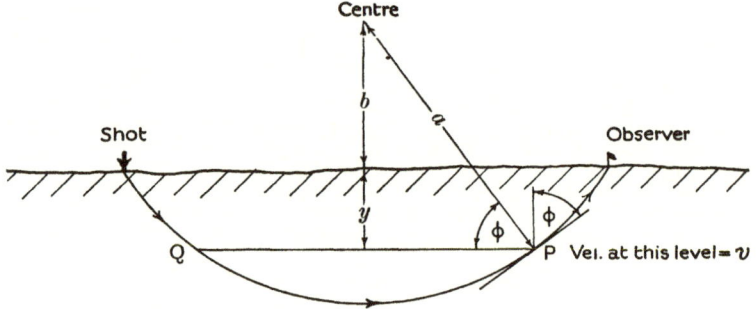

Fig. 160. Diagram for the proof that if the path is circular, the velocity increases linearly with the depth, and the converse.

But the equation of a circle is from Fig. 160

$$\sin \phi = \frac{b+y}{a},$$

where y is the depth of the point considered beneath the ground. From the two expressions above it follows that

$$\frac{v}{v_0} = \frac{b+y}{a},$$

which is a linear relation between velocity and depth, and this completes the proof! Note that a is the radius of the circular arc, and b the height of the centre above ground.

If R is the horizontal range, shot-point to observer, and T is the time of traversing the circular arc, it can be shown that

$$v_0 T = 2a \log \frac{1 + \dfrac{R}{2a}}{1 - \dfrac{R}{2a}}, \text{ exactly}$$

$$= R\left[1 - \frac{1}{3}\left(\frac{R}{2a}\right)^2 + \dots\right].$$

By drawing an experimental 'time-travel' curve, with T abscissa, R ordinate, the deviation from a linear law $v_0 T = R$, having slope v_0, may be studied.

This elegant proof is due to Dr L. V. King who adds that if the velocity increases exponentially with depth, as some suppose, then the path would be a *catenary of equal strength*, but information is lacking as to the actual path.

BIBLIOGRAPHY

The list of references given below is mainly to leading books and articles sufficient for most investigators. Those who desire a more exhaustive list of the vast number of publications will do well to refer to Ambronn for early work, and to Jakosky, who has listed the numerous patents at the end of each chapter in *Exploration Geophysics*.

GENERAL

AMBRONN, R. (1926). *Methoden der angewandten Geophysik*. Leipzig: Steinkopff.
AMBRONN, R. (1928). *Elements of Geophysics*. Tr. M. C. Cobb. New York: McGraw-Hill.
ANGENHEISTER, G. (1930). *Angewandte Geophysik*. Leipzig.
BROUGHTON EDGE and LABY (1931). *Principles of Geophysical Prospecting*. Cambridge University Press.
DOBRIN, M. M. (1952). *Introduction to Geophysical Prospecting*. New York: McGraw-Hill.
Documents 1–4 (1949–50). Schlumberger Well Surveying Corp.
EVE, A. S. and KEYS, D. A. (1929). Geophysical methods of prospecting. *Tech. Pap. Bur. Min., Wash.*, no. 420. Government Printing Office.
EVE, A. S. and KEYS, D. A. (1929, 1934). *Geophysical Prospecting*. New York: A.I.M.E.
Geophysical Abstracts, U.S. Bureau of Mines, Information Circulars (1929–52). Washington: Government Printing Office.
Geophysical methods of prospecting, principles and recent successes. *Quarterly Colorado School of Mines*, **24**, 1929.
Geophysics (1936–53). Society of Exploration Geophysicists.
GUTENBERG, B. (1951). *Physics of the Earth*. New York: Dover.
GUTENBERG, G. (1927). *Lehrbuch der Geophysik*. Berlin: Bornträger.
HADDOCK, M. H. (1926). *The Location of Mineral Fields*. Crosby, Lockwood and Son.
HAMMER, S. (1952). Geophysical exploration comes of age. *Bull. Amer. Ass. Petrol. Geol.* **36**, 1318–22.
Handbuch der Experimental Physik, Wien, W., Harms, F., **25**, *Geophysik*. Leipzig (1930).
HEILAND, C. A. (1926). Instruments and methods for the discovery of useful mineral deposits. *Engng Min. J.* (*-Press*), **121**, 9 Jan.
HEILAND, C. A. (1940). *Geophysical Exploration*. New York: Prentice Hall.
JAKOSKY, J. J. (1950). *Exploration Geophysics*. Los Angeles: Trija.
JEFFREYS, H. (1952). *The Earth*, 3rd ed. Cambridge University Press.

KELLY, S. F. (1935). The magnetic, gravitational and electrical methods of geophysical prospecting. *Explos. Engr*, **13**, 263–70, 303–12.

KRAHMANN, R. (1926). *Die Anwendbarkeit der geophysikalischen Lagerstätters-Untersuchungs-Verfahren u.s.w.* Halle.

LANDSBERG, H. E. (1952). *Advances in Geophysics*, **1**. New York: Academic Press.

MABY, J. C. and FRANKLIN, T. B. (1939). *The Physics of the Divining Rod.* London: Bell.

MATHIAS, E. (1924). *Traité d'Électricité Atmosphérique et Tellurique.* Paris.

MISENER, A. D. (1949). Temperature gradients in the Canadian Shield. *Trans. Canad. Min. Inst. (Inst. Min. Metall.)*, **52**, 125–32.

MOULTON, G. F. (1950). Is American petroleum the cause or result of the greatest prosperity known in history. *Petrol Engineering*, **22**.

NETTLETON, L. L. (1940). *Geophysical Prospecting for Oil.* New York: McGraw-Hill.

PAUTSCH, E. (1927). *Methods of Applied Geophysics.* Houston, Texas.

ROTHÉ, E. (1930). *Les Méthodes de Prospection du Sous-sol.* Paris: Gauthier Villars.

RUSSELL, W. L. (1951). *Principles of Petroleum Geology.* New York: McGraw-Hill.

Science of Petroleum (1938). Oxford University Press.

SHAW, H. (1931). *Applied Geophysics.* London: H.M. Stationery Office.

Trans. Amer. Inst. Min. (Metall.) Engrs, **81** (1929), **97** (1932), **110** (1934), **138** (1940).

TROMP, S. W. (1949). *A Scientific Analysis of Dowsing, Radiesthesis and Kindred Divining Phenomena.* London: Cleaver-Hume.

MAGNETIC METHODS

BAILEY, R. (1948). Canadian aerial magnetic surveys. *Canad. J. Res.* **26**, 523–39.

BALSLEY, J. R. (1946). The airborne magnetometer. *U.S. Dept. of the Interior Geophysical Investigations, Preliminary Report* (3). Washington: Government Printing Office.

BAUER, L. A., PETERS, W. J. and FLEMING, J. A. (1926). The compass variometer. *Publ. Carnegie Instn*, no. 175.

CHAPMAN, S. and BARTELS, J. (1940). *Geomagnetism.* Oxford.

DORSEY, N. E. (1913). Theory of the earth-inductor as inclinator. *Terr. Magn. Atmos. Elect.* **18**, 1.

Engng Min. J. (-Press), **150**, Sept. 1949, 114, Editorial note.

EVE, A. S. (1931). *A Magnetic Method of Estimating the Height of Some Buried Magnetic Bodies.* A.I.M.E.

EVE, A. S., KEYS, D. A. and others (1932). *Geological Survey of Canada Memoir*, nos. 165, 170, Ottawa, 1931. Ottawa: Queen's Printer.

EWING, J. A. (1900). *Magnetic Induction in Iron and Other Metals.* London (n.d.).

GLAZEBROOK, R. (1923). *Dictionary of Applied Physics*, **2**, 528–61. London: Macmillan.

HAALCK, H. (1934). *Lehrbuch der angewandten Geophysik*. Berlin: Bornträger.

HAANEL, E. (1904). *On the Location and Examination of Magnetic Ore Deposits by Magnetometer Measurements*. Dept. of Interior, Ottawa.

HEILAND, C. A. (1929). *Theory of Adolf Schmidt's Horizontal Field Balance*. A.I.M.E.

JENNY, W. P. (1932). Magnetic vector study of regional and local geological structure in the principal oil states. *Bull. Amer. Ass. Petrol. Geol.* **16**, 1177–203.

JENSEN, H. (1951). Airomagnetic survey helps find new Pennsylvania iron orebody. *Engng Min. J.* (*-Press*), **152**, 56–9.

JOYCE, J. W. (1937). *Manual Geophysical Prospecting with the Magnetometer, U.S. Bureau of Mines*. Washington: Government Printing Office.

KEYS, D. A. (1937). A survey of methods of determining depth of magnetic ore bodies. *Tech. Publ. Amer. Inst. Min. Engrs*, no. 830.

KOULOMZINE, T. and BROSSARS, L. (1947). The use of geophysics in prospecting for gold and base metals in Canada. *Geophysics*, **12**, 651–62.

LaCOUR (1930). *Magnetic Communications of the Danish Meteorological Institute*, no. 8.

RUMBAUGH, L. H. and ALLDREDGE, L. R. (1949). Airborne equipment for geo-magnetic measurements. *Trans. Amer. Geophys. Un.* **30**, 836–48.

STEARN, N. H. (1930). Practical geomagnetic exploration with the Hotchkiss Superdip. *Tech. Publ. Amer. Inst. Min. Engrs*, no. 370.

THOMSON, J. J. (1920). *Elements of Electricity and Magnetism*. Cambridge University Press.

ELECTRICAL AND ELECTROMAGNETIC METHODS

ALTY, T. and ALTY, S. (1930). The detection of rock salt by the methods of electrical surveying. *Canad. J. Res.* **3**, 521–6.

AMBRONN, R. (1926). *Methoden der angewandten Geophysik*. Leipzig: Steinkopff.

AMBRONN, R. (1928a). Elektrische Bodenforschung mittels Wechselströmen. *Beitr. Geophys.* **19**, part I.

AMBRONN, R. (1928b). Elektrische Bodenforschung mittels Wechselströmen. *Beit. Geophys.* **19**, part I, 5–58.

BIELER, E. S. (1928). The electromagnetic method, geophysical prospecting. *Canad. Min. Metall. Bull.* **21**, 631–44.

BOWEN, A. E. and GILKESON, C. L. (1930). Mutual impedances of ground-return currents. *Bell Syst. Tech. J.* **9**, 628.

CROSBY, I. B. and KELLY, S. F. (1929). Electrical subsoil exploration and the civil engineer. *Engng News Rec.* 14 Feb.

CROSBY, I. B. and LEONARDEN, E. G. (1928). Electrical prospecting applied to foundation problems. *Tech. Publ. Amer. Inst. Min. Engrs*, no. 131.

DELLINGER, J. H. (1922). *Principles of Radio Transmission and Reception in the Antenna and Coil Aerials.* U.S. Bureau Standards, 15.

DEUSSEN, A. and LEONARDON, E. G. (1935). Electrical exploration of drill holes. *Drill. Prod. Pract.*

DRYSDALE, C. V. and JOLLEY, A. C. (1924). *Electrical Measuring Instruments.* London: Benn.

EVE, A. S. (1930). Absorption of electro-magnetic induction and radiation by rocks. *Tech. Publ. Amer. Inst. Min. Engrs*, no. 316.

EVE, A. S. and KEYS, D. A. (1928). Geophysical prospecting; some electrical methods. *Tech. Pap. Bur. Min., Wash.*, no. 434.

EVE, A. S., KEYS, D. A. and DENNY, E. H. (1927). Penetration of radio waves. *Nature*, 120, 406.

EVE, A. S., STEEL, W. A. and others (1929). Reception experiments in Mount Royal Tunnel. *Proc. Inst. Radio Engrs, N.Y.*, 17, 347–75.

FLEMING, J. A. (1906). *The Principles of Electric Wave Telegraphy.* London: Longmans, Green and Co.

FOX, R. W. (1820). On the electromagnetic properties of metalliferous veins in the mines of Cornwall. *Phil. Trans.* 130, 399.

FRIEDL, K. (1927). Über die jüngsten Erdölforschungen im Wiener Becken. *Z. Petroleum*, Vol. 23, 189–240.

GELLA, N. (1921). Technische Blätter. *Wochenschrift zur Deutschen Bergwerkszeitung*, 8, 10.

GELLA, N. (1925). *Z. int. Ver. Bohring.* 33, 167–8.

GILCHRIST, L. (1931). Measurements of resistivity by the central electrode method at the Abana Mine, North-western Quebec, Canada. *Tech. Publ. Amer. Inst. Min. Engrs*, no. 386.

GILCHRIST, L. (1950). Distribution of potential in a two-layered medium due to an internal source and sink and the determination of the approximate average resistivity of the medium. *Canad. J. Res.* 28, 1–28.

GILCHRIST, L., CLARK, A. R. and BERNHOLTZ, B. (1950). Methods of determination of the average resistivity of a two-layered medium. *Canad. Min. J. (-Press)*, 71, 55–64.

GISH, O. H. (1923). General description of the earth-current measuring system at the Watheroo Magnetic Observatory. *Terr. Magn. Atmos. Elect.* 28, 89–108.

GISH, O. H. and ROONEY, W. J. (1925a). Measurement of resistivity of large masses of undisturbed earth. *Terr. Magn. Atmos. Elect.* 30, 161–3.

GISH, O. H. and ROONEY, W. J. (1925b). Measurement of the resistivity of large volumes of undisturbed earth. *Phys. Rev.* 30, 254.

HAYCOCK, O. C., MADSEN, E. C. and HURST, S. R. (1949). Properties of electromagnetic waves in the earth. *Geophysics*, **14**, 162–71.

HEDSTROM, H. (1937). Phase measurements in electrical prospecting. *Tech. Publ. Amer. Inst. Min. Engrs*, no. 827.

HEILAND, C. A. (1926). Instruments and methods for the discovery of useful mineral deposits. *Engng Min. J.* (*-Press*), **121**, 9 Jan.

HOTCHKISS, W. O., ROONEY, W. J. and FISHER, JAS. (1928). Earth-resistivity measurements in the Lake Superior Copper Country. *Tech. Publ. Amer. Inst. Min. Engrs*, no. 825.

HUMMEL, J. N. (1931). A theoretical study of apparent resistivity in surface potential methods. *Tech. Publ. Amer. Inst. Min. Engrs*, no. 418.

JAKOSKY, J. J. (1928*a*). Electrical methods of geophysical prospecting. *Engng Min. J.* (*-Press*), **125**, 238, 293.

JAKOSKY, J. J. (1928*b*). Operating principles of inductive geophysical processes. *Tech. Publ. Amer. Inst. Min. Engrs*, no. 134.

JEANS, J. H. (1925). *Electricity and Magnetism.* Cambridge University Press.

JOYCE, J. W. (1931). Electromagnetic absorption by rocks. *Tech. Pap. Bur. Min., Wash.*, no. 497. Government Printing Office.

KING, L. V. (1933). On the flow of electric current in semi-infinite stratified media. *Proc. Roy. Soc.* A. **139**, 237–77.

KOENIGSBERGER, J. (1928). Field observations of electrical resistivity and their practical application. *Tech. Publ. Amer. Inst. Min. Engrs*, no. 125.

KOENIGSBERGER, J. (1930). Über geoelektrische Methoden mit direkter Stromzuleitung. *Beitr. Geophys.* **1**, 23–107.

KRAHMANN, R. (1926). *Die anwendbarkeit der geophysikalischen hager-stättenuntersuchungsverfahren.* Halle.

LEE, F. W. (1928). Measuring the variation of ground resistivity with a Megger. *Tech. Pap. Bur. Min., Wash.*, no. 440. Government Printing Office.

LEE, F. W. (1929). Some earth resistivity measurements. *U.S. Bureau of Mines Information Circular* no. 6171. Washington: Government Printing Office.

LEE, F. W. (1930). Comparative advantages of applying several geophysical methods of prospecting to the same territory. *Inform. Circ. U.S. Bur. Min.* no. 6235.

LEE, F. W. (1936). Geophysical prospecting for underground waters in desert places. *Inform. Circ. U.S. Dep. Interior*, no. 6899.

LEE, F. W., JOYCE, J. W. and BOYER, P. (1929). Some earth resistivity measurements. *Inform. Circ. U.S. Bur. Min.* no. 6171. Washington: Government Printing Office.

LEONARDON, E. G. and KELLY, S. F. (1928). Some applications of potential methods to structural studies. *Tech. Publ. Amer. Inst. Min. Engrs*, no. 115.

LIVENS, G. H. (1918). *The Theory of Electricity.* Cambridge University Press.

LUNDBERG, H. (1922). Practical experience in electrical prospecting. *Geological Survey of Sweden Year Book.*

LUNDBERG, H. (1928). Recent results in electrical prospecting for ore. *Tech. Publ. Amer. Inst. Min. Engrs,* no. 98.

LUNDBERG, H., ZUSCHLAG, T. and KIHLSTEDT, F. H. (1931). Expansion and progress of electrical prospecting. *Canad. Min. Metall. Bull.* no. 232, 932–62.

MASON, MAX (1927). Geophysical exploration for ores. *Tech. Publ. Amer. Inst. Min. Engrs,* no. 45.

MAXWELL, J. CLERK (1904). *Electricity and Magnetism.* Oxford: Clarendon Press.

MOORE, R. W. (1952). Geophysical methods adapted to highway engineering problems. *Geophysics,* 17, 505–30.

MORECROFT, J. H. (1921). *Principles of Radio Communication.* New York: Wiley.

OLLENDORFF, FRANZ (1928). *Erdströme.* Springer.

PAVER, G. L. (1945). The application of the electrical resistivity method of geophysical surveying to the location of underground water, with examples from the Middle East. *Proc. Geol. Soc. Lond.,* no. 1407, 56–61.

PETERS, L. J. and BARDEEN, J. (1932). Some aspects of electrical prospecting applied in locating oil structures. *Physics,* 2, 103–22.

PIERCE, G. W. (1920). *Electric Waves and Oscillations.* New York: McGraw-Hill.

ROMAN, I. (1934). Some interpretation of earth resistivity data. *Trans. Amer. Inst. Min. (Metall.) Engrs,* 110, 183–201.

ROMAN, I. (1941). Superposition in the interpretation of two-layer earth-resistivity curves. *Bull. U.S. Geol. Surv.* no. 927–A.

ROONEY, W. J. (1927). Earth resistivity measurements in the Copper Country, Michigan. *Terr. Magn. Atmos. Elect.* 32, 97.

ROONEY, W. J. and GISH, O. H. (1927). Results of earth-resistivity surveys near Watheroo, Western Australia and at Erbo, Spain. *Terr. Magn. Atmos. Elect.* 32, 50.

RUEDY, R. (1945). The use of cumulative resistance in earth resistivity surveys. *Canad. J. Res.* 23, 57–72.

SCHLUMBERGER, C. (1920; 2nd ed. 1930). *Étude sur la Prospection Électrique du Sous-sol.* Paris: Gauthier-Villars.

SHEPARD, E. R. (1935). Subsurface exploration by earth-resistivity and seismic methods. *Pub. Rds., Wash.,* 16, 58–68.

SOMMERFELD, A. (1904). Über das Wechselfeld und der Wechselstrom-widerstand von Spulen und Rollen. *Ann. Phys., Lpz.,* 14, 673–708.

STIPE, C. G. and KELLY, S. F. (1937). Geophysical methods aid construction work. *Civil Engng.* 7, 264.

E & K

370 BIBLIOGRAPHY

Studies of geophysical methods. *Mem. Geol. Surv. Can.* nos. 165 (1931) and 170 (1932).

SUNDBERG, K., LUNDBERG, H. and EKLUND, J. (1925). *Electrical Prospecting in Sweden.* Stockholm.

SUNDBERG, K. and NORDSTROM, A. (1928). Electrical prospecting for molybdenite at Questa. *Tech. Publ. Amer. Inst. Min. Engrs,* no. 122.

SWARTZ, J. H. (1931). Resistivity measurements upon artificial beds. *Inform. Circ. U.S. Bur. Min.* no. 6445.

SWARTZ, J. H. (1937). Resistivity studies of some salt-water loundaires in the Haiwaian Islands. *Trans. Amer. Geophys. Un.* 18, 387–93.

SWARTZ, J. H. (1939). Geophysical investigations in the Hawaiian Islands. *Trans. Amer. Geophys. Un.* 20, 292–8.

TAGG, G. F. (1931). Practical investigations of the earth resistivity method of geophysical surveying. *Proc. Phys. Soc.* 43, 305–23.

TAGG, G. F. (1934). Interpretation of resistivity measurements. *Trans. Amer. Inst. Min. (Metall.) Engrs,* 110, 135.

TAGG, G. F. (1937). Interpretation of resistivity measurements. *Tech. Publ. Amer. Inst. Min. Engrs,* no. 755.

THOMSON, J. J. (1920). *Elements of Electricity and Magnetism.* Cambridge University Press.

WAIT, J. R. (1952). A conducting permeable sphere in the presence of a coil carrying an oscillating current. *Can. Journ. Research,* 31, 670–8, 1953. (At meeting of Society of Exploration Geophysicists, Toronto, Oct. 1952.)

WATSON, H. G. I. (1931). The Bieler-Watson Method. *Mem. Geol. Surv. Can.* 165, 144–60.

WEAVER, W. (1928). Certain applications of the surface potential methods. *Tech. Publ. Amer. Inst. Min. Engrs,* no. 120.

WENNER, F. (1916). A method of measuring earth resistivity. *Bull. U.S. Bur. Stand., Wash.,* no. 258.

WETZEL, W. W. and McMURRAY, H. V. (1937). A set of curves to assist in the interpretation of the three-layer resistivity problem. *Geophysics,* 2, 320–41.

ZENNECK, J. A. (1915). *Wireless Telegraphy.* New York: McGraw-Hill.

GRAVITATIONAL METHODS

BARTON, C. E. (1928). The Eötvös torsion balance method of mapping geologic structure. *Tech. Publ. Amer. Inst. Min. Engrs,* no. 50.

BARTON, D. C. (1931). Gravity measurements with the Eötvös torsion balance. *Bull. Nat. Res. Coun., Wash.,* no. 7, pp. 167–90.

BARTON, D. C. (1945). Case histories and quantitative calculations in gravimetric prospecting. *Trans. Amer. Inst. Min. (Metall.) Engrs,* 164, 17–65.

ECKHARDT, E. A. (1952). Geophysical activity in 1951. *Geophysics,* 17, 441–51.

EDDINGTON, A. (1928). *The Nature of the Physical World.* Cambridge University Press.

EÖTVÖS, R. VON (1896). *Ann. Phys., Lpz.,* **59,** 354–400.

GEORGE, P. W. (1928). Experiments with Eötvös torsion balance in the Tristate Zinc and Lead District. *Tech. Publ. Amer. Inst. Min. Engrs,* no. 65.

HAMMER, S. (1939). Terrain corrections for gravimeter stations. *Geophysics,* **4,** 184–94.

HAMMER, S. and ECKHARDT, E. A. (1952). *Geophysics,* **17,** 435–51.

HEILAND, C. A. (1928). A cartographic correction for the Eötvös torsion balance. *Tech. Publ. Amer. Inst. Min. Engrs.* no. 52.

HEILAND, C. A. (1929). A new graphical method for torsion balance, topographic corrections and interpretations. *Bull. Amer. Ass. Petrol. Geol.* **12,** 39.

HEILAND, C. A. (1932). *Directions for the Use of the Askania Torsion Balance.* Houston, Texas: American Askania Corporation.

INNES, M. J. S. (1947). An investigation of the applicability of gravimetric and magnetometric methods of geophysical prospecting. *Dom. Obs. Repr., Ottawa,* **11,** 355–61.

LACOSTE, L. J. B. (1934). A new type long-period vertical seismograph. *Physics,* **5,** 178–80.

LANCASTER-JONES, E. (1928). The computation of Eötvös gravity effects. *Tech. Publ. Amer. Inst. Min. Engrs,* no. 75.

McLINTOCK, W. F. P. and PHEMISTER, J. (1928). The use of the torsion balance in the investigation of geological structure in south-west Persia. *Summary of Progress of the Geological Survey of Great Britain for the Year* 1926. Appendix XII with bibliography. London: H.M. Stationery Office.

McLINTOCK, W. F. P. and PHEMISTER, J. (1927). *Min. Mag., Lond.,* **37,** 363–5.

ROGERS, G. R. (1952). Subsurface gravity measurements. *Geophysics,* **17,** 365–71.

SHAW, H. (1928). Gravity surveying in Great Britain. *Tech. Publ. Amer. Inst. Min. Engrs,* no. 74.

SHAW, H. (1929). Interpretation of gravitational anomalies. *Tech. Publ. Amer. Inst. Min. Engrs,* no. 178.

SHAW, H. and LANCASTER-JONES, E. (1923). The Eötvös torsion balance. *Proc. Phys. Soc.* **35,** 151–65 and 204–12.

SHAW, H. and LANCASTER-JONES, E. (n.d.). *The Eötvös Torsion Balance.* London: L. Oertling, Ltd.

VON THYSSEN, S. (1938). Ueber das Wessen des Seismische-Elektrischen Effektes. *Beit. angew. Geophys.* **7,** 218–29.

WOOLLARD, G. P. (1950). The gravity meter as a geodetic instrument. *Geophysics.* **15,** 1–29.

WYCKOFF, R. D. (1941). The Gulf gravimeter. *Geophysics,* **7,** 13–33.

SEISMIC METHODS

ADAMS, F. D. and COKER, E. G. (1906). *Investigations into the Elastic Constants of Rocks.* Washington: Carnegie Inst.

BARTON, D. C. (1928). Seismic methods of mapping geological structures (at meeting of A.I.M.E. Aug. 1928).

BARTON, D. C. (1929). *Geophysical Prospecting,* pp. 572–624. A.I.M.E.

BLAU, L. W. (1928). Experimental investigation of forced vibrations. *J. Franklin Inst.* **206**, 3, 355–78.

BYERLY, P. (1942). *Seismology.* New York: Prentice Hall.

Dictionary of Applied Physics. Theory of Elasticity (Southwell, R.V.), Elastic Constants (Batson, R. G.).

DIX, C. H. (1952). *Seismic Prospecting for Oil.* New York: Harper.

DRYSDALE, C. V:, LAMB, H. and others (1923). *The Mechanical Properties of Fluids.* London.

Encyclopaedia Britannica (1911). Seismometer.

EWING, M. and LEET, D. (1932). Comparison of two methods for interpretation of seismic time-distance graphs which are smooth curves. *Trans. Amer. Inst. Min. (Metall.) Engrs,* **81**, 263–70.

GABRIEL, V. G. (1936). Seismic prospecting in exploration for oil. *Louisiana Conserv. Rev.* **5**, 4, 4–8.

GUTENBERG, B. and RICHTER, C. F. (1949). *Seismicity of the Earth and Associated Phenomena.* Princeton University Press.

HEILAND, C. A. (1935). The seismic method of geophysical prospecting. *Explos. Engr,* **13**, 359–72.

KNOTT, C. G. (1908). *The Physics of Earthquake Phenomena.* Oxford.

LEET, D. (1930). Earth vibrations from dynamic blasts. *Trans. Amer. Geophys. Un.* **11**, 49–62.

LOVE, A. E. H. (1920). *Mathematical Theory of Elasticity.* Cambridge University Press.

SERVICE, J. H. (1928). The transmission of sound through sea water. *J. Franklin Inst.* **206**, 779–807.

SHEPARD, E. R. and HAINES, R. M. (1942). Seismic sub-surface exploration on the St Lawrence River Project. *Proc. Amer. Soc. Civ. Engrs,* **68**, 1743–67.

WALKER, G. W. (1913). *Modern Seismology.* London: Longmans, Green and Co.

RADIOACTIVE METHODS

BROWNELL, G. M. (1950a). Radiation surveys with a scintillation counter. *Econ. Geol.* **45**, 167–74.

BROWNELL, G. M. (1950b). Prospecting with a scintillometer. *Precambrian,* **23** (3), 23–9.

CARMICHEAL, H. (1946). Design of the Chalk River Ion chambers, Types TPA, TPJ and TPD. *Can. Nat. Res. Counc. Rep. C.R. Tech.* no. 276.

GEIGER, H. and MÜLLER, W. (1928). Mode of action and construction of a counter tube. *Phys. Z.* **29**, 839–41.

GEIGER, H. and MÜLLER, W. (1929). Experimental investigation of the electron counter. *Phys. Z.* **30**, 489–93.

FRANKLIN, E. and HARDWICK, J. (1951). The design of portable gamma and beta radiation measuring instruments. *Journ. British Inst. Radio Eng.* **11**, 417–34.

GEITEL, H. (1920). Die Radioaktivität der Erde und der Atmosphäre. *Handbuch der Radiologie,* **1**. Leipzig.

GOCKEL, A. (1908). *Die Luftelektrizität.* Hirzel.

HESS, V. F. (1928). *The Electrical Conductivity of the Atmosphere and its Causes,* tr. by L. W. Codd. London: Constable.

HILLBORN, J. W. (1951). Equilibrium assaying of uranium ore. *Can. Dep. Mines, Radioactivity Division, Topical Report* no. 92/51. Ottawa: King's Printer.

HORWOOD, J. L. and McMAHON, C. (1950). Use of a high-pressure ionization chamber in assaying uncrushed ore samples. *Can. Dep. Mines, Radioactivity Division, Memorandum* no. 106. Ottawa: King's Printer.

JOLY, J. (1909). *Radioactivity and Geology.* London: Constable.

LAPOINTE, G. (1950). Equilibrium corrections in Geiger analysis. *Can. Dep. Mines, Radioactivity Division, Document* no. SR-31/50. Ottawa: King's Printer.

LAPOINTE, C. and WILLIAMSON, W. R. (1952). The use of Geiger-Müller tubes for the state of equilibrium of the radio-elements in uranium ore. *Can. Dep. Mines, Radioactivity Division, Document* no. TR-101/52. Ottawa: Queen's Printer.

LAPP, H. E. and ANDREWS, H. L. (1950). *Nuclear Radiation Physics.* New York: Prentice Hall.

LUNDBERG, H. (1950). Airborne electrical surveys for regional studies in oil and ore prospecting. *Trans. Canad. Min. Inst. (Inst. Min. Metall.),* **53**, 130–2.

LUNDBERG, H. (1952). Prospecting for oil and gas. *Precambrian,* **25**, 9–21.

LUNDBERG, H. and ISFORD, G. (1953). Oil prospecting with the radioactive method. (At meeting of A.I.M.E., Feb. 1953.)

LUNDBERG, H., ROULSTON, K. I., PRINGLE, R. W. and BROWNELL, G. M. (1952). Oil exploration with airborne scintillation counters. *Oil in Canada,* **4**, 40–6.

MAKOWER, W. and GEIGER, H. (1912). *Practical Measurements in Radioactivity.* London: Longmans, Green and Co.

MEYER, S. and SCHWEIDLER, E. (1927). *Radioaktivität.* Berlin: Teubner.

PRINGLE, R. W., ROULSTON, K. I., LUNDBERG, H. and BROWNELL, G. M. (1952). The scintillation counter in the search for oil (at meeting of A.I.M.E., Feb. 1952).

PRINGLE, R. W., ROULSTON, K. I. and TAYLOR, H. W. (1950). The scintillation counter as a low resolution gamma-ray spectrometer. *Rev. Sci. Instrum.* **21,** 216.

Prospecting for Uranium (1951). Washington: Government Printing Office.

Aerial prospecting for radioactive materials. *Reports MR*–17 *and CRR*–495 (1952). Nat. Research Council of Canada. Ottawa: Queen's Printer.

RUSSELL, W. L. (1941). Well logging by radioactivity. *Bull. Amer. Ass. Petrol. Geol.* **25,** 1768–88.

RUTHERFORD, E. (1915). *Radioactive Substances.* Cambridge University Press.

RUTHERFORD, E., CHADWICK, J. and ELLIS, C. D. (1930). *Radiations from Radioactive Substances.* Cambridge University Press.

SHRUM, G. M. and SMITH, R. (1934). A portable Geiger-Müller tube counter as a detector for radioactive ores. *Canad. J. Res.* **11,** 652–7.

WILMOT, R. D. and McMAHON, C. (1951). Radioactivity of uranium ore with Geiger type equilibrium counter. *Can. Dep. Mines, Radioactivity Division, Memorandum* no. 115. Ottawa: King's Printer.

WELL-LOGGING METHODS

DI GIOVANNI, H. J., GRAVESON, R. T. and FOLI, A. H. (1953). Scintillation unit for drill-hole logging. *Nucleonics,* **11,** 34–9.

DOLL, H. G. (1949). Introduction to induction logging and application to logging of wells drilled with oil base mud. *J. Petrol. Technology, Tech. Paper,* no. 2641.

DOLL, H. G. (1951). The Laterlog: a new resistivity logging method with electrodes using an automatic focusing system. *Trans. Amer. Inst. Min. Metal.* **192,** 305–16.

FEARON, R. E. (1949a). Neutron well logging. *Nucleonics,* **4,** 30–42.

FEARON, R. E. (1949b). Gamma-ray well logging. *Nucleonics,* **4,** 67–75.

JACKSON, W. J. and CAMPBELL, J. L. P. (1945). Some practical aspects of radioactivity well-logging. *Tech. Publ. Amer. Inst. Min. Engrs,* no. 1923.

MISENER, A. D. and THOMPSON, L. G. C. (1950). Temperature gradients in Ontario and Quebec. *Trans. Canad. Min. Inst. (Inst. Min. Metall.),* **53,** 1–4.

PONTECORVO, B. (1941). Neutron well logging. *Oil Gas J.* **40,** Sept. 11, 32–3.

SUMMERS, G. C. and BRODING, R. A. (1952). Continuous velocity logging. *Geophysics,* **17,** 598–614.

WILSON, J. T. (1949). Some major structures of the Canadian Shield. *Trans. Canad. Min. Inst. (Inst. Min. Metall.),* **52,** 231–42.

OTHER METHODS AND APPLICATIONS

BANKS, R. T. (1929). Diamond drilling for oil. *Trans. Canad. Min. Inst. (Inst. Min. Metall.)*, **32**, 335–43.

BEVIER, G. M. (1925). The Damon Mound oil field, Texas. *Bull. Amer. Ass. Petrol. Geol.* **9**, 505.

BULLARD, E. C. (1945). Thermal history of the earth. *Nature*, 35–6.

DOLL, H. G. (1943). The S.P. dipmeter. *Tech. Publ. Amer. Inst. Min. Engrs*, no. 1547.

DOLL, H. G. and SCHWEDE, H. F. (1947). Radioactive markers in oil field practice. *Tech. Publ. Amer. Inst. Min. Engrs*, no. 2261.

KELLEY, P. K. (1925). The sulphur salt dome, Louisiana. *Bull. Amer. Ass. Petrol. Geol.* **9**, 479.

PEELE, R. (1927). *Mining Engineers' Handbook*, 2nd ed. New York: John Wiley.

PETTERSON, H. (1953). Radium and the deep sea. *Amer. Scientist*, **41**, 245–55.

POSTON, R. H. (1928). Prospect drilling in south-east Missouri. *Engng Min J. (-Press)*, **125**, 14–16.

THEODORSEN, T. (1930). Instrument for detecting metallic bodies buried in the earth. *J. Franklin Inst.* **210**, 311–26.

WRIGHT, L. B. (1928). Borehole surveying at the Homestake. *Engng Min. J. (-Press)*, **125**, 57–58.

INDEX

For EU product safety concerns, contact us at Calle de José Abascal, 56–1°,
28003 Madrid, Spain or eugpsr@cambridge.org.

www.ingramcontent.com/pod-product-compliance
Ingram Content Group UK Ltd.
Pitfield, Milton Keynes, MK11 3LW, UK
UKHW010852090126
466816UK00011B/178